LASERS IN METALLURGY

LASERS IN METALLURGY

Proceedings of a symposium sponsored by the Physical Metallurgy and Solidification Committees of The Metallurgical Society of AIME, held at the 110th AIME Annual Meeting, Chicago, Illinois, February 22-26, 1981.

Edited by
K. MUKHERJEE
Michigan State University
East Lansing, Michigan 48824

and

J. MAZUMDER
University of Illinois at
Urbana-Champaign
Urbana, Illinois 61801

A Publication of The Metallurgical Society of AIME

A Publication of The Metallurgical Society of AIME
P.O. Box 430
420 Commonwealth Drive
Warrendale, Pa. 15086
(412) 776-9000

Printed in the United States of America.
Library of Congress Card Catalogue Number 81-85419)
ISBN Number 0-89520-385-5 7

Foreword

Since its initial development, the laser has been hailed as a power-ful tool for a variety of potential applications. The theoretical concept of the laser was first put forward by Schawlow and Towns in 1958, and in 1960 Maiman developed the first working ruby laser. A variety of laser systems have been developed since, but the development of multikilowatt continuous CO_2 lasers, around 1970, has greatly increased the commercial feasibility of lasers in metallurgy and materials processing. The availability of high power density, $\geq 10^6$ watts/cm^2, is a primary factor in establishing the laser as a useful tool in metal and materials industries.

During the past ten years, both the increasing demand for advanced materials and the availability of these high power sources have stimulated considerable interest in research and development related to laser applications. The repertoire of metallurgical applications of lasers has grown considerably during this period. The more obvious applications such as laser cutting, drilling, welding and heat treatment are already finding their way to industrial production lines. Lasers offer an easily maneuvered, chemically clean heat source which can produce a narrow heat-affected zone and low distortion. The use of lasers for such applications as surface alloying, cladding, glazing, and annealing of semiconductors offer the possibility of producing new materials with superior quality. Laser shock hardening is a process which allows hardening of otherwise unhardenable materials. The inherent rapid rates of heating and cooling also open up the possibility for producing novel materials. Theoretical and experimental investigations and mathematical modeling of thermal and shock wave phe-nomena have already been initiated by several researchers. Much more remains to be done.

In view of the foregoing, the organizers felt that it was timely and desirable to convene a symposium on lasers in metallurgy. This symposium and its proceedings are aimed at stimulating further research in this impor-tant and interesting area. Hopefully, these proceedings will serve as a handy reference book for graduate students and researchers entering this relatively new field of research.

The symposium organizers wish to express their appreciation to the members of the Physical Metallurgy and Solidification Committees of The Metallurgical Society of AIME for their support and encouragement. Finally, our sincere thanks to the session chairmen and the authors for making this symposium successful.

K. Mukherjee
Michigan State University
East Lansing, Michigan 48824

J. Mazumder
University of Illinois
 at Urbana-Champaign
Urbana, Illinois 61801

October 5, 1981

Table of Contents

SOLIDIFICATION MICROSTRUCTURE OF LASER PROCESSED ALLOYS
AND ITS IMPACT ON SOME PROPERTIES*

Theo Z. Kattamis
Department of Metallurgy
University of Connecticut
Storrs,Connecticut 06268

The solidification microstructure of laser or electron beam beads was
studied in a Ni-Al-Cr alloy and in the binary model alloy systems,Al-Cu
and Mg-Zn. The study on Ni-Al-Cr indicated that in extrapolating dendrite
arm spacing versus local cooling rate curves the cellular dendrite spacing
should be reported on the secondary,not the primary dendrite arm spacing
curve. In laser beads deposited on grain-refined nondendritic Mg-5.3%Zn-
0.6%Zr alloy the microstructure is finely dendritic. Despite the high
cooling rate prevailing during solidification of high energy beam deposited
beads microsegregation is high,because the corresponding growth rate is
low and equal to a large fraction of the beam travel speed. In high
energy beam deposited spots both cooling and growth rates are high and the
material is very homogeneous. The microhardness of a laser beam spot is
roughly equal to that of a solutionized and aged bead. Preliminary corrosion
studies have indicated that corrosion rate of a rapidly solidified Mg-Zn
alloy immersed in a 3.5%NaCl aqueous solution is higher than in a slowly
solidified specimen. However,after solutionizing the finely distributed
interdendritic nonequilibrium Mg_7Zn_3 phase,the corrosion rate is substantially reduced below that of the slowly solidified material. It may be
reduced further by polishing the specimen surface,thus reducing the
probability of operation of crevice corrosion.

*   This work was partially supported by the Air Force Office of Scientific
Research through Grant No. 77-3344.

The solidification microstructure of laser or electron beam re-
melted and rapidly solidified Ni-Al-Ta and Ni-Al-Cr alloys was previously studied (1,2). Further studies were recently conducted on Ni-Al-Cr
alloy ,as well as on the binary model alloy systems,Al-Cu and Mg-Zn.
Results are reported herein. Following a brief consideration of the as-
cast microstructure of high energy beam deposited beads solute microsegregation is examined. Results are given on the microhardness of rapidly
solidified Al-Cu alloy beads and spots,and on the corrosion behavior of
rapidly solidified Mg-Zn alloy.

Dendritic Microstructure of a Ni-Al-Cr Alloy

In laser or electron beam beads columnar dendrites are often adjacent to cellular dendrites. Comparing locations A and B,Figure 1,which
correspond to the same distance from the bottom of the bead,hence to the
same average cooling rate,it can be seen that the cellular dendrite spacing at B is equal to the secondary,not the primary dendrite arm spacing
at A. This observation is general and has been made for a variety of
rapidly solidified alloys,atomized,splat-cooled or extracted from the
liquid. It would indicate that coarsening,which is known to control
secondary dendrite arm spacing in columnar or equiaxed dendrites,controls
also cellular dendrite spacing in cellular dendrites. The equality was
further confirmed by solidifying dendritic monocrystals of Ni-18.0at%Al-
4.1at%Cr under various growth conditions (3),Figure 2. Monocrystal A was
grown under a gradient $G=8x10^3K/m$ and at a growth rate R=0.25m/h and
monocrystal B under $G=20x10^3K/m$ and at R=0.10m/h. For both monocrystals
the corresponding cooling rate $\varepsilon=G\cdot R=0.55K/s$. Secondary dendrite arm
spacing in the columnar monocrystal A was found to be equal to cellular
dendrite spacing in the cellular dendritic monocrystal B (3). The dependence of primary and secondary dendrite arm spacing and of cellular
dendrite spacing on growth velocity for two gradients: $8x10^3$ and $20x10^3$
K/m is illustrated in Figure 3. For each gradient cellular dendrite
spacings and secondary dendrite arm spacings fall on the same curve. The
critical growth rates R_1 and R_2 at which the cellular dendritic to columnar dendritic transition occurs for gradients G_1 and G_2,respectively,
are indicated in the same figure and were deduced from Figure 2. The

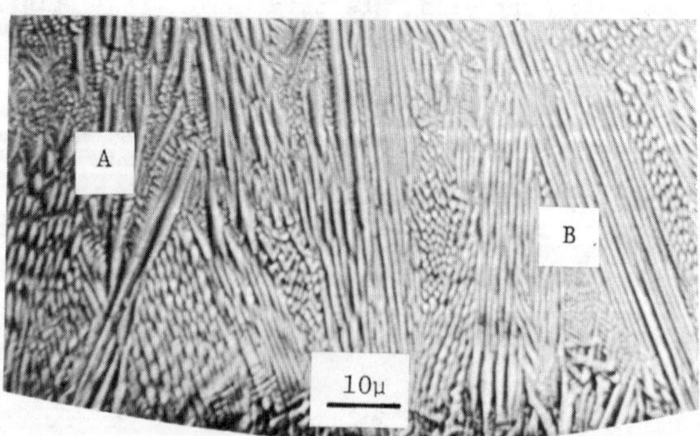

Fig. 1- Photomicrograph of part of a laser bead deposited on Ni-18.0
at%Al-4.1at%Cr alloy plate at 0.05 m/s.

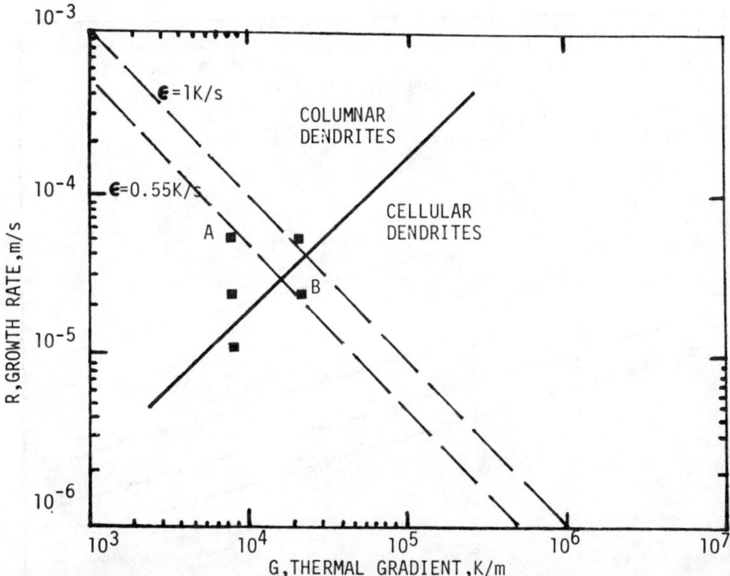

Fig. 2- G-R-ε diagram for Ni-18.0at%Al-4.1at%Cr indicating the cellular
dendritic-to-columnar dendritic transition.

data points of Figure 3 were reported in Figure 4 versus local cooling
rate. Again,secondary dendrite arm spacings and cellular dendrite spacings
fall on one and the same curve,and primary spacings on another. By extrapolating these curves to very fine microstructures it is possible to
approximately determine the corresponding local cooling rate. In laser
beads the microstructure is usually cellular dendritic. If measured
spacings are reported on the secondary dendrite arm spacing curve,as in
Figure 4,the local cooling rate within the bead is found to vary between
1.5 and 6.5×10^5K/s. Had the cellular dendrite spacings been reported on
the primary spacings curve,as often and erroneously done,the deduced
local cooling rates would have been at least one order of magnitude
higher.

Dendritic Microstructure of a Mg-Zn-Zr Alloy

The microstructure of bulk solidified Mg-5.3wt%Zn-0.6wt%Zr alloy is
nondendritic,consisting of fine grains of Mg-rich α-phase within which
the distribution of zinc is spherical (4,5). The intergranular spaces are
occupied by a nonequilibrium Mg_7Zn_3 intermetallic phase. Inoculation of
the melt with zirconium causes the formation of $ZrZn_2$ intermetallic particles on which the α-phase nucleates heterogeneously,most likely through
a peritectic reaction mechanism (6). In splat-cooled foils the microstructure remains nondendritic with very fine grains. However laser or electron beam beads solidify with a normal dendritic microstructure,Figure 5,
consisting of hexagonal dendrites of α-phase surrounded by nonequilibrium
Mg_7Zn_3. The formation of this microstructure may be attributed to the
lower undercooling required for epitaxial growth on the unmelted alloy,

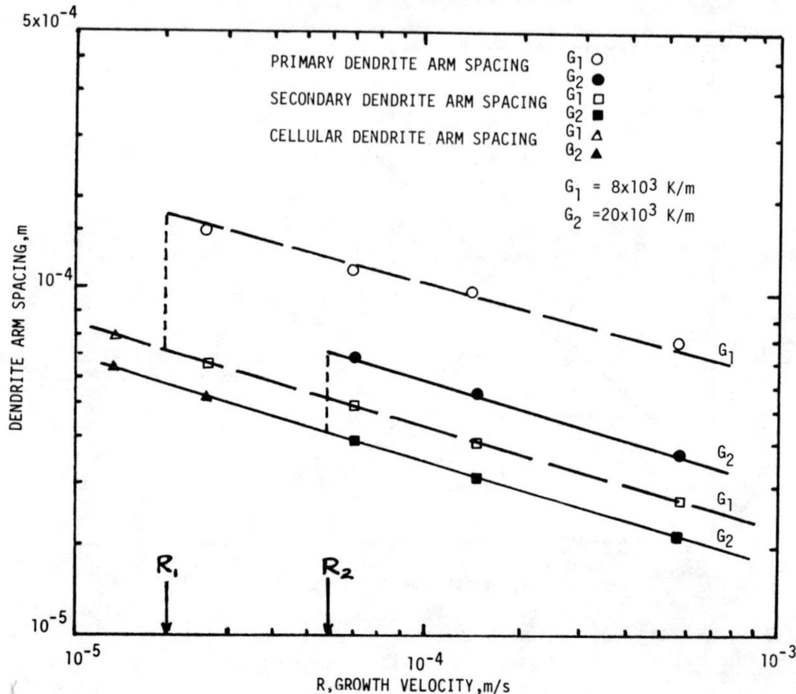

Fig. 3- Dendrite arm spacing versus growth velocity and thermal gradient. Ni-18.0at%Al-4.1at%Cr dendritic monocrystals.

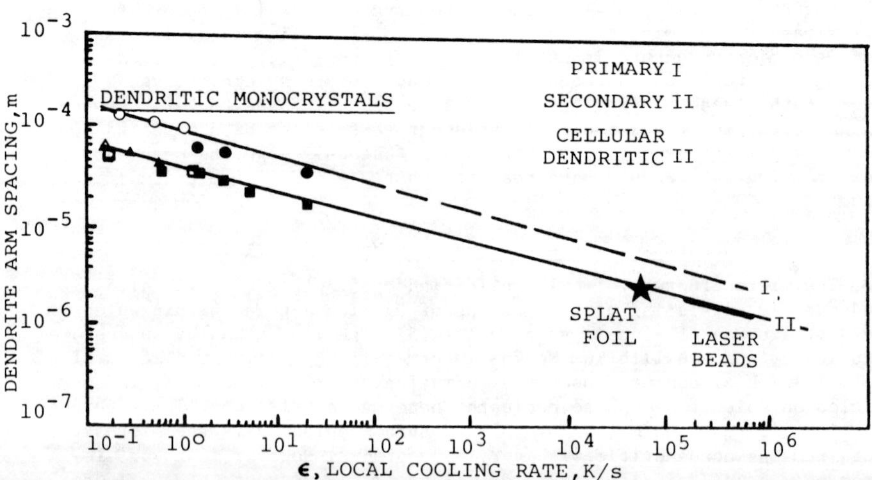

Fig. 4- Dendrite arm spacing versus local cooling rate. Ni-18.0at%Al-4.1at%Cr dendritic monocrystals,splat-cooled foil and laser beads.

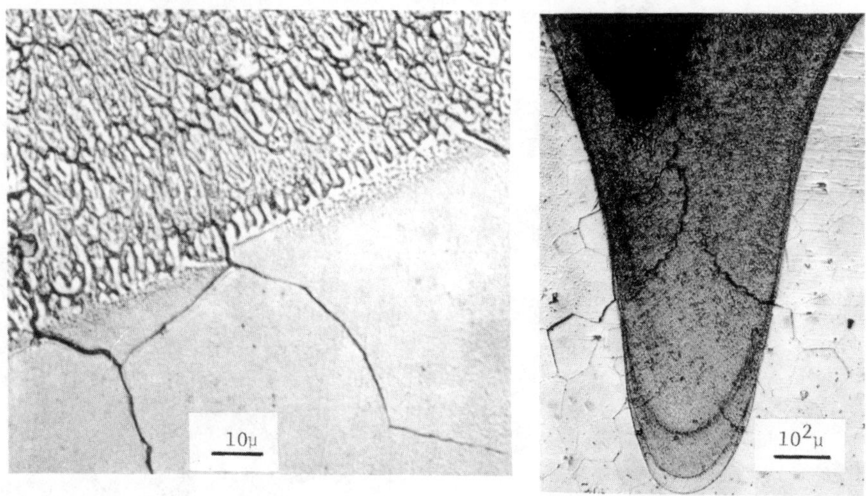

Fig. 5 - Photomicrographs of a laser bead deposited on a Mg-5.3wt%Zn-0.6 wt%Zr plate at 0.10m/s.

rather than for heterogeneous nucleation of the α-phase on the $ZrZn_2$ particles.

Dendritic Microsegregation

Dendritic microsegregation of a two-phase alloy, such as Al-4.5wt%Cu, is best expressed by the volume fraction of interdendritic nonequilibrium eutectic or θ-Al_2Cu secondary phase which results from divorce of the eutectic. In electron beam beads deposited at 0.05, 0.10 and 0.25 m/s the volume fraction of θ phase, f_θ, was measured as described previously (1,7), using X-ray diffraction analysis. The microstructure in beads deposited at 0.05 and 0.10m/s was coarse enough, allowing determination of f_θ by quantitative metallography. The average volume fraction, f_θ, was found to be 0.04 in the base metal, and 0.043, 0.045 and 0.055, respectively in the three beads corresponding to travel speeds of 0.05, 0.10 and 0.25m/s. Figure 6 exhibits the microstructure of a cross-section of the bead deposited at 0.15m/s. Cellular dendrite spacings in the three beads were: $1.3x10^{-7}$, $1.1x10^{-7}$ and $0.8x10^{-7}$ m, respectively, corresponding to local cooling rates of: $1x10^5$, $2x10^5$ and $4x10^5$K/s (8). For this range of cooling rates segregation increases with increasing local cooling rate, presumably because then the time available for back-diffusion of copper in the solid during solididication decreases (9). At these high cooling rates the corresponding growth rate is low and the thermal gradient high.

In high energy beam linear remelting and resolidification growth rate, R, is simply related to travel speed, V, Figure 7. Assume two consecutive positions of the trailing edge of the puddle, at time t and t+1 s, hence V m apart. During this unit time interval the solid has grown by R m in a direction normal to the solid-liquid interface. It can be seen

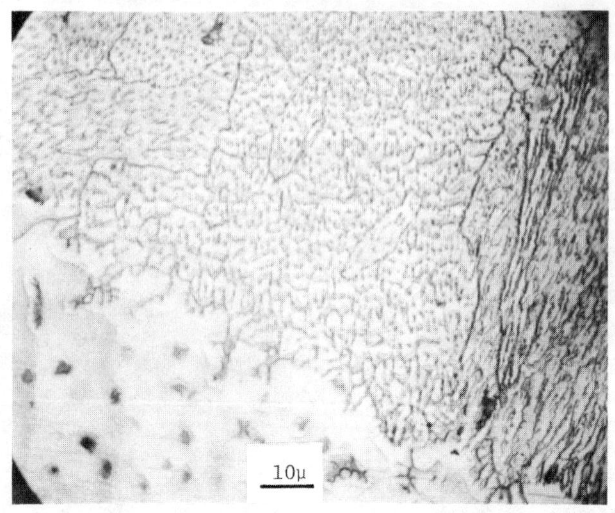

Fig. 6- Photomicrograph of a transverse section of an electron beam
bead deposited on an Al-4.5wt%Cu plate at 0.15m/s.

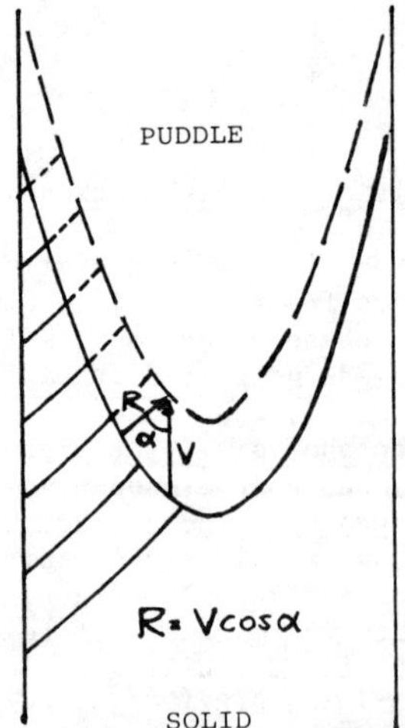

Fig. 7- Relationship between beam travel speed,V and growth rate
of the solid,R .

that approximately R=Vcosα, where the angle α depends on travel speed.
Thus, as V increases the width of the bead decreases and α increases. For
usual travel speeds R is a large fraction of V. At higher growth rates
solute trapping in the solid is expected to operate yielding a smaller
volume fraction f_θ . Work is now in progress to determine the critical
travel speed or growth rate at which the nonequilibrium interdendritic θ-
phase disappears completely. In a preliminary experiment electron beam
and laser beam spots were deposited. The transverse section of such a
spot is illustrated in Figure 8. In this specimen no interdendritic θ-
phase has been detected, instead a very fine and uniform precipitation
occurred in the solid. A rough calculation(10) has shown that the average
growth rate,R,within the spot is in excess of 3m/s. At this stage of
the investigation it is speculated that this rate is higher than the critical growth rate at which θ disappears and that a supersaturated α –phase
solidifies,within which solid state precipitation occurs during cooling.

Microhardness of Al-4.5%Cu Laser Beads and Spots

Microhardness was measured in the base metal,the laser beads both in
the as-solidified and the solutionized-and-aged conditions,as well as in
laser spots in the as-solidified condition. The average microhardness
measured in the base metal was 93 VHN. In the laser bead deposited at
0.10m/s the microhardness varied between 99 and 85 VHN in the as-cast
condition with an average of 94 VHN. Following a solution treatment at
798K for 1 h with subsequent aging at 428K for 3 h the measured average
microhardness was 128 VHN. In the as-solidified laser spot the microhardness was very uniform and averaged 123 VHN.

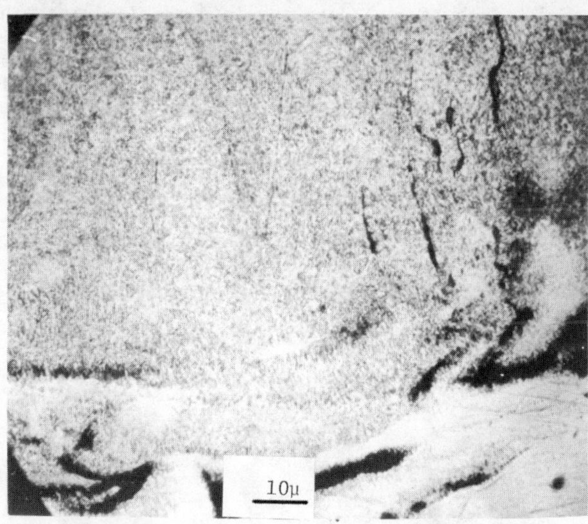

Fig. 8- Photomicrograph of a transverse section of an electron beam
spot deposited on an Al-4.5wt%Cu plate.

<u>Corrosion Behavior of Rapidly Solidified Mg-6.1%Zn Alloy</u>

Preliminary results are given herein of the study of the corrosion
behavior of rapidly solidified Mg-6.1wt%Zn alloy immersed in a 3.5wt%NaCl
aqueous solution. The work was first focused on electron beam beads deposited on an as-cast plate of the same alloy,Figure 9. The beads contained
a large amount of gas porosity which may be attributed to the high vapor
pressure of Mg and the prevailing vacuum conditions during melting.
Corrosion of this rapidly solidified material was so rapid that the beads
completely dissolved in about 24 h,Figure 9. In a similar test conducted
on laser beads which were processed under inert atmosphere and exhibited
significantly less porosity the dissolution time increased appreciably.
To further investigate this rapid dissolution specimens were sectioned at
an early stage of corrosion and metallographically examined. It became
evident that corrosion occurs within the Mg-rich α-phase at its interface
with the interdendritic Mg_7Zn_3 phase. Galvanic corrosion operates between
the electropositive α-phase and the electronegative Mg_7Zn_3 phase. In the
rapidly solidified beads the amount of interdendritic Mg_7Zn_3 is higher
than in the base metal and it is very finely distributed. Thus,the effect
of galvanic corrosion is maximized. These qualitative findings were
further explored by measuring cumulative weight loss versus time in rapidly solidified bulk material ,such as splat-cooled foils. Figure 10 illustrates weight loss curves for as-cast bulk material,as-splat-cooled foil,
solutionized and homogenized foil, and surface polished solutionized and
homogenized foil. Corrosion of the as-splat-cooled foil is slightly faster
than that of the bulk material,because of the larger amount and the finer
dispersion of interdendritic Mg_7Zn_3 present. Following dissolution of
this phase there is a substantial decrease in corrosion rate. The corrosion
behavior is improved by finely polishing the foil,thus eliminating potential sites for crevice corrosion which is extensively observed at the
rough surface of as-processed splat foils. It is expected that in specimens
processed at a growth rate high enough that the interdendritic phase vanishes corrosion behavior will be further improved.

Fig. 9- SEM micrograph of a Mg-6.1wt%Zn alloy specimen immersed in a
3.5wt%NaCl aqueous solution for 24 h. Complete dissolution of
electron beam beads.

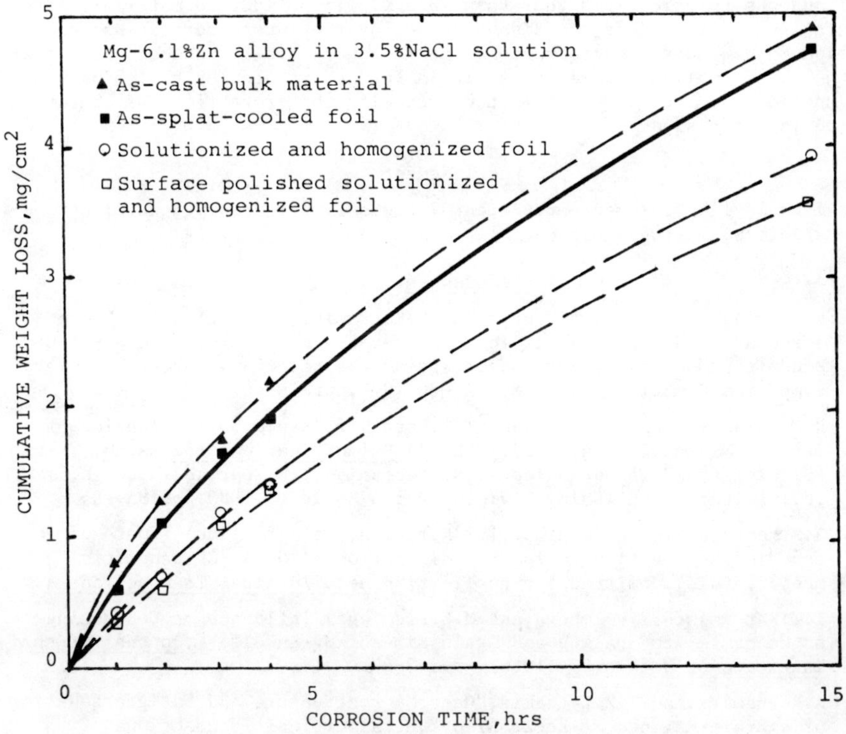

Fig. 10- Cumulative weight loss versus corrosion time. Mg-6.1wt%Zn
         alloy immersed in a 3.5wt%NaCl aqueous solution.Bulk cast
         material and rapidly solidified material in various conditions.

Conclusions

1) In extrapolating dendrite arm spacing versus local cooling rate curves
for the purpose of determining cooling rates within a laser or electron
beam bead cellular dendrite spacings should be reported on the secondary,
not the primary dendrite arm spacing curve.

2) Laser beads or electron beam beads deposited on grain-refined Mg-5.3wt%
Zn-0.6wt%Zr alloy,which is nondendritic, are finely dendritic.

3) Growth rate in high energy beam deposited beads is a large fraction
of the beam travel speed.

4) In laser and electron beam beads deposited on Al-4.5%Cu the interdendritic volume fraction of nonequilibrium θ-phase is higher than in slowly
solidified alloys. In high energy beam deposited spots on the same alloy
the interdendritic θ-phase vanishes.

5) The microhardness of a laser beam spot is roughly equal to that of a
solutionized and aged bead.

6) The corrosion rate of a rapidly solidified Mg-Zn alloy immersed in a

NaCl aqueous solution is higher than in a slowly solidified material. After a solution treatment and the disappearance of the interdendritic nonequilibrium Mg_7Zn_3 phase,the corrosion rate is substantially reduced below that of the slowly solidified material. It is further reduced by polishing the specimen surface very finely,thus reducing the probability of operation of crevice corrosion.

Acknowledgment

I would like to thank Dr. Brian G. Lewis for processing the electron beam specimens, using equipment acquired by Dr. P. R. Strutt.

References

1) R.Kadalbal,J.J.Montoya-Cruz,and T.Z.Kattamis,"Solidification Microstructure and Microsegregation of Laser-Glazed Ni-Al-Ta and Ni-Al-Cr Dendritic Monocrystals",Proceedings of the Materials Research Society Symposium,Cambridge,Mass.,Nov.27-30,1979,pp.740-46.

2) R.Kadalbal,J.Montoya-Cruz,and T.Z.Kattamis,"Rapid Solidification of Ternary Nickel-Base Alloys",in Rapid Solidification Processing, Principles and Technologies II,R. Mehrabian,B.H.Kear,and M.Cohen, ed.;Claitor's Publishing Division,Baton Rouge,La,1980,pp.195-205.

3) J.J.Montoya-Cruz,R.Kadalbal,T.Z.Kattamis,and A.F.Giamei,"Coarsening and Microsegregation During Solidification of Ni-Al-Cr Dendritic Monocrystals",submitted for publication,Metallurgical Transactions A .

4) T.Z.Kattamis,U.T.Holmberg,and M.C.Flemings,"Influence of Coarsening on Dendrite Arm Spacing and Grain Size of Mg-Zn Alloy",Journal of the Institute of Metals,95 (1966),pp.343-47.

5) A.K.Bhambri,and T.Z.Kattamis,"Cast Microstructure and Fatigue Behavior of a Grain-Refined Mg-Zn-Zr Alloy",Metallurgical Transactions, 2 (1971),pp.1869-74.

6) E.F.Emley,Principles of Magnesium Technology,Pergamon Press,1966.

7) T.Z.Kattamis,Y.V.Murty,and J.A.Reffner,"Microstructure and Segregation in Splat-Cooled Al-Cu Alloy",Journal of Crystal Growth, 19, (1973),p.237.

8) H.S.Cairnie, Quantitative Determination of Nonequilibrium θ-Phase in Rapidly Solidified Al-Cu,M.S.Thesis,Department of Materials Science and Engineering,M.I.T.,December 1980.

9) H.D.Brody,and M.C.Flemings,"Solute Redistribution in Dendritic Solidification",Transactions TMS-AIME, 236 (1966),pp.615-23.

10) M.C.Flemings,Solidification Processing ,p.18; McGraw-Hill,Inc., New York,N.Y.,1974.

SOLIDIFICATION OF CONTINUOUS LASER-MELTED TRAILS

S. M. Copley, D. G. Beck, O. Esquivel, and M. Bass
Department of Materials Science
University of Southern California
Los Angeles, California 90007

By moving a metallic substrate under the beam of a continuous wave
carbon dioxide laser, it is possible to melt and solidify a continuous melt
trail. The solidification of such trails often involves marked undercooling
of the solid-liquid interface and considerable convective flow. The microstructure and topography of such trails in Udimet 700 and Ag-Cu alloys have
been examined by a variety of metallographic and analytical techniques.
Experimental results concerning shape, surface relief, epitaxy and metastable
phase formation in such trails are presented.

Introduction

The carbon dioxide laser is a source of intense radiant energy, which can be employed to melt the surface of a metal or alloy. By moving a specimen under a laser beam it is possible to produce a shallow, continuous melt trail that is quenched by conduction of the absorbed heat to the surrounding substrate. This paper describes the results of an investigation of the constitution, microstructure and topography of such continuous laser-melted trails. It is possible to understand these features in terms of mass transport, heat flow and the nucleation and growth of new phases and grains.

Experimental

The relationship of the laser beam, alloy specimen and laser melted trail in our experiment is illustrated in Fig. 1. The laser beam is focussed

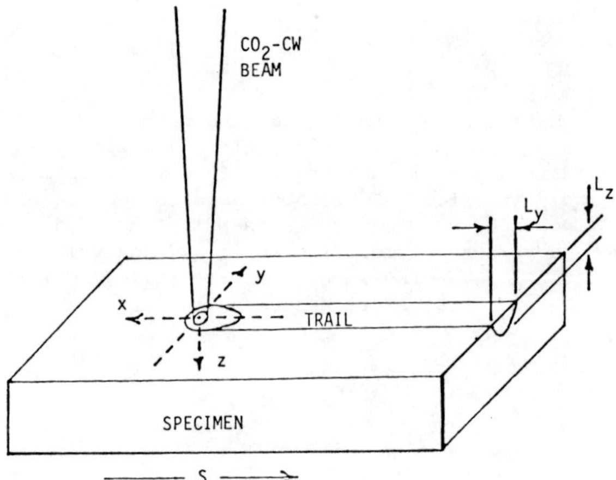

Fig. 1: Experimental arrangement for laser melting.

to a radius R and is stationary; the specimen moves at a constant speed S in the negative x direction. A fraction α of the incident beam power P is absorbed melting a trail of width L_y and depth L_z. By suitably adjusting P, R and S, it is possible to melt a trail with small L_y and L_z. In this case, heat is extracted very rapidly from the melt pool and considerable undercooling of the melt may occur near the melt-solid interface, which may lead to the nucleation and growth of new grains or phases.

In order to ensure the reproducibility of a laser melting experiment it is necessary to measure the beam power and the spatial distribution of intensity in the focal plane. The beam power is normally measured with commercially available calorimetric units. The spatial distribution of intensity is measured in our laboratory by positioning a pyroelectric detector behind a pin hole that scans the laser beam in the focal plane. The condition of the surface should also be characterized because this influences the absorption of the laser beam.

Results and Discussion

Udimet 700

We began our investigation by laser melting single crystals of Udimet 700[†]. This alloy was chosen for study because: (i) it has a cored dendritic structure that can readily be revealed by etching; (ii) the equilibrium gamma-prime precipitates have a cuboidal morphology, which gives a visual indication of the crystallographic orientation of the unmelted substrate as well as the extent of the melted and solidified trail; and (iii) its microstructure in a variety of heat treated conditions is well established (1). An important objective of our initial experiments was to determine if solidification was epitaxial or involved the nucleation of new grains.

Examination of a trail lying along < 100 > on a {001} surface produced by a 200 W beam at a specimen speed of 0.5 cm s^{-1} suggested that solidification was entirely epitaxial occurring by the growth of cellular dendrites along the < 100 > most nearly parallel to the direction of maximum heat flow, see Fig. 2 (2).

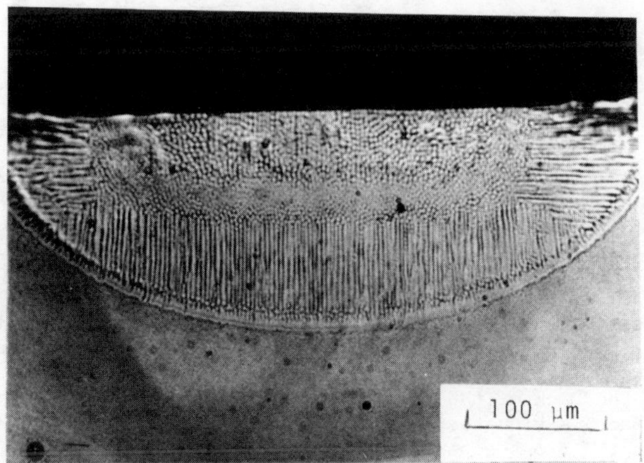

Fig. 2: Transverse section of laser melted
region in Udimet 700.

A more detailed investigation showed, however, that near the center line and top surface of the melt trail misoriented grains were often present, which extended their boundaries through growth of cellular dendrites following the laser beam (3). We studied the origin of such misoriented grains and concluded they nucleated at "ripples", undulations in surface elevation, observed in melt trails (4).

A mechanism for ripple formation based on surface tension driven (Marangoni-type) convection was proposed, which also explained the nucleation of the misoriented grains in terms of an overflowing of the melt at the back edge of the melt pool rather than undercooling of the melt near the melt-solid interface (5). Such overflowing causes the melt to contact the oxidized, recently solidified edge resulting in heterogeneous nucleation of the misoriented grains.

[†]Udimet 700:	Ni	Co	Cr	Ti	Al	Mo	C	B
	Bal	18.5	15.0	3.5	4.3	5.3	0.08	0.030

13

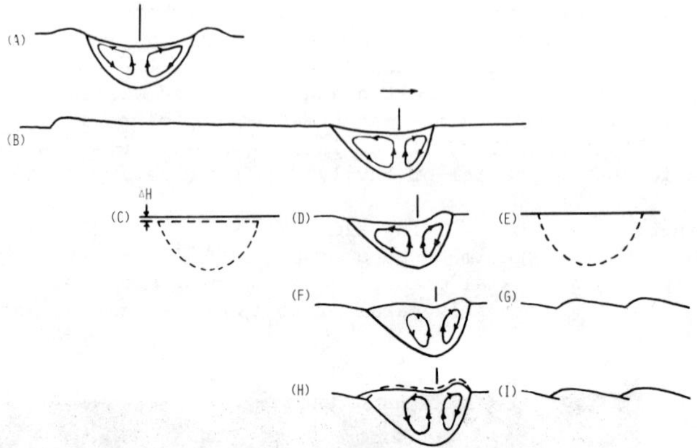

Fig. 3: Mechanism of ripple formation: (a) velocity equals zero;
(b) and (d) velocity not equal to zero; projected area of
back side of pool (c) less than that at front edge of pool
(e); (f) and (g) ripples without back flow; (h) and (i)
ripples with back flow.

Figure 3 illustrates the proposed mechanism. In the case V = 0, convection currents in the melt cause an overflowing of the melt pool to occur;
the shape of the melt pool in the vicinity of the impingement point of the
laser is concave if viewed looking down into the pool. In the case V ≠ 0,
a hill marks the back of the initial melt pool. As the SV interface replaces
the LV interface, a situation develops where the back edge of the melt pool
is lower than the front edge. In this case, the liquid added to the pool due
to the melting at the front edge exceeds that subtracted from the pool due
to solidification at the back edge. Nevertheless, the LV interface at the
back edge of the pool continues to grow in a downwards direction with respect
to the horizontal due to the convection.

As solidification continues, a tipping upward of the LV interface directly in front of the back edge of the melt pool must occur. At this point,
either: (i) the SV interface will continue to replace the LV interface,
reducing and eventually eliminating the imbalance of liquid flowing in to and
out of the melt pool; or (ii) the melt will flow back over the back surface.
The condition for the flow back to occur is $A(\gamma_{SV} - \gamma_{SL} - \gamma_{LV}) < 0.5\ gm\Delta h^2$
where A is the SV interface area covered by the back flow and γ_{SV}, γ_{SL} and
γ_{LV} are the specific energies of the SV, SL and LV interfaces, respectively,
g is the gravitational constant, m is the mass of the liquid and Δh is the
lowering of the center of mass of the liquid meniscus due to the back flow.
This equation indicates that back flow occurs if the accompanying increase
in interface energy is less than the decrease in gravitational potential
energy of the liquid.

The essential requirement for rippling to occur is that the LV interface
be tipped from horizontal. This may occur for reasons other than convection
such as gradients in pressure over the melt due to evaporation. Also, non-
uniform melting of a coarse dendritic structure, bubble formations and

mechanical or power fluctuations may cause ripples. A movie showing convection in melt pools of camphene-tricyclene, a transparent analog system, was made to demonstrate the proposed mechanism (6).

Anthony and Cline had previously suggested the occurrence of surface tension driven convection in laser melting and had explained a feature of surface topography in laser melted trails, which they also called "ripples", in terms of this phenomenon (7). The features explained by Anthony and Cline were the hump marking the back edge of the initial melt pool, Fig. 3(B) and the centerline ridge of the trail illustrated by the profilometer trace perpendicular to the trail, Fig. 4. They are different features of laser melted trails than those we have called ripples although both can be explained on the basis of convection.

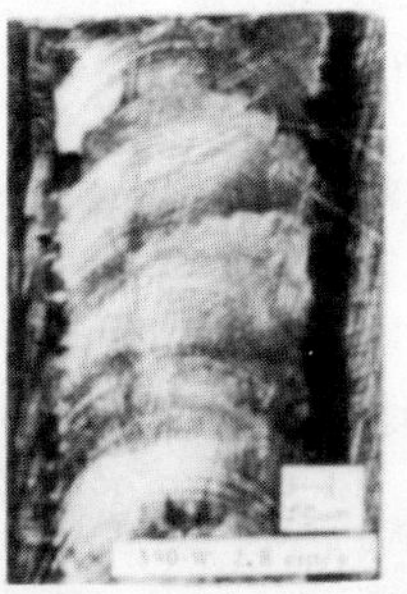

Fig. 4: Fine and coarse ripples in Udimet 700.

Recently, we presented preliminary results of an investigation of the size, shape and surface topography of laser melted trails in Udimet 700 obtained for various beam powers and scan velocities (8). Two types of ripples, coarse and fine, were observed as shown in Fig. 4. The fine ripples were attributed to the quasi-steady-state convective flow model previously proposed (5). It seems more likely that coarse ripples are formed by the mechanism shown in Fig. 5. Surface tension driven flow from under the beam decreases the elevation of the melt vapor interface as the beam moves forward. Beam heating occurs primarily along bc; little occurs along ab due to the glancing angle of the beam. Thus the velocity of ab, the part of the interface along which melting is occurring that lies near the surface, becomes progressively slower until it is overtaken by the beam. As the beam moves ahead of ab, it can couple to the unmelted surface beyond point (a), which is perpendicular to its path. This produces an increase in the flow of liquid to the back edge of the melt pool forming another coarse ripple.

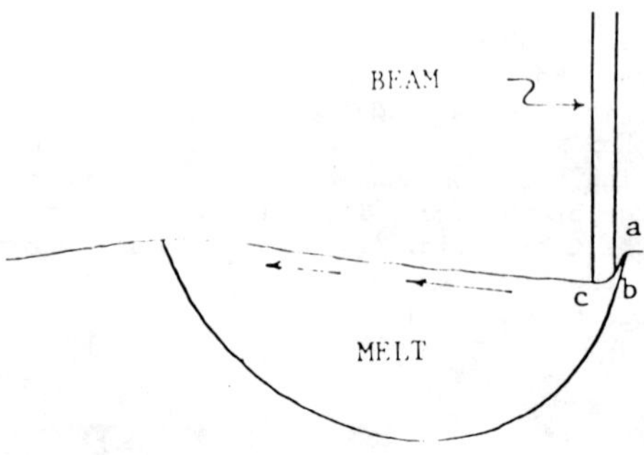

Fig. 5: Longitudinal section along centerline
of melt pool showing surface tension
driven convection.

Silver Copper Alloys

In addition to our research on melt trails in Udimet 700, we have been
investigating the possibility of using laser melting to produce a metastable
crystalline phase in Ag-Cu alloys. The Ag-Cu system is a simple binary
eutectic. Over a wide range of compositions cooling from the melt yields
an equilibrium mixture of Cu-rich and Ag-rich face centered cubic solid solutions. It has been demonstrated that a metastable extended solid solution
with the same composition as the melt can be produced by "splat-quenching"
(9). This result has been duplicated by self-substrate quenching small
volumes melted with a pulsed Nd:glass laser beam (10). Our objective is to
utilize undercooling of the melt near the melt solid interface to produce
the metastable extended solution in a continuous laser melted trail.

Our initial effort to produce the metastable extended solid solution
in Ag-Cu alloys was unsuccessful (11). Apparently, the 0.7 cm s^{-1} scan velocity used did not provide sufficient undercooling. Very fine eutectic plates
were observed instead. Higher scan velocities were not investigated because
of laser power limitations. At higher scan velocities, the trails produced
with the available power were too small for examination.

With the installation of a 1400W CW-CO$_2$ laser in November 1978, it became possible to produce trails in Ag-Cu of sufficient size for examination
at scan velocities greater than 0.7 cm s^{-1}. Positive identification of the
metastable extended solid solution was reported in Ag-25 at pct Cu, Ag-50 at
pct Cu and Ag-75 at pct Cu alloys in trails produced at a beam scan velocity
of 10 cm s^{-1} (12). A comparison of lattice parameters of the metastable
extended solid solutions with those obtained by Dewez through use of the splat
cooling technique is given in Fig. 6.

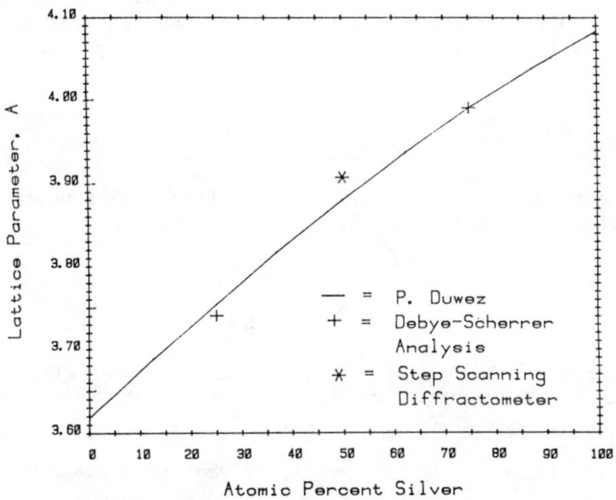

Fig. 6: Lattice parameters of metastable
extended solid solutions.

An important feature in the melt trails of the Ag-Cu alloys were bands
made visible by etching that appear to mark locations of the melt-solid
interface during solidification. Heavily etched bands in a Cu-50 at pct Ag
alloy are shown in Fig. 7.

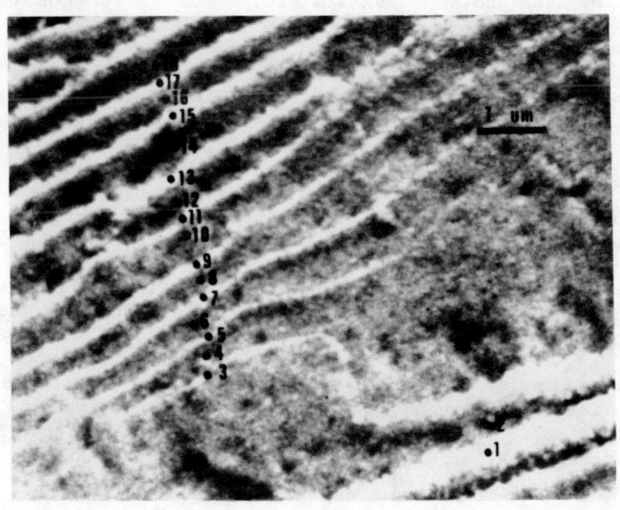

Fig. 7: Heavily etched bands in transverse
section of melt trail in Cu 50 at. pct. Ag.

17

A more detailed investigation has confirmed the presence of the metastable extended solid solution in laser melted trails in Ag-Cu alloys and has demonstrated that the previously observed bands correspond to approximately 1 pct variations in composition (13). The origin of these bands is not understood.

Acknowledgement

The authors are pleased to acknowledge financial support of this research by the National Science Foundation, Metallurgy Program, Grant No. DMR 78-07532.

References

1. E. H. Van Der Molen, J. M. Oblak and O. H. Kriege, "Control of γ' Particle Size and Volume Fraction in the High Temperature Superalloy Udimet 700", Met. Trans. 2 (6) (1971) pp. 1627-33.

2. S. L. Narasimhan, S. M. Copley, E. W. VanStryland and M. Bass, "Epitaxial Solidification on Laser Melted Superalloys", IEEE J. Quantum Electron, QE-13 (9) (1977) p. 20.

3. S. L. Narasimhan, S. M. Copley, E. W. Van Stryland and M. Bass, "Solidification of a Laser Melted Nickel-Base Superalloy", Met. Trans. 10A (5) (1979) pp. 654-55.

4. D. Beck, S. M. Copley, M. Bass and E. W. Van Stryland, "Solidification of Superalloy Melts Produced by Laser Irradiation", pp. 104-11 in Rapid Solidification Processing - Principles and Technologies, R. Mehrabian, B. H. Kear and M. Cohen, Editors, Claitor's Publishing Co., Baton Rouge, LA, 1978.

5. S. M. Copley, D. Beck, O. Esquivel and M. Bass, "Laser Melt Quenching and Alloying", pp. 161-72 in Laser-Solid Interactions and Laser Processing - 1978, S. D. Ferris, H. J. Leamy and J. M. Poate, Editors, Conf. Proceedings No. 50, American Institute of Physics, New York, NY, 1979.

6. O. Esquivel, S. M. Copley and M. Bass, "Convective Processes in Continuously Melted Zones", to borrow copy of film write S. M. Copley, Dept. of Materials Science, USC, Los Angeles, CA 90007.

7. T. R. Anthony and H. E. Cline, "Surface Rippling Induced by Surface-Tension Gradients during Laser Surface Melting and Alloying", J. Appl. Phys., 48 (9) (1977) pp. 3888-94.

8. O. Esquivel, J. Mazumder, M. Bass and S. M. Copley, "Shape and Surface Relief of Continuous Laser-Melted Trails in Udimet 700", pp. 180-88 in Rapid Solidification Processing, Principles and Technologies, II, R. Mehrabian, B. H. Kear and M. Cohen, Eds., Claitor's Publishing Div. Baton Rouge, LA, 1980.

9. P. Duwez, "Structure and Properties of Alloys Rapidly Quenched from the Liquid State", Trans. ASM, 60 (4) (1967) pp. 607-33.

10. W. A. Elliott, F. P. Gagliano and G. Krauss, "Metastable Phases Produced by Laser Melt Quenching", Met. Trans. 4 (9) (1973) pp. 2031-37.

11. S. M. Copley, M. Bass, E. W. Van Stryland, D. G. Beck and O. Esquivel, "Microstructures of Surface Alloyed Ag-Cu Films Produced by Laser Melt Quenching", pp. 147-50 in Rapidly Quenched Metals III, vol. 1, Brian Cantor, editor, The Metals Society, London, 1978.

12. D. G. Beck, S. M. Copley and M. Bass, "Constitution and Microstructure of Ag-Cu Alloys Produced by Continuous Laser Melt Quenching", pp. 734-39 in Laser and Electron Beam Processing of Materials, C. W. White and P. S. Peercy, Eds., Academic Press, New York (1980).

13. D. G. Beck, S. M. Copley and M. Bass, "The Microstructure of Metastable Phases in Ag-Cu Alloys Generated by Continuous Laser Melt Quenching", Met. Trans., 12A (9) (1981) pp. 0000-00.

THE EFFECTS OF LASER SURFACE MELTING

ON COPPER ALLOYS

C. W. Draper

Laser Studies Group
Western Electric Engineering Research Center
P. O. Box 900
Princeton, NJ   08540

The laser surface melting and associated rapid self quenching of a
wide variety of copper alloys has been studied. The quenched surfaces
have been examined metallographically with particular interest in alloy
systems demonstrating high temperature phase metastability. Results
showing enhanced surface sensitive behavior for laser quenched copper
alloys in both erosion and corrosion testing are summarized. Laser
irradiated surfaces have been characterized with optical and scanning
electron microscopy, energy dispersive x-ray analysis and Auger electron
spectroscopy with depth profiling.

Introduction

We have undertaken a generic study of the effects of laser surface melting and rapid self quenching on a wide variety of copper alloys. Our interests include understanding the coupling phenomena, characterizing the surface topography and melt/quenched layer microstructure and determining the effects of laser quenching on the surface sensitive behavior of copper alloys in corrosion and erosion environments.

Although we have affected laser surface melting (LSM) with Q-switched Nd-YAG (1), both fundamental and frequency doubled, the majority of the work has been done using relative motion cw CO_2 laser scanning. This laser processing method is widely used and will not be described here. Details may be found in reference 2.

Table I gives the UNS designation and chemical composition of the alloys studies. Samples are cut in 1.9 cm squares, polished to a random grit finish, and finally ultrasonically cleaned in detergent, purified water and spec grade methanol.

Coupling 10.6 µm laser radiation to copper alloy surfaces

It was not obvious that it would be relatively easy to couple cw CO_2 radiation into copper alloys. After all copper mirrors are frequently used as reflectors for 10.6 µm radiation in high power laser applications. We have observed a very large variation in this coupling factor. For the purposes of our discussion the coupling is simply a measure of the ease (generally equated to laser power) with which laser surface melting is accomplished.

The pure coppers C10100 through C18200 do not laser surface melt. At a high, critical laser power, surface vaporization will occur, but controllable melting cannot be achieved. The zinc bearing alloys C23000, C26000, C44300 and C74500 will laser surface melt, but surface vaporization of zinc makes uniform processing difficult. The boiling point of elemental zinc lies below the melting point of copper and the other alloy constituents. In fact zinc can be made to boil off without any surface melting whatsoever. For this reason, the zinc bearing alloys have a very narrow range over which LSM processing is quasi-controllable. The remaining alloys all have a wide power range over which LSM processing can be done without surface vaporization inhibiting the quality of the quenched surface.

TABLE I

UNS Designation and Chemical Analysis of Copper Alloys

UNS Designation	Common Name	Elemental Composition (Wt. %)
C10100+	OFHC	Cu-99.9
C11400+	Oxygen Free Copper with Ag	Cu-99.92, 0-0.04, Ag-0.03
C12200+	Phosphorus Deoxidized Copper	Cu-99.90, P-0.02
C18200	Chromium Copper	Cu-99.21, Cr-0.68, Fe-0.07
C23000	Red Brass	Cu-85.38, Zn*-14.61,
C26000	Cartridge Brass	Cu-70.54, Zn-29.46
C44300	Inihibited Admiralty	Cu-71.81, Zn*-27.04, Sn-1.08 As-0.065
C51000	Phosphor Bronze A	Cu-95.43, Sn-4.42, Zn-0.063, P-0.060
C52100	Phosphor Bronze C	Cu-91.95, Sn-7.71, P-0.13, Zn-0.10
C52400	Phosphor Bronze D	Cu-89.86, Sn-9.71, P-0.23, Zn-0.20
C60800	Aluminum Bronze	Cu-94.52, Al*-5.46, Zn-0.10, Fe-0.02
C61400	Aluminum Bronze D	Cu-90.01, Al-6.30, Fe-3.68
C62400	Ampco 18	Cu-85.61, Al-10.43, Fe-3.64
C62500	Ampco 21	Cu-81.94, Al-12.54, Fe-4.09, Mn-1.10
C70600	Copper Nickel 10%	Cu-86.65, Ni-10.91, Fe-1.59, Zn-0.18
C71500	Copper Nickel 30%	Cu-68.78, Ni-29.93, Fe-0.48, Zn-0.40,
C72500+	Tin Modified Copper-Nickel	Cu-88.2, Ni-9.5, Sn-2.3
C74500	Nickel Silver 65-10	Cu-65.27, Zn*-24.61, Ni-9.89, Mn-0.23

+ Nominal compositions

* By difference

In order to achieve a relative ranking, a series of samples were meticulously mounted on the wheel so that their exposed surfaces were the same, $\pm$ 0.0076 cm, distance from the focusing lens. For a 6.35 cm focusing lens the optical depth of focus (for 5% change in focused spot radius) is approximately 0.1 cm. We are, therefore, confident that each surface has been exposed to <u>identical</u> incident power densities. Laser power was increased from 200 to 400 watts in increments of 50 watts, and then to 1200 watts in 100 watt steps.

We have compared this experimentally determined ranking with calculated thermal diffusivities, melt temperatures and experimentally measured normal spectral (10.6 µm) reflectances (3). Although the correlation is not absolute the general trends can best be explained on the basis of relative thermal diffusivities and <u>not</u> changes in the reflectance. Both the laser power required and thermal diffusivities change by an order of magnitude, while the melt temperature and reflectance change by only 20 and 34% respectively.

Surface Topography

The appearance of a LSM processed copper alloy is like that of an instantly frozen ocean. Hills and valleys (long wavelength roughness) due to rippling and melt stripe overlap, and ultrasmooth (short wavelength roughness) walls in between. The causes for these features have been described in detail elsewhere (4,5) and will not be repeated here. Figure 1, a composite of low and high magnification photomicrographs, clearly shows these differences relative to a 400 grit random finish on C72500. Table II summarizes the change in long wavelength roughness as a function of material velocity. Long wavelength roughness decreases with increasing processing speed. It has been shown in many published reports that the degree of long range roughness is also sensitive to cover gas, preirradiation surface finish and, of course, the thermophysical parameters of the alloy and its elemental constituents.

The important point to be made is that the surface remaining following the laser quench will, in general, be rougher than most standard metal finishes. From a surface wear application this will in many cases prove detrimental.

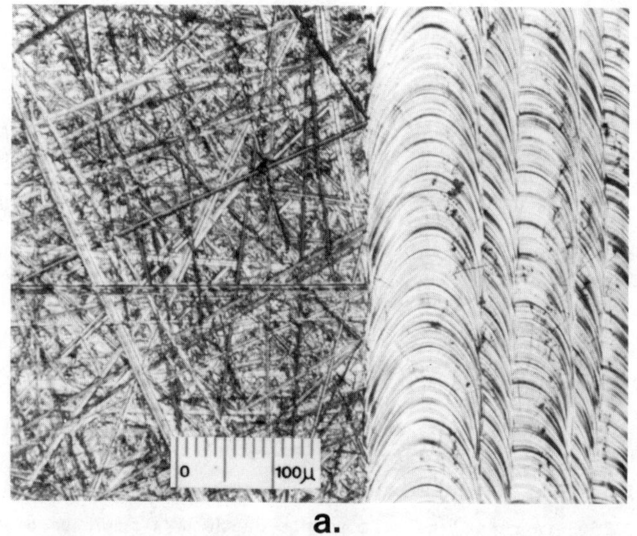

a.

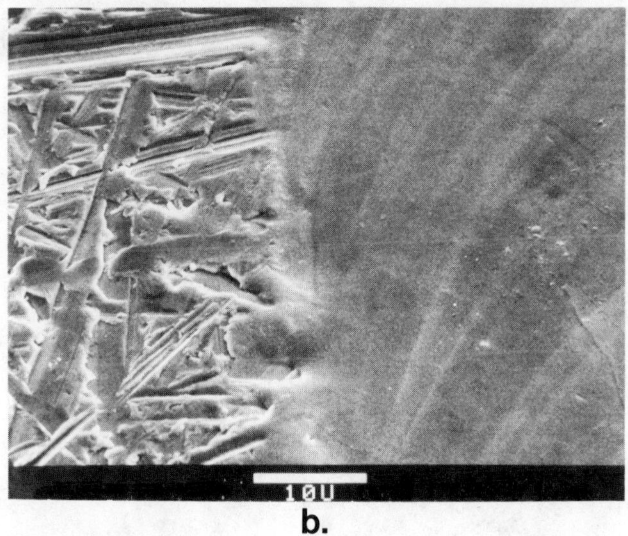

b.

Figure 1.

Laser surface melting of C72500, Sn modified Cu-Ni.
Preirradiation finish is 400 grit random. Processing parameters
were 800 W and 60 cm sec^{-1}. a. Optical micrograph - note
overlapped melt stripes and rippling within each stripe.
b. SEM micrograph of non-LSM/LSM border.

TABLE II

Mean Roughness Height for C51000, C52100, C60800 and C70600 as a
Function of Sample Transverse Velocity

		MEAN ROUGHNESS HEIGHT AT	
ALLOY	LASER POWER (Watts)	50.5 cm/sec	152 cm/sec
C51000	450	0.15 μm	0.12 μm
C52100	300	1.7	0.35
C60800	400	2.0	1.2
C70600	300	3.9	2.4

The Melt Region

Figures 2 and 3 illustrate some of the quantitative information obtained on C72500 in regard to melt puddle depth as a function of processing parameters: laser power and material velocity. Other copper alloys studied to this detail exhibit similar behavior.

In general, the following observations have been made:

(i)   The melt puddle depth, width and long wavelength roughness are sensitive functions of both the laser power and processing speed.

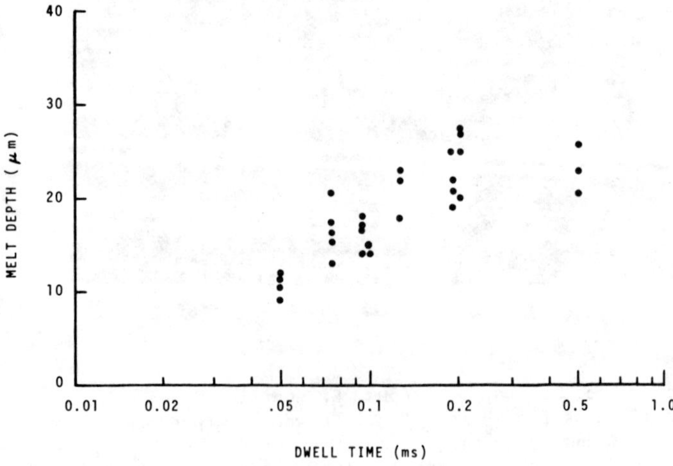

Figure 2.

Plot of melt depths (determined by metallographic cross sectioning) in C72500 alloy as a function of calculated dwell times.

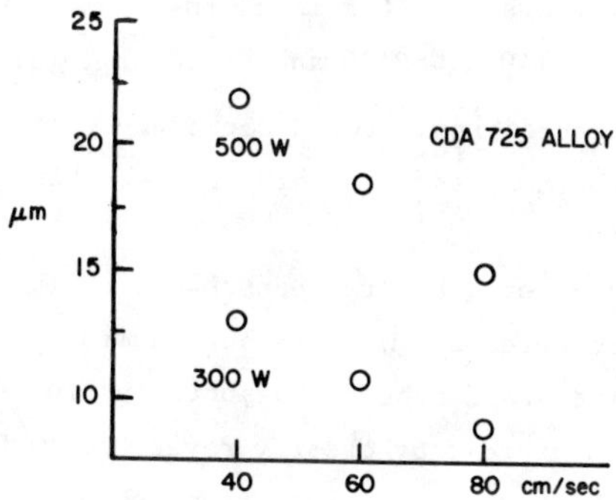

Figure 3.

Melt depth versus laser power and sample transverse velocity for C72500 alloy.

(ii)   The melt puddle width is not as sensitive as the melt puddle depth - this is a measure of the near one dimensionality of the heat flow.

(iii) Workpiece - focal point window tolerances (2) of approximately 50 μm are required for the wide range of laser powers, focusing optics, and material velocities used here.

Of the eighteen copper alloys listed in Table I, eight were chosen for metallographic cross sectioning. Those alloys are the phosphor bronzes A and C, tin modified copper-nickel, 10% copper nickel, and the four aluminum bronzes. Samples were polished to a 0.25 μm paste finish. The aluminum bronzes were etched with ammonium persulfate (ASTM No. 31) the rest with potassium dichromate (ASTM No. 29). The Cu-Sn and Cu-Ni-Sn phase diagrams indicate multiphase room temperature systems for the phosphor bronzes and tin modified copper nickel, but discernible second phases do not form in normal production and fabrication procedures. The aluminum bronze D and Ampco alloys are the only multiphase alloys studied.

As a result of microscopic inspection of the metallographic cross sections several observations can be made. These are summarized below.

(i)   The depth advance of the melt front is readily visible and sharply defined in C51000, C52100 and C72500, but not for C70600, C60800, C61400, C62400 and C62500. The Cu-Sn distribution coefficient is significantly different from unity in contrast to either Cu-Ni or Cu-Al, and melt front segregation (6) is undoubtedly responsible for these differences.

(ii)   The rapid quench has resulted in a single phase LSM region in
the multiphase C61400, C62400 and C62500 alloys.

(iii) A melt front-segregation related secondary phase has been
deposited at the melt edge in C51000. The phase has not been
identified.

(iv) Different degrees of grain reorientation, refinement and
elimination are evident. We have not examined the quenched
regions from the standpoint of measuring grain refinement.
Optical and SEM microscopy clearly reveal regrowth orientation
controlled by heat flow dynamics irrespective of original grain
orientation.

Detailed results on the microstructural changes found in the aluminum
bronzes have been (7, 8) and are to be (9, 10) published elsewhere.
Figure 4 is a composite of the Cu-Al-Fe phase diagram and of optical and
scanning Auger micrographs, which clearly shows that, following laser
quenching, the Fe-rich phase has been eliminated from the surface region
to a depth of about 10 μm.

In both the C62400 and C62500 alloys a single phase, presumably
metastable β, results from LSM processing of complex multiphase
structures. Inspection of the phase diagram in Figure 4 shows that for
Al compositions between about 10 and 15 weight percent a single phase β
region is first encountered when cooling from the melt. The quenching
rates used in these alloy systems was estimated to be on the order of
10^6 K sec^{-1}.

Effects on Surface Sensitive Behavior

The corrosion and erosion results discussed below represent a summary
of work published in detail elsewhere (7, 10, 11 .

Enhanced resistance to dealuminification has been shown (7) to result
from LSM processing of C61400 (aluminum bronze D). Samples irradiated
over 50% of their surface area were exposed at elevated temperatures, to
aerated 3% NaCl solutions (pH 4.0) for up to one day. The corrosion film
on the unirradiated region was noticeably thicker and continuous as
compared with the patchy film on the laser quenched surface.
Differential SEM/EDAX analysis was used to show that the films on the
unirradiated surface were measurably thicker and Al and Cl rich.

28

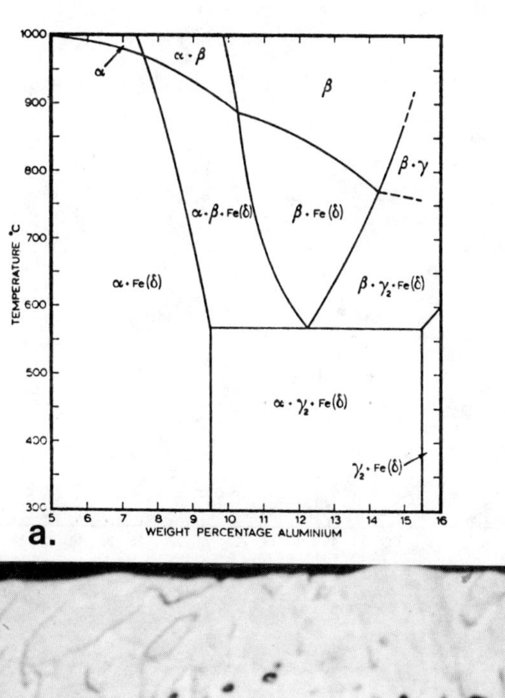

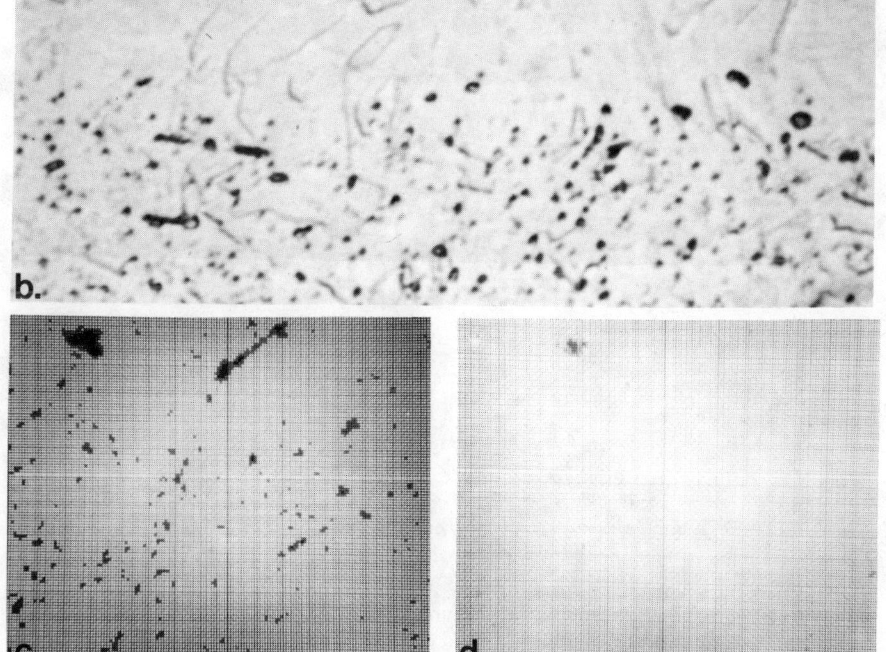

Figure 4.

Laser surface melting of C61400, Fe-aluminum bronze.
a. Vertical section of Cu-Al-Fe phase diagram at 3 wt. % Fe.
b. Optical micrograph of metallurgical cross section. The melt
depth is about 10 μm. c. SAM micrograph of preirradiation
surface. d. SAM micrograph of LSM treated surface. Note the
elimination of the Fe-rich δ phase in micrographs b. and d.

29

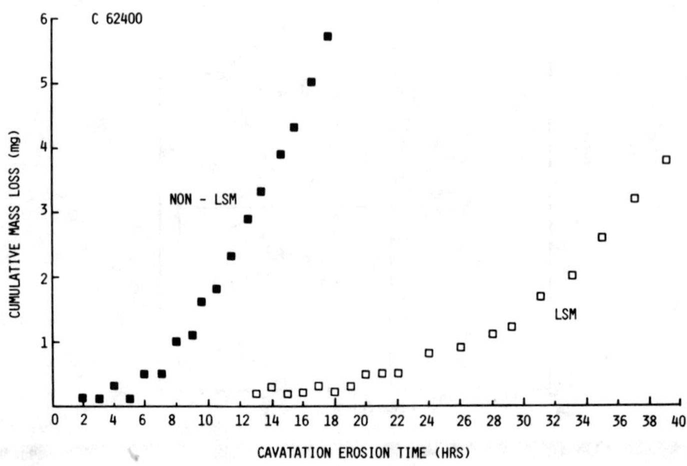

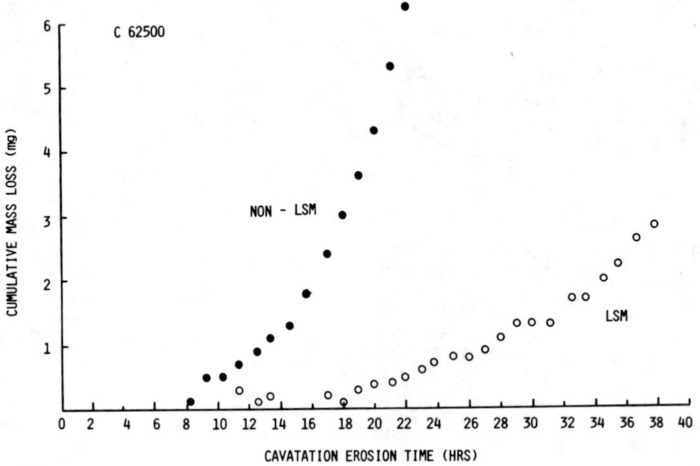

Figure 5:
Cumulative mass loss versus exposure time to cavitation erosion
for Fe-aluminum bronzes C62400 and C62500. Both Non-LSM and LSM
surfaces. Laser quenched surfaces exhibit increased incubation
periods and decreased erosion rates relative to the untreated
conventionally prepared microstructures.

Samples of C72500 prepared in the same manner as above, that is 50%
irradiated, were exposed to aggressive polutant vapor testing (11).
Separate sets of samples were exposed for 24 hours to 5 ppm H_2S and 80%
relative humidity, and 4 days to 1.8 ppm Cl_2 also at 80% relative
humidity. The thickness of the corrosion films was measured by Auger
depth profiling. The corrosion film thickness is reduced by 20% in the
H_2S environment for the LSM processed region and reduced by 75% in the
Cl_2 environment.

The results on the cavitation erosion (10)   of C62500 and C62400
Fe-aluminum bronzes, both laser quenched and unirradiated, are presented
in Figure 5. It is well known that the γ_2 phase (see phase diagram in
Figure 4) is deleterious to material performance in both corrosion and
erosion environments. The elimination of this phase in the laser
quenched surface markedly improves the material performance. The
incubation period (time to measureable mass loss) has been increased
significantly and the erosion rate has been reduced. This improvement is
seen <u>in spite of</u> the increased surface roughness, which accompanies LSM
processing, and is recognized as detrimental in cavitation erosion.

Summary

It has been shown that the high reflectance of copper and its alloys
to laser radiation at 10.6 μm does not prohibit relatively easy surface
melting of many copper alloys. Considerable refinement of microstructure
is possible. In the multiphase Fe-aluminum bronzes laser quenching
results in single phase metastable surfaces with enhanced corrosion and
erosion properties. This may be of some practical significance as the
aluminum bronzes are widely utilized in marine service.

References

1. L. Buene, J. M. Poate, D. C. Jacobson, C. W. Draper and J. K. Hirvonen, Appl. Phys. Lett. **37** (1980) 385.

2. C. W. Draper and J. W. Benko, Appl. Opt. **18** (1979) 3205.

3. R. Webb, private communication.

4. T. R. Anthony and H. E. Cline, J. Appl. Phys. **48** (1977) 3888.

5. S. L. Narasimhan, S. M. Copely, E. W. Van Stryland and M. Bass, Metall. Trans. **10A** (1979) 654.

6. See for example, B. Chalmers, "Physical Metallurgy", John Wiley & Sons, New York (1959) p. 249.

7. C. W. Draper, R. E. Woods, and L. S. Meyer, Corrosion, NACE **36** (1980) 405.

8. C. M. Preece, J. M. Poate and C. W. Draper, in Synthesis and Properties of Metastable Phases. E. S. Machlin and T. J. Rowland, Editors (T.M.S.-A.I.M.E., New York, 1981), pp. 151-164.

9. C. W. Draper, "The Use of Laser Surface Melting to Homogenize Fe-Aluminum Bronzes". Submitted to J. Mat. Sci.

10. M. W. Gabriel, C. M. Preece, A. Staudinger and C. W. Draper, "Cavitation Erosion of Laser Quenched Fe-Aluminum Bronze". Submitted to I.E.E.E. J. Quantum Electron.

11. C. W. Draper and S. P. Sharma, "The Effects of Laser Surface Melting on Tin Modified Copper-Nickel". Submitted to Thin Solid Films.

ELECTRON BEAM SURFACE MELTING TECHNIQUE AND

ITS APPLICATION TO A HIGH CARBON-LOW ALLOY STEEL

Mohan Kurup, B. G. Lewis and P. R. Strutt

Metallurgy Department
University of Connecticut
Storrs, Connecticut 06268

1. Electron Beam Scanning Technique

The intriguing metallurgical structures produced by laser and electron
beam glazing (1-6) create considerable interest in developing a convenient
method for glazing the entire surface of a substrate material. In this
respect electron beam glazing is admirably suitable since the beam may be
rapidly deflected over the surface area by electromagnetic beam deflection
coils as shown in Fig. 1. Following previously reported electron beam surface scanning experiments (7-9) the present paper will show that a beam
scanning mode employed for generalized surface heating (10) may also be used
for producing well controlled surface glazing. With the degree of precision
control attainable with electronic circuitry now available, the electron
beam scanning technique may be used for obtaining high quality smooth surfaces on a M2 tool steel substrate.

The type of scanning pattern used is obtained by applying two nearly
equal frequency triangular waveform signals to the X and Y axis beam deflection coils of the electron beam surface melting unit. The succession of
rectangular Lissajous figures traced out in this scanning mode are shown in
Fig. 2, where X and Y axis deflection of the spot on the substrate has maximum value of a and b respectively. The rectangular path results from the
progressive amount of phase lag of the Y-axis signal behind the X-axis signal. As the beam sweeps diagonally across from A (where the phase difference
is zero) the path intercepts BC when the X-axis signal attains its maximum
amplitude, the intersection point is at a distance $b(\Delta f/f)/(1+\Delta f/f)$ below C.
Following this, the path next intercepts CD when the Y-axis signal attains
its maximum value. The distance of this intercept point to the left of C is
a $(\Delta f/f)$. In proceeding with this simple analysis it is readily apparent
how the successive rectangular type of scanning pattern is obtained.

Simple consideration shows that the angle θ in Fig. 2 is given by Eqn. 1

$$\tan\theta = \frac{b}{a} \cdot \frac{1}{(1 + \Delta f/f)} \; ; \quad \text{------------------------------------(1)}$$

Also by geometrical construction the spacing d between two adjacent diagonal
traces is given by Eqn. 2. It should be noted at this juncture that no
mention has been made of the actual beam spot diameter (w). When considering

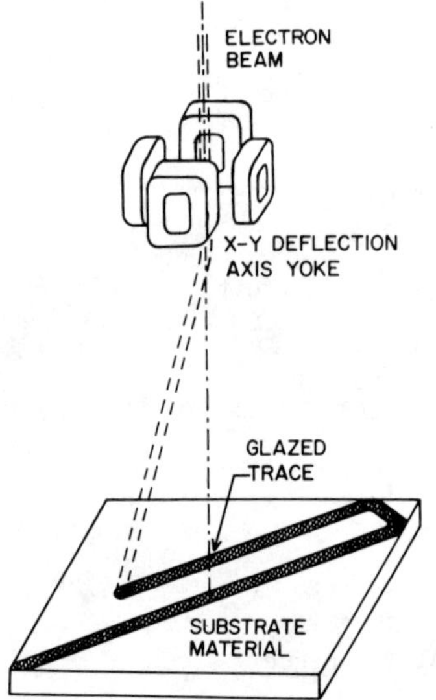

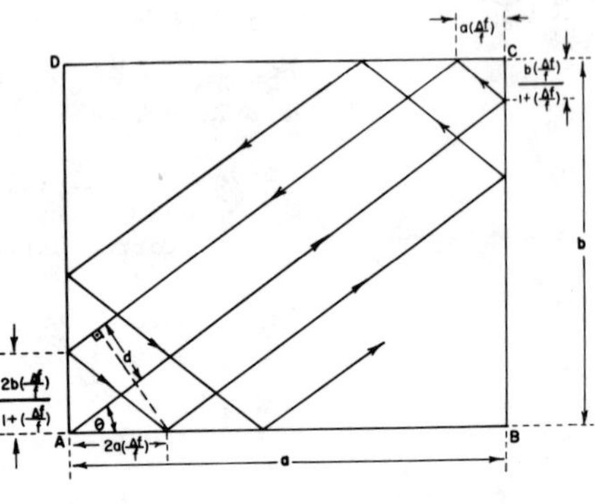

Fig. 1 – Electromagnetic deflection of an electron beam during surface melting.

Fig. 2 – Schematic diagram of the geometry of scanning mode.

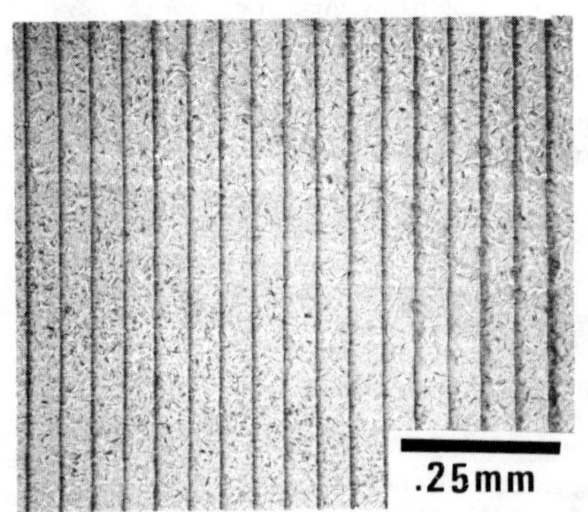

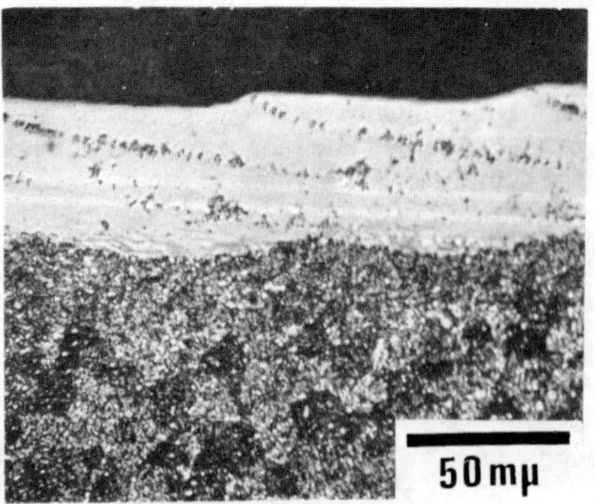

Fig. 3 – (a) Plan view and (b) cross-section of as melted surface formed by beam scanning of an M2 tool steel substrate.

a beam of finite width the diagonal lines in the pattern will represent the path traced out by the center of the spot.

$$d = \frac{2ab(\Delta f/f)}{\sqrt{a^2(1 + \Delta f/f)^2 + b^2}} \; ; \text{------------------(2)}$$

In considering the nth cycle, the beam path intercept on BC is a distance y_n away from C, see Eqn. 3.

$$y_n = \frac{(2n - 1)\cdot(\Delta f/f)\cdot b}{(1 + \Delta f/f)}; \; \text{------------------(3)}$$

Equation 3 may be used to find n, the number of cycles required to sweep out the entire ABCD, simply by noting that y_n will then be equal (or about equal) to b. The precise condition is given by Eqn. 4.

$$n \geq 1/2(\Delta f/f) > (n - 1) \; ; \text{------------------(4)}$$

It should be noted that although the time to complete a frame is (n/f) the line spacing is frequency independent and depends on the ratio $(\Delta f/f)$.

2. Topography and Microstructure of Rapidly Solidified Surfaces

Fig. 3(a) shows portion of a 2.5cm x 2.0 cm surface melted area on M2 tool steel produced when f is 60 hertz and $(\Delta f/f)$ is 1/450, the corresponding beam velocity and frame time are 385 cm.s^{-1} and 3.75 s. The high quality surface finish without the familiar surface "ripple" effect is clearly evident in the micrograph in Fig. 3(a).

The overlapping of glazed zones is clearly seen in cross-sectional view, see Fig. 3(b). It would seem reasonable that the rather dramatic increase in surface uniformity achieved by repetitive melting and solidification arises from the dispersion or evaporation of surface contaminants. This cleansing promotes progressively better melting of the solid surface by molten liquid during each successive cycle. By examination of the micrographs it is evident that it is possible to eliminate the "ripple structure" and obtain a surface with minimal periodic topographical variation, see Fig. 3(b). The surface relief effects (i.e. martensite platelets and slip lines) results from the solid state transformations and thermal stresses produced during cooling.

A graphic illustration of the refined microstructure obtainable with electron beam surface melting is shown in Fig. 4(b) where the material under investigation is a commercial "Ferro-TiC" alloys. In conventionally processed material, see Fig. 4(a),   large titanium carbide particles are embedded in a steel matrix, which may be either martensitic, maraging, or tool steel; the volume fraction of TiC varies between 36 to 45%. After electron beam melting, however, an entirely new highly refined dendritic carbide microstructure is formed, see Fig. 4(b). Thus the combined turbulent mixing and heating of the molten liquid as it flows around the incident beam is highly effective in dissolving the high melting point carbide particles.

3. Surface Melting and Post Treatment of a High Carbon-Low Alloy Steel

High carbon-low alloy steels represent an interesting class of materials, since for a carbon content of > 1wt% the M_s temperature is below room temperature. In such a material, therefore, it is intriguing to determine the dependency of microhardness on cooling rate in laser and electron beam surface melting experiments, since grain size and carbide refinement may considerably modify the M_S temperature. Particular interest in studying grain refinement in a high carbon-low alloy steel arises from the work of Wadsworth and Sherby (11) who, by thermomechanical processing produced fine grained material containing a dispersion of relatively fine cementite particles. By controlling the austenite grain size before quenching into liquid nitrogen these workers were able to vary the scale of the martensitic structure to a considerable degree.

The alloy used in the present investigation had the following composition 1.3wt%C, 1.5wt%Cr, 0.86wt%Mn, 0.16wt%Si, 0.08wt%Ni, 0.06wt%Al, balance Fe; the M_S temperature is 212°K (12). The as-received material used in this study had been thermomechanically processed with the type of microstructure just discussed (11). Laser surface melting was carried out using a continuous CO_2 laser at United Technologies Research Center, East Hartford and electron beam surface melting at the Metallurgy Department, University of Connecticut. The irradiation beam diameter was ∿0.5mm in each case, however, owing to differences in coupling efficiency (13) the electron beam power was adjusted to give the same depth of melting as the laser at a given beam velocity. Hence the beam power of the laser and electron beam were 5.1 kW and 375 W respectively and the beam velocities ranged from 0.5 to 125 cm.s^{-1}. In one particular case a wide beam (1mm), produced by rapid oscillation, was used instead of a circular spot; the beam velocity relative to the melting direction was 3.5cm.s^{-1}.

Quite surprisingly the microhardness of the rapidly solidified high carbon-low alloy steel was, in general, appreciably different following laser and electron beam irradiation. Fig. 5 shows microhardness as a function of beam velocity (v). For the laser the microhardness corresponding to the maximum value of v is in excess of 900 V.H.N. and then progressively decreases as v decreases. In contrast, the microhardness following electron beam surface melting is low in all cases and, in fact, only increases slowly with decreasing v, see Fig. 5.

In addition to the evaluation of microhardness and microstructural characteristics of as-melted material the effect of post thermal treatment was studied. Table I shows that the microhardness of electron beam melted specimens is, in some instances, substantially increased by quenching into liquid nitrogen. Apart from the specimens with a large melt zone (the first three in Table I) the hardness following the subsequent quench into liquid nitrogen does not vary in systematic way with beam velocity. The highest microhardness value obtained (1050 to 1100 V.H.N.) (see Table I) was in the specimen melted using a wide beam (1mm) whose microhardness was only 335 V.H.N. in the as-melted condition. The microstructure corresponding to the hardened state in this specimen is seen in Fig. 6 where a dense, relatively fine scale martensitic structure is clearly evident. It should be pointed out that evidence of small scale martensitic formation was observed in all cases where quenching into liquid nitrogen increased the microhardness to ≳ 500 V.H.N., see the microstructure in Fig. 7 where the beam velocity v is 75 cm.s^{-1}. The corresponding structure typical of all as-melted electron beam processed specimens when v ≳ 30 cm.s^{-1} is seen in the highly etched scanning electron micrograph in Fig. 8. Within the cellular regions there is evidence of a filamentary carbide structure and also a few larger carbide particles; apart from these discernable features it is difficult to resolve any other details of the microstructure.

Considering the effect of quenching laser melted material Table I shows that the microhardness of the specimen with the deep melt zone (corresponding to v equals 5 cm.s^{-1}) increases from 350 to 760 V.H.N. Again, microscopic examination reveals that the well defined martensitic structure, similar to that in Fig. 6, is produced by post treatment quenching. In all cases where v ≳ 25 cm.s^{-1}, however, subsequent quenching into liquid nitrogen does not

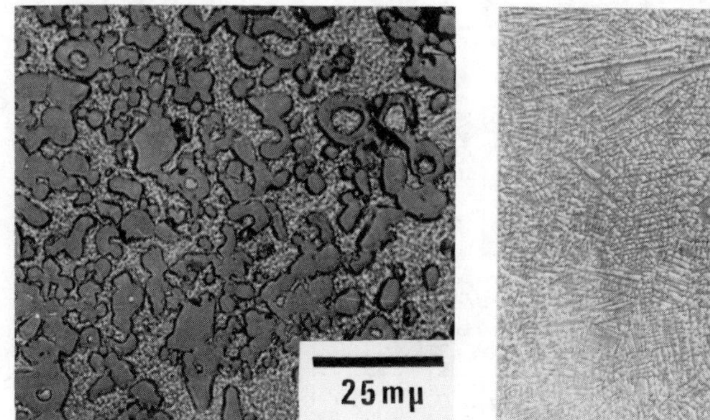

Fig. 4 - Microstructure of a cemented carbide material in which TiC
particles are embedded in a steel matrix, (a) structure before
processing and (b) the as-surface melted structure.

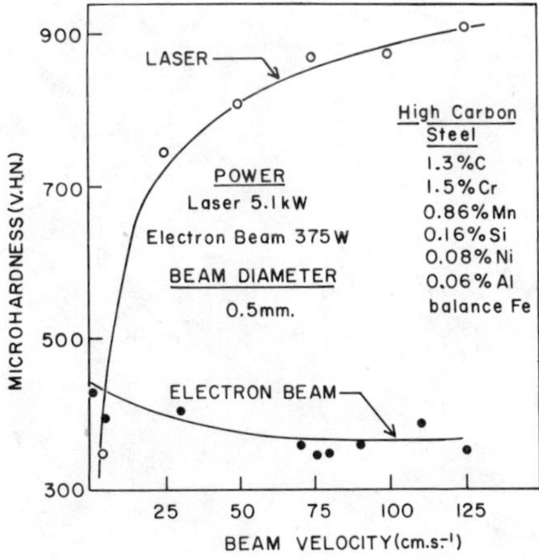

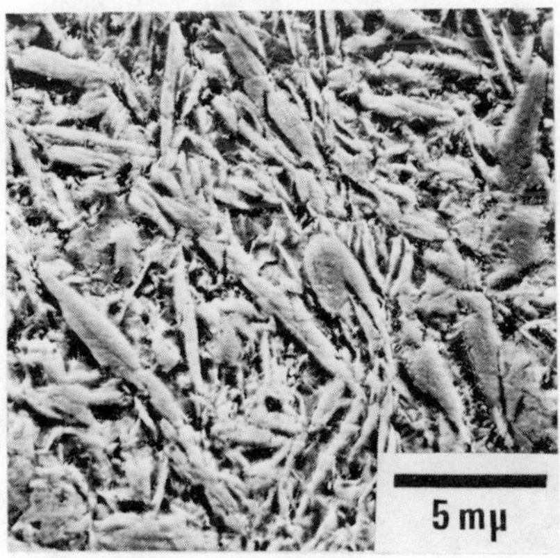

Fig. 5 - Microhardness as a function
of beam velocity for a high
carbon-low alloy steel using
(i) laser and (ii) electron
beam surface melting.

Fig. 6 - Martensitic structure
following liquid nitrogen
quenching of melted zone
formed by a line scanning
electron beam.

TABLE I

Electron Beam

Beam Velocity (cm.s^{-1})	0.5	3.5*	5.0	30	60	70	75	80	90	110	125
Hardness (V.H.N.)	430	355	395	405	355	360	345	350	360	390	350
After Liquid N$_2$ Quench	750	1075	680	425	505	500	505	600	400	800	460

Laser

Beam Velocity (cm.s^{-1})	5.0	25	50	75	100	125
Hardness (V.H.N.)	350	745	810	870	875	910
After Liquid N$_2$ Quench	760	775	820	830	860	920

MICROHARDNESS OF ELECTRON BEAM AND LASER MELTED SPECIMENS
(BEFORE AND AFTER LIQUID NITROGEN QUENCH)

* Melted using 1mm wide oscillated beam.

statistically change the as-melted microhardness. The hardened condition
of as-melted material when v $\gtrsim$ 25 cm.s^{-1} is particularly curious since the
normal M_s temperature is 212°K (12). Figure 9 shows the as-melted microstructure for v equal to 50 cm.s^{-1}, it appears to consist of irregularly
shaped martensite plates together with a filamentary carbide network in the
original intercellular regions of the solidification structure. A higher
magnification micrograph is seen in the offset of Fig. 9. In comparing
Figs. 6 and 9 there is a resemblance between the irregular martensite plates
in Fig. 9 and some of those in the more discernable martensitic structure in
Fig. 6. With increasing beam velocity and correspondingly higher cooling
rate the scale of the microstructure in Fig. 9 reduced without a discernable
change in character.

4. Discussion

The most surprising aspect of the findings in the preceeding section is
the difference in properties obtained by laser and electron beam surface
melting at higher beam velocities (i.e. shorter interaction times). At, and
below a beam velocity of 5 cm.s^{-1} (using a circular source) the beam penetrates into the substrate as in the deep penetration mode (13). Thus the
mode of energy transfer between beam and material, and the coupling efficiency is rather similar for the two types of energy source. With this
situation, nearly identical materials properties are found after melting
(and subsequent quenching into liquid nitrogen) for both the laser and electron beam (see the microhardness values in Table I). Similar results are
also found using a rapidly oscillated beam which produces a large cross-
section melt zone (1mm x 0.02 mm) without actual beam penetration. In these
large melt zones the cooling rate is relatively low (10^3 to 10^4°K.s^{-1}),
nevertheless the scale of the microstructure is sufficiently refined that
small martensite plates nucleate homogeneously throughout the region upon
quenching into liquid nitrogen, see Fig. 6. In comparison to the large
martensite platestypical of a high carbon steel the scale in Fig. 6 is particularly fine. Furthermore, the homogeneous structure in Fig. 6 results in
a particularly high microhardness (1050 to 1100 V.H.N.) as compared with an
average value of $\sim$750 V.H.N. in a conventionally processed and quenched
material.

At higher cooling rates more complex materials properties are obtained
as a result of laser and electron beam processing and post treatment. Mean
cooling rates throughout the melt zone, as calculated from the simple semi-
infinite plate model of Greenwald et al (14) range from 10^5 to $5x10^6$°K.s^{-1}
as the beam velocity increases from 25 to 125 cm.s^{-1}. In considering electron beam melted material the erratic response to quenching in liquid nitrogen when v $\gtrsim$ 30 cm.s^{-1} obviously reflects underlying phase stabilities.
Related behavior has been previously noted in recent studies of M2 tool steel
where it was found that the preponderance of δ-ferrite, austenite, and martensite could fluctuate somewhat irregularly. Interestingly, the phase
regions appeared to be associated with convective and turbulent flow patterns.
In Table I the specimen melted at a beam velocity of 110 cm.s^{-1} with the
electron beam obviously contains a high proportion of austenite which transforms to martensite upon quenching into liquid nitrogen. Curiously, the two
specimens melted at v equal to 90 and 125 cm.s^{-1} show an inappreciable change
in hardness upon being quenched. One can only conclude therefore that the
as-melted microstructure of these specimens consists of austenite sufficiently well stabilized that it resists the martensitic transformation. Clearly,
this is an area for quantitative study including X-ray diffraction analysis
and detailed transmission electron microscopy.

Finally, a most surprising finding is the high microhardness following
laser melting when v $\gtrsim$ 25 cm.s^{-1} in a material whose normal M_s temperature

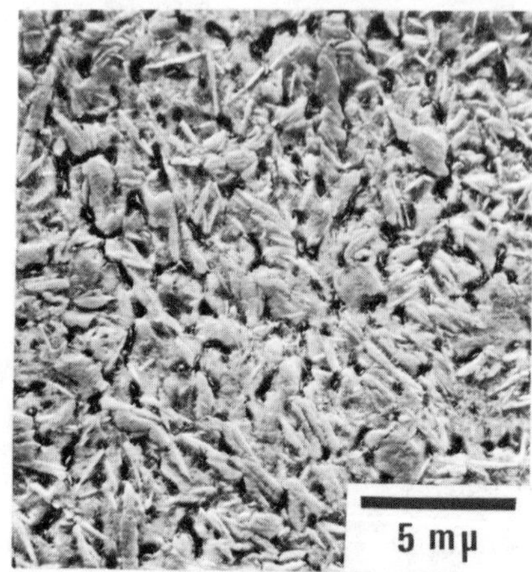

Fig. 7 - Martensitic structure in liquid nitrogen quenched zone formed by electron beam melting of a velocity of 80 cm.s^{-1}.

Fig. 8 - Microstructure formed after liquid nitrogen quenching a zone formed by electron beam melting at a velocity of 30 cm.s^{-1}, note there is no evidence of martensite.

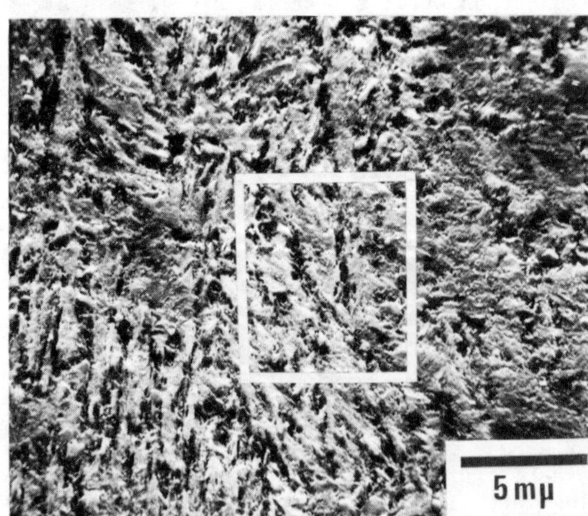

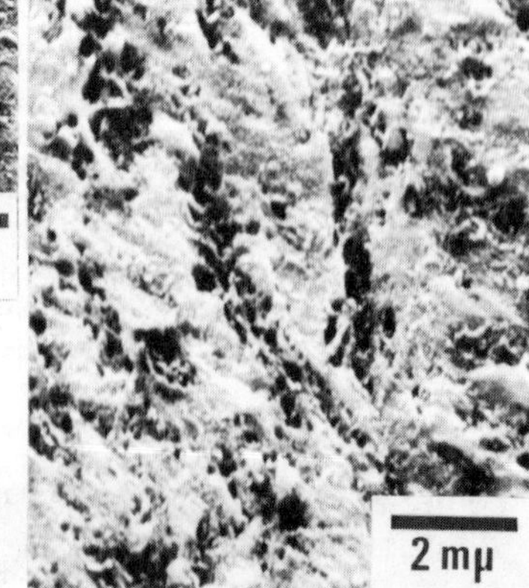

Fig. 9 - Irregular martensitic structure in as-laser melted material at a beam velocity of 50 cm.s^{-1}. The higher magnification offset shows the filamentary carbide structure.

is 212°K. Considering only the results for the laser experiments in Fig. 5
it appears that a high cooling rate results in a significant increase in
the M_s temperature and, furthermore, that M_s increases with increasing
cooling rate. That **rapid** cooling, in itself, is insufficient to produce
the hardening effect is shown by the results obtained by using the electron
beam where even at the higher velocity (125 cm.s^{-1}) the microhardenss was
only 350 V.H.N. It is interesting to note that the laser melting results
show that a particularly high cooling rate is not required to produce the
effect, since significant hardening occurs when v is 25 cm.s^{-1}; the
corresponding cooling rate at this velocity is $\sim 2 \times 10^5$°K.s^{-1}. Thus other
effects, such as the type of processing environment and nature of the fluid
flow (i.e. the degree of turbulence and magnetic stirring) play an important
role.

In considering the effect of environment, it should be noted that the
protective helium atmosphere used in laser processing, whilst greatly
restricting exposure to oxygen, does not entirely eliminate oxide formation.
On the basis of models for martensite nucleation (15,16,17) the small oxide
particles formed during the melting process may serve as martensite nucleation sites during cooling in the solid state. Since oxidation is dramatically reduced during melting in the electron beam unit (the vacuum is
$\sim 10^{-5}$ to 10^{-4} Torr), martensite nucleation at oxide particles will be insig-
nificant and hence the microhardness is low. If other types of micro-inclu-
sions and small precipitates are capable of acting as martensite nucleation
sites in the case of electron beam melting then electromagnetic stirring
may be effective in retarding the formation of a martensitic structure.
Certainly, the effects of convective flow, turbulence and electromagnetic
stirring (using the electron beam) are important in determining the nature
of the microstructures observed in many instances.

5. Acknowledgement

The authors with to acknowledge the support of this investigation by
the Office of Naval Research, contract number N00014-78-C-0580.

References

1. E. M. Brienen, B. H. Kear, C. M. Banas and L. E. Greenwald; pp.435 in Superalloys: Metallurgy and Manufacture, B. H. Kear, ed; Claitor, Baton Rouge, LA 1976.

2. P. R. Strutt, H. Nowotny, M. Tuli and B. H. Kear; Mat. Sci. and Engr., (1978) pp.217.

3. Young-Won Kim, P. R. Strutt and H. Nowotny; Met. Trans. A, (10), (1979) pp.881.

4. S. M. Copley, D. Beck, O. Esquivel, and M. Bass; pp.161 in Laser-Solid Interactions and Laser Processing, S. D. Ferris, H. J. Leamy, and J. M. Poate, ed; Amer. Inst. of Physics, New York, NY, 1979.

5. R. Mehrabian, S. Kou, S. C. Hse, A. Munitz; pp.129 in Laser-Solid Interactions and Laser Processing, S. D. Ferris, H. J. Leamy and J. M. Poate, ed; Amer. Inst. of Physics, New York, NY, 1979.

6. W. M. Steen; pp.189 in Rapid Solidification Processing, Principles & Technologies II, R. Mehrabian, B. H. Kear and M. Cohen, ed; Claitor, New Orleans, LA, 1980.

7. W. Gruhl, B. Grzemba, G. Ibe and W. Hiller; Aluminum, (53), (1977), pp.177.

8. T. R. Tucker, J. D. Ayers and R. J. Schaefer; pp.760 in Laser and Electron Beam Processing of Materials, C. W. White and P. S. Peercy, ed; Academic Press, New York, 1980.

9. B. G. Lewis, D. A. Gilbert and P. R. Strutt; pp.221 in Rapid Solidification Processing, Principles & Technologies II, Claitor, New Orleans, LA, 1980.

10. R. Mayer, W. Dietrich, and D. Sundermeyer; Welding Journal, (56), (1977), pp.35.

11. J. Wadsworth and O. D. Sherby, J. Mech. Working Technol., (2), (1978), pp.53.

12. E. R. Petty, Martensite, Longmans Press, London, 1970.

13. P. R. Strutt, J. of Mat. Sci. and Engr., (44), (1980), pp.239.

14. L. E. Greenwald, E. M. Breinan, and B. H. Kear, pp.189 in Laser-Solid Interactions and Laser Processing, S. D. Ferris, H. J. Leamy, and J. M. Poate, ed; Amer. Inst. of Physics, New York, NY, 1979.

15. A. R. Entwisle, Met. Trans., (2), (1971), pp.2395.

16. J. M. Galligan and T. J. Garosshen, Nature, (274), (1978), pp.664.

17. M. Suezawa and H. E. Cook, Acta Met., (28), (1980), pp.423.

CONTROLLED RAPID SOLIDIFICATION BY ELECTRON BEAM SURFACE MELTING

R. J. Schaefer, W. J. Boettinger, F. S. Biancaniello, and S. R. Coriell
National Bureau of Standards
Washington, D.C.

The use of a modified electron beam welding apparatus to produce controlled rapid solidification is described. For research purposes, electron beam heating has several advantages over laser heating, most significantly that the absorbed power is known better and the beam can be deflected electronically at high speed. Experiments can thus be carried out in which local solidification velocities can be determined from the known thermal input by means of heat flow calculations. We report here several methods of utilizing electron beam heating, and some observations of cellular substructures in surface melted Al-Ag alloys.

Introduction

The melting of thin layers on metal surfaces by directed energy sources
such as lasers or electron beams can result in rapid resolidification as
heat is conducted into the cooler substrate. Such surface layers can have
any of the structural characteristics which have been observed in rapidly
solidified metals, such as amorphous or highly supersaturated crystalline
phases, or highly refined microstructures. For studies attempting to
correlate these structures to the thermal conditions prevailing during
solidification, surface melting has an advantage over other rapid solidification processes such as spray atomization or melt spinning in that no
coefficient for the transfer of heat from the metal to an external cooling
medium need be estimated. Calculations of the melting and freezing of
metals subjected to surface heating by directed energy sources (1-4) can
thus be used with relative confidence, provided that the thermal input of
the energy source can be adequately characterized.

The objective of the research reported here has been to relate the
microstructure of surface melted layers to the solidification conditions,
and to use the observed relations to test and improve upon theories of
rapid solidification (5, 6). Experimental data of this type are needed to
better define, for example, the possible dependence of solute distribution
coefficients on solidification velocity.

For the purpose of producing well-characterized thermal inputs for
these experiments, electron beam heating has certain advantages, to be
discussed below. We seek especially modes of heating which will produce
solidification conditions which are homogeneous over sufficiently large
areas that the microstructural analysis does not depend on a knowledge of
the exact location of the sample within the melt.

Energy Input to Metal Surfaces

Although lasers have several attractive characteristics as instruments
for surface heating, it is difficult to predict how much of the energy in
an incident laser beam will be absorbed by a metal surface, especially at
high power densities. The variation of absorptivity with temperature and
wavelength can be reasonably well accounted for, but the dependence of the
energy absorption on sample surface condition is a rather unpredictable
factor which may change greatly when the surface oxidizes or a surface
layer becomes molten. Therefore the melting and freezing of the surface
can be more reliably calculated if the thermal energy input is provided by
an electron beam.

When a beam of high-energy (10-100 keV) electrons impinges upon a
metal surface, a fraction of the electrons are elastically backscattered
with little transfer of energy to the metal. The backscattered energy
increases with larger atomic number targets, so that for 30 keV electrons
one would expect elastic backscattering to produce a 7% energy loss for an
aluminum target but almost a 35% loss for a uranium target (7). Although
x-rays are generated at intensities which demand safety measures such as
the use of leaded glass in viewing ports, the energy loss in this form is
minute.

The electron beam current actually entering the target can readily be
monitored by observing the voltage across a resistor by which the target is
connected to ground. By means of such a device, one can observe that as
the beam power density is increased the current entering the sample suddenly

44

falls sharply (about 20-30 percent in a typical case) when a plasma is
ignited above the target surface. This is clearly a situation to be avoided
in quantitative melting experiments, for in the presence of the plasma one
does not know how much energy remains in those electrons which do reach the
target surface, or how much of the energy which is given up to the plasma
is subsequently transferred to the target by secondary processes. McKay
and Schriempf (8) have carried out a calculation of the energy transferred
to a metal surface by the thermal emission (mostly ultraviolet) from a
laser-induced plasma above the metal surface, but it has yet to be shown
that such a calculation can attain high accuracy.

Equally as important as a knowledge of the total energy entering the
target is a knowledge of the spatial distribution of the energy. We have
used several devices for measuring beam profiles, based on modifications of
devices reported in the literature (9). With either lasers or electron
beams, it is usually easier to measure the beam profile at much lower beam
powers than those used for the actual surface melting, or at some plane
other than the target plane, where the beam is expanded to a much lower
power density. We have sought methods which allow beam profile measurements in the target plane at fairly high beam currents.

An easily constructed profile-measuring device for use with electron
beams consists of an array of parallel fine tungsten wires (Fig. 1a). The
electron beam is scanned rapidly across the array and the current entering
the wires is recorded on an oscilloscope. If the wire diameter is small
compared to the beam diameter, the peaks seen on the oscilloscope show the
beam profile. The multiple wires facilitate the determination of the beam
scanning velocity. This device can be used only at relatively low power
levels if the wires are to survive.

Another device, which is shown in Fig. 1b, is similar in concept to
that shown in Fig. 1a, but in this case the signal on the oscilloscope
represents the current which passes through the narrow slit between the two
knife edges as the electron beam is scanned across it. The advantage is
that the knife edges, especially if they are water cooled, can tolerate a
considerably higher beam current than could the fine wires. In the device
as shown, the beam current passing through the slit is captured by a
Faraday cup, which collects most of the backscattered electrons as well as
those that are directly absorbed by the bottom of the cup. The device
could readily be modified to count only the directly absorbed electrons,
but this would change only the amplitude, and not the shape, of the recorded signal.

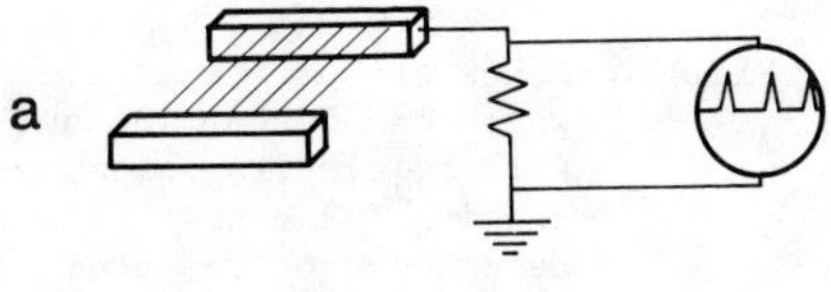

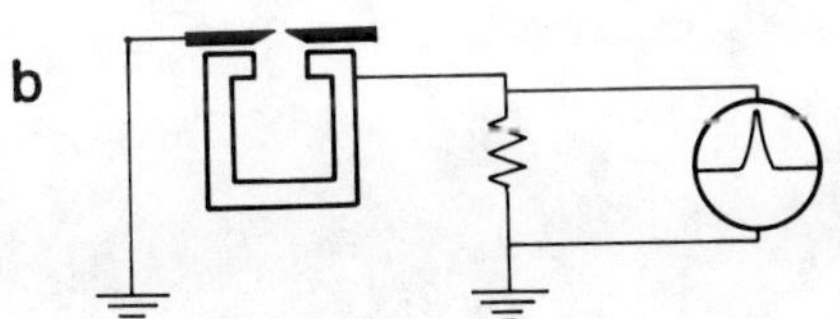

Fig. 1. Devices for measuring
electron beam current distributions. (a) Multiple
wire device (b) Slit and
Faraday cup device.

Both the wire and the slit device measure the current in a narrow
strip passing through the electron beam. Except for a simple gaussian-
shaped beam, it is difficult to deconvolute this information into an exact
description of the current as a function of position, and impossible if the
beam is not radially symmetric. This information can be obtained by re-
placing the slit of Fig. 1b by a pinhole and scanning the beam over the
pinhole repetitively in one direction while scanning at a slower rate in
the perpendicular direction. One thereby obtains a useful oscilloscope
trace which represents a series of cross-sectional slices through the beam
profile.

Our measurements of beam profiles, using the devices shown in Fig. 1,
have shown that the most sharply focused beam is approximately gaussian in
profile with a FWHM of approximately 10^{-4} m. At focusing current greater
than that required for sharp focus, the beam becomes wider but with a sharp
central peak, whereas at lower focusing current the beam becomes wider with
a flat top or with a higher power around the perimeter.

Useful Heating Modes

Fig. 2 illustrates schematically some of the modes of surface heating
which are being used to produce rapidly solidified metals. The line and
strip are steady state modes in a moving reference frame, whereas the spot,
patch and pedestal are transient or pulsed modes.

A full heat-flow analysis of the spot mode has been published (2), but
this is a situation in which the solidification parameters vary from point
to point with only radial symmetry. It is thus not ideally suited for
microstructural studies.

In the line mode, a beam sweeps at constant velocity across the target
surface, leaving behind a linear trail. This mode, which has also been
analyzed (3), has the advantage over the spot mode that any cross-sectional
cut perpendicular to the direction of motion should show an identical
configuration.

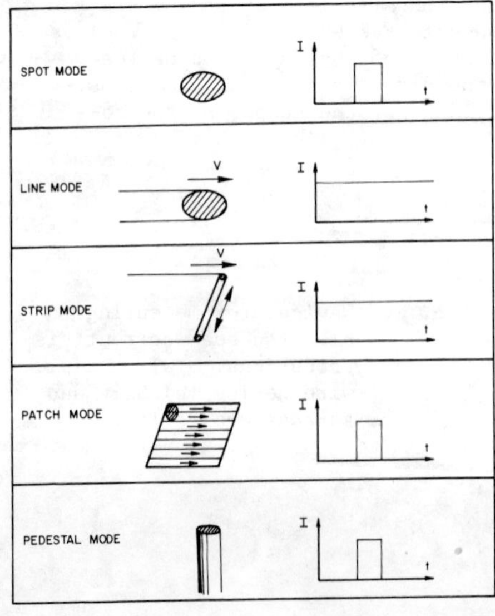

Fig. 2. Modes of surface melting
which can be carried out
with an electron beam,
showing for each the time
dependence of the beam
current.

The strip mode uses a beam which is electronically scanned very
rapidly back and forth to simulate a line source of heat with a width which
is equal to the beam diameter and a length which is determined by the
electronic deflection. The simulation is effective only if the scanning is
sufficiently rapid that the heating appears continuous at a small distance
below the surface. The target is translated perpendicular to the direction
of the electronic scan, leaving behind a wide, shallow melt pass. In this
mode, freezing conditions can be uniform across the width of the strip
except near the edges, so that large areas of sample with identical processing conditions can be obtained. However, the heat flow analysis of
this mode has not yet been carried out.

The patch mode is a method for producing uniform heating over an extended area. Ideally one could simply use a large diameter beam with
uniform intensity, but this is in practice not readily attained. It is
therefore simulated by rapidly rastering a beam over a patch so that the
average power density has the desired value. Once again, the rastering
must be sufficiently rapid to appear continuous at the maximum melt depth.
The objective of this mode is to produce uniform heating over an area
sufficiently large that the heat flow is effectively one-dimensional and
the solidification conditions depend only on the depth below the surface.
In the pedestal mode the patch covers the entire end of a rod-shaped sample,
so that heat flow is approximately one-dimensional even if the pulse time
is rather long.

Fig. 3 illustrates the range of power density and pulse time combinations which can be used with the patch mode for surface melting of
aluminum. The lower boundary of the trapezoid represents the critical
value of $q\tau^{\frac{1}{2}}$ required to get the surface just barely to the melting point
(1), where q is the beam power density and τ is the pulse duration. The
upper boundary is the value of $q\tau^{\frac{1}{2}}$ at which the average metal surface
temperature reaches the boiling point (at atmospheric pressure). The left
side of the trapezoid represents the minimum time required to complete 10
scans of the patch (at 4 kHz) while the right side limit represents the
maximum pulse time for which the machine has sufficient power (10 kW in our
case) to melt an area sufficiently large to attain one-dimensional heat
flow. Because electron beam melting is carried out in a vacuum, it is not
reasonable to expect the surface of the metal to reach the atmospheric
pressure boiling point (at least not without violent side effects), so
operations should be confined to the lower part of the trapezoid. In
addition, the smaller the beam size in relation to the patch size and
the smaller the total number of raster scans during the pulse time, the

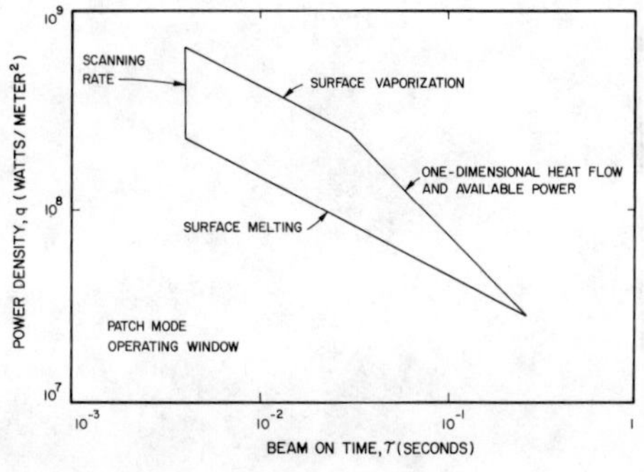

Fig. 3. Operating window for
surface melting of
aluminum in the patch
mode.

hotter is the local transient surface temperature directly under the beam.
To avoid the danger of local overheating, it is best to defocus the beam so
that its diameter is perhaps one half the width of the patch and to scan as
rapidly as possible.

Alloy Solidification

One of the most important characteristics sought in rapidly solidified
metals is chemical homogeneity. Previously published calculations (5, 6)
have indicated that at sufficiently rapid rates of solidification, metals
can solidify with stable solid-liquid interfaces at alloy concentrations
much greater than those predicted for instability by constitutional supercooling theory. This regime of "absolute stability" can be attributed to
the influence of the solid-liquid surface energy in opposing the growth of
perturbations with very short wavelengths.

The previous calculations for Al-Cu and Si alloys, which assumed that
the solute distribution coefficient was that given by the equilibrium phase
diagram, indicated that the transitions from instability to absolute
stability should occur at very low alloy concentrations or at very high
interface velocities. We have chosen to work with the system Al-Ag because
it has a rather high solute distribution coefficient (k = 0.41) which
results in a predicted range of interesting stability-instability transitions at compositions of 0.1-1 wt.%.

In Fig. 4 the curved line shows the behavior of this alloy system as
predicted by morphological stability theory, while the straight lines
indicate the predictions of simple constitutional supercooling and absolute
stability criteria (5, 6). The physical properties used in the calculations
are listed in Table I.

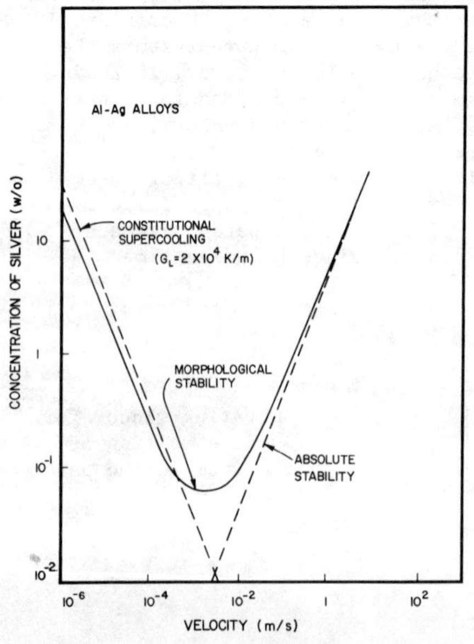

Fig. 4. Stability of planar
interfaces in Al-Ag
alloys.

Table I. Properties Used for Stability Calculations in Al-Ag

Property	Symbol	Value
Diffusion rate in liquid	D	$4.5 \times 10^{-9} \, m^2/s$
Solute distribution coefficient	k_o	0.41
Liquidus slope	m	-1.7 K/wt.%
Surface energy parameter	$T_M \Gamma$	$1.0 \times 10^{-7} \, m \cdot K$
Latent heat of fusion	L	$1.08 \times 10^9 \, J/m^3$
Thermal conductivity of liquid	k_L	$90.4 \, J/s \cdot m \cdot K$
Thermal conductivity of solid	k_s	$211 \, J/s \cdot m \cdot K$
Thermal diffusivity of liquid	κ_L	$3.5 \times 10^{-5} \, m^2/s$
Thermal diffusivity of solid	κ_s	$6.8 \times 10^{-5} \, m^2/s$

The constitutional supercooling line is given by

$$c_\infty = \frac{G_L Dk}{v(k-1)m}$$

where G_L is the temperature gradient in the liquid, D is the interdiffusion rate in the liquid, k is the solute distribution coefficient, v is the velocity of solidification, and m is the slope of the liquidus curve.

The absolute stability criterion predicts

$$c_\infty = \frac{k^2 T_m \Gamma v}{D(k-1)m}$$

where T_m is the melting temperature and Γ is the ratio of the solid-liquid surface energy to the latent heat of fusion per unit volume.

The calculation was carried out assuming that the solute distribution coefficient does not depend on velocity. If there is a dependence, one would expect the distribution coefficient to approach unity at higher velocities, which would lead to additional stabilization of planar inter-faces. The concentrations for instability on the left side of the curve are approximately proportional to the temperature gradient in the liquid (here assumed to be 200 K/cm), but the concentrations on the right side are almost independent of gradient.

There exists a body of experimental experience which demonstrates the type of behavior described by the left side of Fig. 4 but the absolute stability transition on the right side has never been accurately measured, although evidence for this transition has been reported for laser-melted Si (10, 11). Conventional solidification processes cannot attain the required velocities, and in most rapid solidification processes the interface

velocity is not well enough known for a meaningful correlation to the
theory. Well-controlled surface melting can be used to investigate this
region.

Our experiments with Al-Ag alloys have used samples with 0.1, 0.3, and
1.0 wt.% silver. Line mode tracks have been melted using an electron beam
1 mm in diameter sweeping over the sample surface at velocities of 0.05,
0.15, and 0.42 m/s. At these velocities, plane-front solidification is
predicted to be unstable at silver compositions above 0.28, 0.71, and 1.8
wt.% Ag, respectively. If the sweep velocities were equal to the actual
interface velocity, then most of these samples would have been in the
absolute stability regime. However, when the samples were sectioned, and
examined by scanning electron microscopy, it was found that cellular
structures were present throughout all melt passes, except for a very thin
layer near the bottom of each pass which represents an initial transient in
the melt pool solidification. The observed structures are sometimes
elongated cells but in other regions are rather irregular arrays with
occasional very small regions of hexagonal symmetry (Fig. 5).

At steady state in the line mode of melting, local solidification
velocities must everywhere be less tnan the sweep velocity of the beam
across the sample surface, by a factor equal to the cosine of the angle θ
between the normal to the interface and the direction of the sweep.

Some of the curves shown in Ref. (3) were computed for melting parameters close to those used in these experiments, so they may be used to
estimate local interface velocities. Thus within the central regions of
the melt pass, the local solidification velocity should be no less than
about 50% of the sweep velocity. This factor is insufficient to move the
experimental points into the theoretically unstable region of Fig. 4.

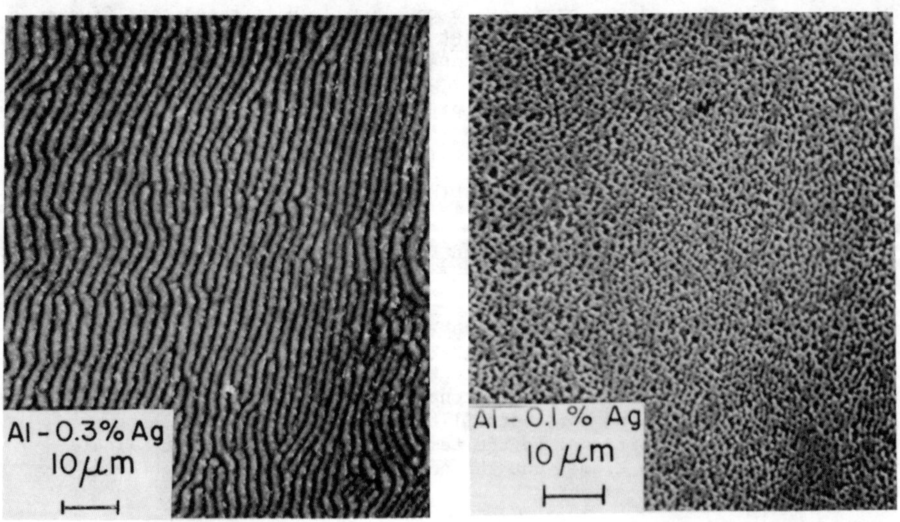

Fig. 5. Examples of well-developed elongated cells and irregular arrays
         in electron-beam melted Al-Ag alloys. Both samples melted in line
         mode with sweep speed 0.05 m/s.

Other factors which could account for the observation of cells when
none were expected include inaccuracy in the parameters used in the theory,
especially the diffusion coefficient D, which was obtained from macro-
segregation experiments (12), and the surface energy term Γ. In addition,
all of the interfaces in this experiment which were predicted to be in the
absolute stability regime had propagated through regions near the bottom of
the melt which would have been in the unstable regime. Once the cellular
structure is developed in this region, we do not know that the conditions
required to restore planar solidification are the same as those required
for a planar interface to break down, nor do we know how long a transient
may be required for the restoration of planar solidification. Thus a
cellular interface may continue to propagate in the cellular mode even in
regions where a planar interface would be stable. This subject requires
further study.

Cell spacings were measured in some of the more regular structures,
and they showed good agreement with the predicted wavelength at the onset
of instability, when plotted as a function of velocity (Fig. 6). However,
some of the measured cellular structures lay well within the predicted
region of absolute stability, so the significance of the agreement is
unclear.

More experiments with these alloys, using different melting modes,
will be required before any firm conclusions can be drawn with respect to
the correct location of the absolute stability limit.

Acknowledgments

The authors thank C. Fenimore for heat flow calculations, C. H. Brady
for metallography and D. B. Ballard for microscopy. This work was supported
by a contract from the Defense Advanced Research Projects Agency.

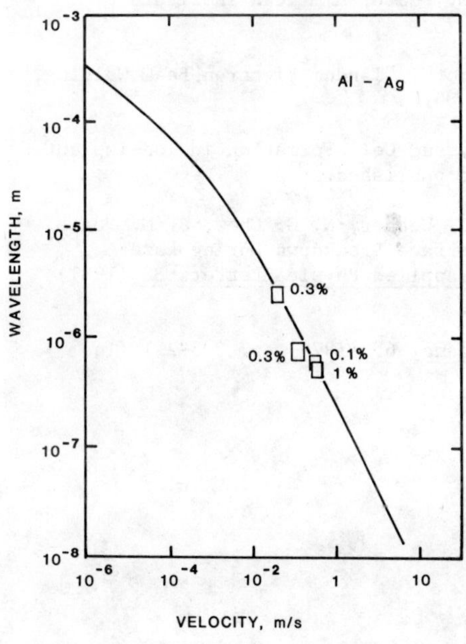

Fig. 6. Wavelength at the onset of
instability as a function
of solidification velocity.
The plotted uncertainty in
the experimental points
represents a variation between 0.5 and 1 of the
cosine of Θ, which is the
angle describing the local
growth direction.

References

1. S. C. Hsu, S. Chakravorty, and R. Mehrabian, "Rapid Melting and Solidification of a Surface Layer," Metallurgical Transactions 9B (1978) pp. 221-229.

2. S. C. Hsu, S. Kou and R. Mehrabian, "Rapid Melting and Solidification of a Surface Due to a Stationary Heat Flux," Metallurgical Transactions 11B (1980) pp. 29-38.

3. S. Kou, S. C. Hsu and R. Mehrabian, "Rapid Melting and Solidification of a Surface Due to a Moving Heat Flux," Metallurgical Transactions 12B (1981) pp. 33-46.

4. R. Mehrabian, S. C. Hsu, S. Kou, and A. Munitz, "Laser Surface Melting and Solidification - Moving Heat Flux," pp. 213-227 in Applications of Lasers in Materials Processing, E. A. Metzbower, ed.; ASM, Metals Park, Ohio, 1979.

5. S. R. Coriell and R. F. Sekerka, "Interface Stability During Rapid Solidification," pp. 35-49 in Rapid Solidification Processing: Principles and Technologies, II, ed. R. Mehrabian, B. H. Kear and M. Cohen, Claitor's Publishing Division, Baton Rouge, La., 1980.

6. J. W. Cahn, S. R. Coriell and W. J. Boettinger, "Rapid Solidification," pp. 89-103 in Laser and Electron Beam Processing of Materials, ed. C. W. White and P. S. Peercy, Academic Press, New York, 1980.

7. M. von Allmen, "Coupling of Beam Energy to Solids," pp. 6-19, Ibid.

8. J. A. McKay and J. T. Schriempf, "Transient Heating of Metals by Microsecond-Duration CO_2 Laser Pulses with Air Plasma Ignition," pp. 55-60 in Laser-Solid Interactions and Laser Processing, ed. S. D. Ferris, H. J. Leamy and J. M. Poate, American Institute of Physics, New York, 1979.

9. Y. Arata, E. Nabegata, and N. Iwamoto, "Tandem Electron Beam Welding," Transactions of JWRI, 7 (1978) p. 85.

10. J. Narayan, "Interface Instability and Cell Formation in Ion-Implanted and Laser-Annealed Silicon," to be published.

11. A. G. Cullis, D. T. J. Hurle, H. C. Webber, N. G. Chew, J. M. Poate, P. Baeri and G. Foti, "Growth Interface Breakdown During Laser Recrystallization from the Melt," Applied Physics Letters 38 (1981) pp. 642-644.

12. S. Asai and I. Muchi, Tetsu to Hagane, 65 (1979) pp. 203-211 (in Japanese).

LASER SURFACE MELTING OF IRON - 4.2 WT. PCT. CARBON ALLOY*

T. R. Tucker, A. H. Clauer, C. T. Walters, S. L. Ream
Battelle Columbus Laboratories
Columbus, OH

B. P. Fairand
Laser Consultants, Inc.
Upper Arlington, OH

Continuous and pulsed CO_2 laser beams have been used to melt shallow surface layers of near-eutectic Fe-C binary alloys in the white cast condition. Broad, relatively slow scan rate laser melts eliminated free carbon inclusions and refined the as-cast microstructure. In addition to grain refinement, significant surface hardening was achieved by the retention of high-carbon martensite. The addition of 1.5 µs laser pulses in vacuum produced thinner melts on top of the homogenized layer. The phase structures observed were equivalent to splat-quenched specimens of the same alloy composition.

*This work was supported by the U.S. Army Research Office.

Introduction

While much interest has been directed at rapid solidification processing of alloys based on the iron-carbon system, relatively little fundamental research has been conducted on laser surface melting of Fe-C binary alloys. Phase structures of the binary alloys produced by splat-quenching techniques have been determined (1,2,3). Laser surface melting of carbon steels, tool steels, and cast irons has also been investigated (4,5). However, the present study seeks to survey the phase and microstructural species attainable in surface-melted layers of Fe-C binary alloys and to determine control parameters for the production of such structures.

Using the earlier rapid quenching results as a guide, the following conditions were assumed. For low-carbon alloy compositions, slight variations of well-known phases were expected in the laser-melted regions. For alloys in the cast iron range of carbon contents, unusual metastable high-carbon phases were possible. Though reference (2) reported the presence of a small amount of amorphous Fe-3.8 w/o C in vacuum splat-cooled specimens, the attainment of a metallic glass layer was considered unlikely. Since most of the novel structures and phases observed to date resulted from alloys near the eutectic composition, we have studied Fe-4.2 w/o C most extensively and have concentrated on that alloy in this paper.

In particular, two process conditions at opposite ends of the laser processing spectrum are discussed. The first process employed a high-power continuous-wave laser to melt a relatively broad region of a slowly translated coupon. The intent of the scanned melt passes was to homogenize the as-cast material and refine the grain size. In the second process, thin surface melt layers (1-10 μm) were produced under irradiation by 1.5 μs CO_2 laser pulses in vacuum.[6] The homogeneous, featureless fusion zones experienced quench rates comparable to those of the splat-quenched specimens in reference (1).

Experiment

CW Laser Melting

High-power continuous-wave CO_2 laser radiation (λ = 10.6 μm) was used to make continuous melt trails on specimens with a variety of surface conditions. In order to understand the energy coupling mechanisms involved and to control the melt process, experiments were conducted on thin-section coupons exposed to a very broad beam, thereby determining laser absorption coefficient as a function of specimen temperature and surface condition. While these results will be reported in the near future, the data relevant to this report are considered here.

Small square coupons 1 mm thick were prepared with various surface conditions from mirror polish to black paint. A thermocouple was installed in the center of the back surface. The coupon was supported by an alumina collar to minimize thermal conduction losses during laser heating. The front surface was protected from oxidation by a flow of helium gas which successfully controlled oxidation everywhere except at the specimen corners. An expanded CW laser beam of about 40 cm^2 area and powers of 1-5 kW flooded the specimen with a nearly uniform flux of 50-250 w/cm^2 (the specimen experienced a beam intensity that was about twice the average flux). By monitoring the specimen temperature changes during laser heating, the absorbed energy could be determined and related to the incident energy to determine the fraction absorbed.

54

To make the scanned laser melts, a different configuration was
employed. 15 x 15 x 3 mm coupons received a SiC surface grit-blasting
treatment before being mounted to a linear translation stage. The
specimens were passed under a CW laser beam in a helium gas shielding
environment. The Air Force Materials Lab 10 kW CO_2 laser produced a
"top-hat" beam (round beam of nearly uniform cross-sectional intensity)
that was focused to a diameter of 11 mm in the plane of the specimen
surface. The measured power at the workpiece was 10 kW.

Surface melting occurred for sample translation speeds in the range
of 2-8 cm/s. Melt depth depended strongly on the process speed with the
most uniform melts achieved at 5.4 cm/s. A network of fine cracks was
observed with some specimens exhibiting larger, through-thickness cracks.
It should be noted that as-cast eutectic iron-carbon is extremely brit-
tle and that no specimen preheat was employed in these tests.

Pulsed Laser Melting
<u>____________________</u>

For the most rapid heating-rate melts, specimens were placed on a
translation stage in a chamber evacuated to approximately 5 mPa ($\sim$4 x
10^{-5} mm Hg). Laser pulses from a 1.5 μs TEA CO_2 laser were incident
over areas of about 0.5 cm^2. The pulse consisted of a gain-switched-
spike with total duration of about 100 ns and intensity about 7 times the
average intensity followed by a linearly decreasing tail of nominally 1.4
μs pulse width. Approximately 38% of the total beam energy was contained
in the spike. Total laser pulse energy reaching the specimen varied up
to a maximum of 50 J.

The pulse-irradiated coupons had one of three surface conditions:
as-machined, polished and "homogenized" (melted by CW laser). Usually, a
thermocouple monitored the specimen temperature in order to obtain
absorption data. In addition, open shutter photography monitored the
presence of luminous plasmas, and a photon drag detector observed the
reflected 10.6 μm radiation. Beam interaction and absorption data have
been reported in more detail elsewhere in this volume[7].

Individual pulse energy was increased until a clean, simply connected
melt shape was observed on the specimen surface. The fluences necessary
to reach this condition were invariably well above plasma initiation
thresholds. Repeated pulsed melts were produced at the determined
fluence level with the specimen being translated about 2 mm between
shots. Thus, a point on the specimen was melted 4 or 5 times. Upon
completion, the coupon contained a rapidly solidifed layer that was about
5 μm deep and extended over a 7 x 15 mm area.

Results

CW Laser Absorption
<u>___________________</u>

The procedure for heating the thin specimens was to irradiate with
the laser for a time necessary to reach a given temperature, turn off the
laser for 2 s, and then turn the laser on again for about 2 s. In this
manner, the specimen cooling curve is immediately adjacent to the heating
curve. A rear surface temperature profile is shown in Figure 1. Calculations indicated that the rear surface temperature was within 5°C of the
front surface temperature for these heating rates. At the point where
the laser was turned on for the second time, there is a negative slope
corresponding to specimen cooling with the beam off and a positive slope
related to the temperature rise with the beam on. The absorbed power at

55

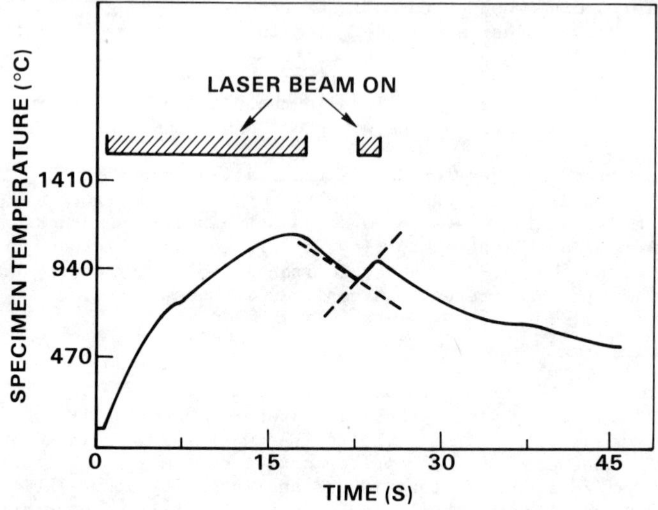

Figure 1. Temperature response of 1 mm-thick grit-blasted Fe-4.2C coupon to an expanded CW CO_2 laser beam.

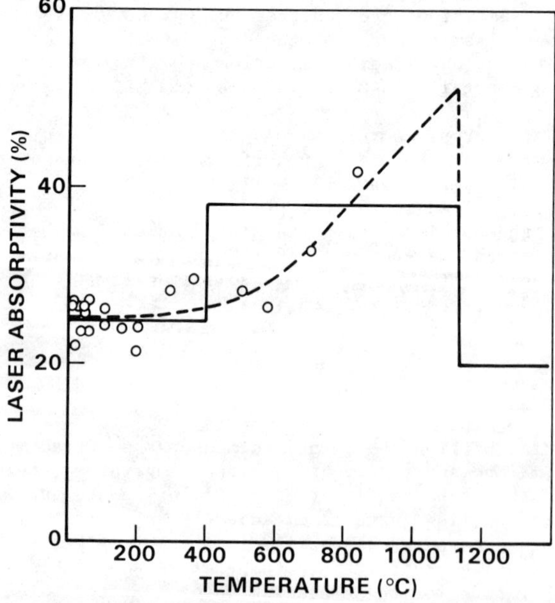

Figure 2. Thermal Coupling coefficient of specimen in Figure 1 to 10.6 µm laser radiation as a function of specimen temperature.

the specimen temperature given by the intersection of the two slopes is
related to the sum of the two slopes by the following formula:

$$Q = \sigma \ell c \; \frac{dT}{dt}$$

where Q is absorbed power density in W/cm^2, σ is density in g/cm^3,
ℓ is specimen thickness in cm, c the specific heat in $J/g°C$, and dT/dt
is the specimen heating rate in °C/s. Dividing Q by the incident power
density yields the absorption coefficient.

In Figure 1, thermal arrests due to the austenitic transformation at
723°C were observed. By the formula above, the absorption coefficient
depends on c which is also a function of temperature. For these experiments, c was taken to be an increasing function of temperature from 0.538
J/g °c at room temperature to 0.835 at the transformation point. Above
that point, specific heat data show more scatter. It was assumed to hold
constant at 0.835 from the transformation temperature up to the melting
temperature (1130°C) in our analysis.

Figure 2 details laser absorption coefficient versus temperature for
a grit-blasted surface as determined by the procedure outlined above.
There is an increasing trend with temperature as shown by the dashed
line. It should be noted that the room temperature average value of 25%
is three times the absorptivity for a highly polished surface of the same
material. The solid curve is explained in the next section.

"Homogenizing" Laser Melt Passes

Application of the scanned CW laser melting procedure to a grit-
blasted coupon resulted in the structures shown in Figure 3. The 15 x 15
x 3 mm specimen was translated under the 10 kW beam at 5.4 cm/s. The
macro photograph of the specimen's irradiated surface reveals a glazed
fusion zone that extends over more than half of the surface area. In
cross-section, a dish-shaped darkened region proved to be the heat
affected zone which reached a depth of 500 μm. A shallow fusion zone
about 7 mm wide and 50 μm thick can be seen clearly in the lower right-
hand photo of Figure 3. Substantial grain refinement has taken place in
the fusion zone microstructure.

Pulsed Laser Melts

An overlapping pulsed-laser-melted cross section is presented in
Figure 4. This micrograph is taken from work which was previously
reported in Reference (6). The as-cast microstructure shows the familiar
white cast structure consisting of the light Fe_3C phase and a darker
portion corresponding to prior austenitic grains. Free carbon is present
in two forms: as small particles within the darker phase network and as
larger, mostly spherical inclusions.

The laser melted layer is seen as a thin featureless surface zone.
The laser pulse interaction with carbon resulted in selective vaporization and particle ejection from the surface leaving a cratered morphology. A large crater appears in the center of Figure 4. These craters
tended to be smoothed by the multiple pulse melts; their presence also
increased the incidence of surface cracks. Crater formation during the
laser pulse interaction was observed by open shutter photography as isolated luminous specks.

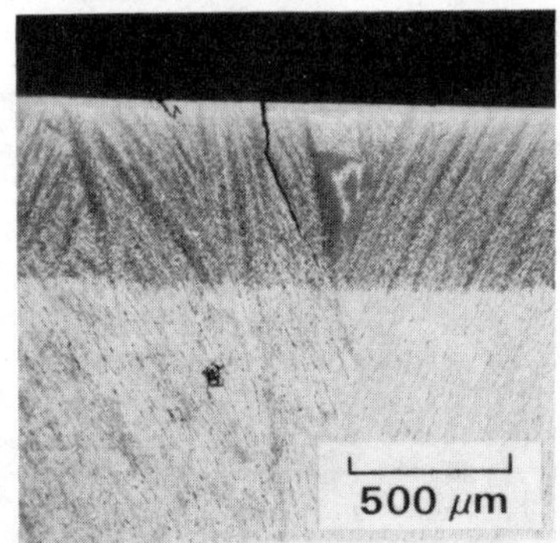

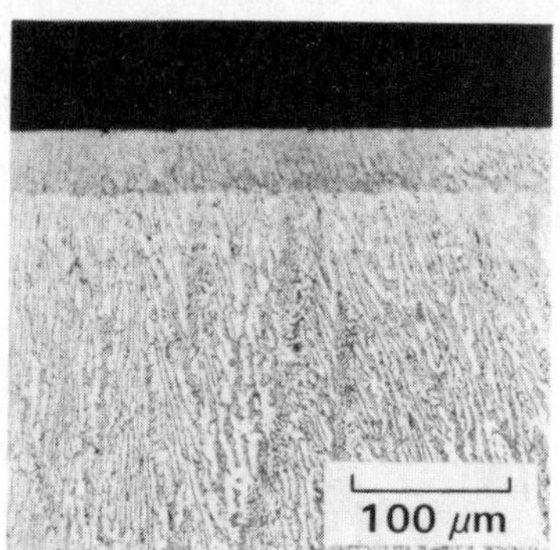

Figure 3.   CW laser surface melted sample.   The 15 x 15 x 3 mm grit-blasted coupon was subjected to a single 10 kW beam pass at 5.4 cm/s.

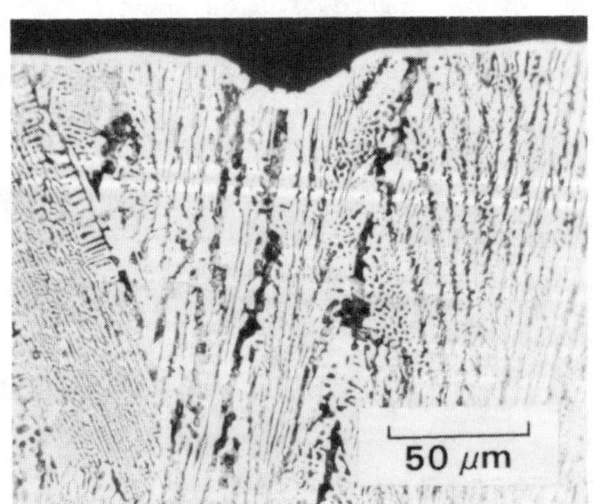

Figure 4.   Overlapping TEA laser pulse surface melts on as-cast Fe-4.2C.

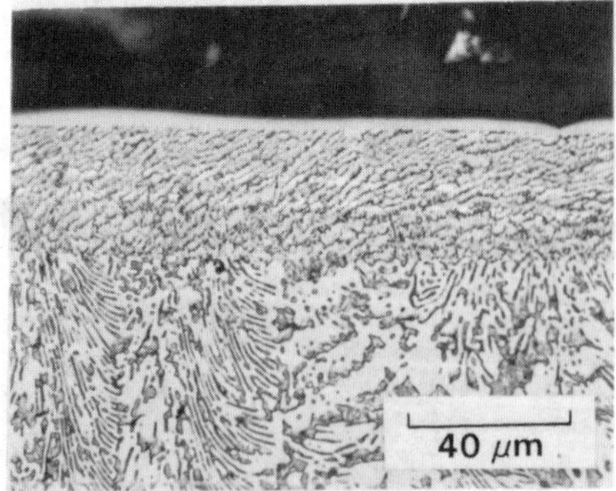

Figure 5.   Overlapping TEA laser pulse melts on an homogenized sample. The 30 μm grain-refined area arises from the scanned CW laser pass.

<u>Doubly Processed Surfaces</u>

Overlapping laser pulses on homogenized coupons produced more uniform surface melts than were achieved with the original cast samples. This point is illustrated in Figure 5. There are three horizontal strata in the figure. The upper layer is the pulsed laser melted zone atop a grain-refined layer produced by the CW laser beam. The lower region is part of the heat affected zone from the scan melt pass. The homogenizing melt zone microstructure resembles the underlying structure as far as grain shape is concerned. This correspondence was observed to be true generally even for different regions of the same specimen. Thus, a strong orientation relationship between the resolidified microstructure and the substructure was inferred. This orientation appeared to take precedence over the presumed heat flow direction for a broad range of scan rates.

The pulsed laser melts on previously melted surfaces showed significantly less cratering than was observed for unhomogenized samples. This was determined by the appearance of the processed surfaces and the absence of specks in the open-shutter photographs. Surface cracking was reduced, but not eliminated entirely. The surface melt in Figure 5 has a smoother and more uniform appearance than that of Figure 4.

Properties of the various processed regions of the sample in Figures 3 and 5 are presented in Table I and Figure 6. Microhardness indents were made with a Knoop indenter under a load of either 25 g or 500 g. The indents were produced on polished cross sections with the exception of the pulsed melt data which were taken from indents of the as-melted surface. Phase analysis was accomplished by X-ray diffractometry on rectangular samples about 0.25 cm^2 in area. Filtered CuKα X-radiation which is strongly absorbed by iron was employed. The four scanning electron micrographs of Figure 6 came from approximately the same vertical section of the sample. The photos are all at the same magnification and preserve the same orientation.

Table I. Doubly Processed Fe - 4.2C

Region	Knoop Hardness 25 g	500 g	Phases Present
As-cast	900	700	Fe$_3$C (Ortho.) α (BCC) α' (BCT) 0.35 w/o C
HAZ	1550	1100	
Scan melt	1590	--	Fe$_3$C (O) α' (BCT) 0.8-1.6 w/o C γ(FCC) 0.4-1.9 w/o C
Pulsed melt	1025*	--	ϵ (HCP) 4.34 w/o C γ (FCC) 1.20 w/o C Fe$_3$C (distorted)

* Surface reading.

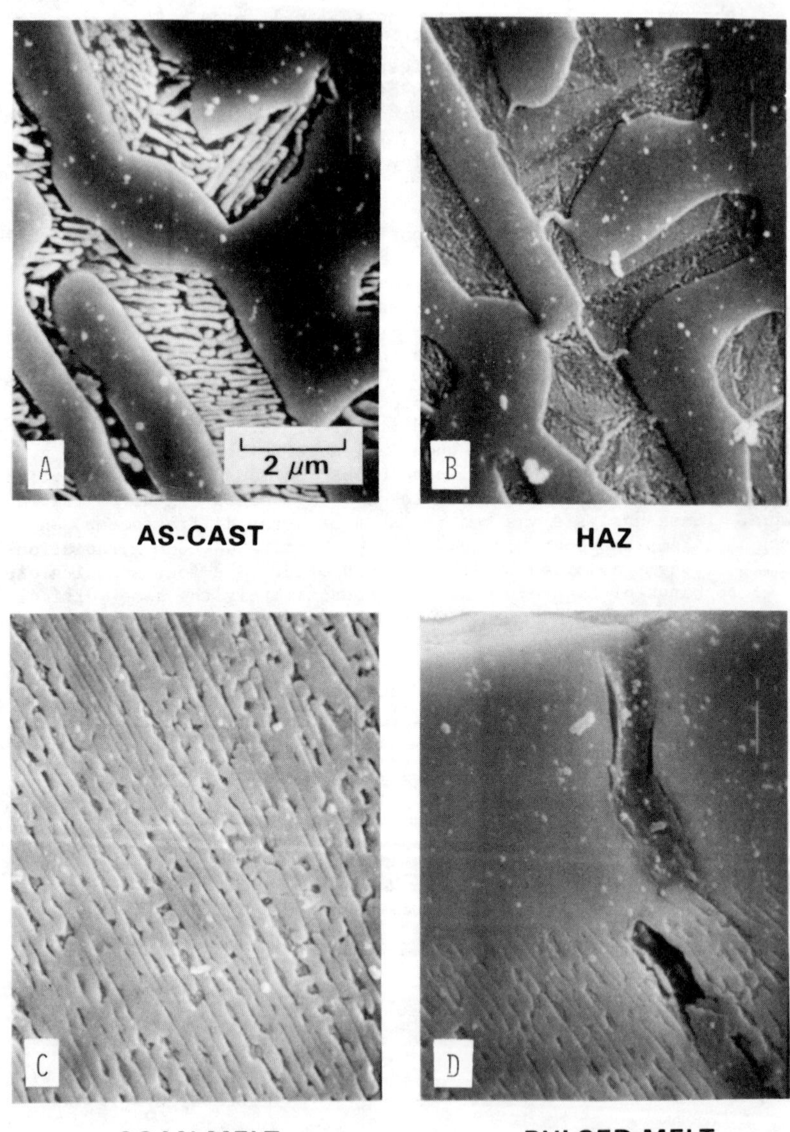

Figure 6.   SEM micrographs of various regions in a doubly
processed specimen. All photos are taken at the
same magnification and orientation to the sample
surface.

The as-cast structure in Figure 6A shows the lightly etched eutectic carbide phase and a deeply etched pearlitic structure. The phase determination of orthorhombic Fe_3C, ferrite, and low-carbon martensite as well as the hardness value of 700 KHN are consistent with the observed microstructure. Significant hardening was measured in the heat affected zone. This increase of more than 50% resulted in a higher hardness reading than for Fe_3C alone. The HAZ microstructure of Figure 6B shows wedge-shaped grains in the prior austenite regions. Their shape and the etching behavior suggest the presence of high-carbon martensite.

Data from the scan melt region confirmed the martensite determination. The hardness in the homogenized zone was nearly identical to that in the HAZ. Moreover, the phases detected were carbide, high-carbon martensite, and a small amount of retained austenite. The average carbon content of the martensitic phase (1.2 wt. %) correlates well to the measured microhardness value of 1100 KHN.

The microstructure of Figure 6C shows the same general appearance as the underlying structure except that the grain sizes are smaller by a factor of about five. The slanted growth direction of the scan melt microstructure extrapolates directly to the orientation of the underlying primary carbide plates in the HAZ. The scale of the homogenized layer microstructure is too small to visually determine the structure of the prior austenite grains.

For the pulsed fusion zone, a different structure was observed. The layer is featureless on the 1 μm scale of Figure 6D. Microhardness data indicate a softer structure than the underlying material, though it is probably harder than the as-cast microstructure. Phase analysis revealed the presence of the hexagonal ε phase, retained austenite, and distorted carbides. The carbon content of the ε phase was determined by application of the formula from Reference (1) to our measured lattice parameter. The austenite exhibited a fairly narrow carbon concentration as opposed to the variations observed in the homogenizing layers. In addition, a strong orientation of the γ phase was observed with the (1,1,1) direction normal to the specimen surface. The measured value of 1.2 w/o C in the phase corresponds closely to the value of 1.18 w/o C reported by Ruhl and Cohen (1) for splat quenched Fe-4.27C.

<u>Discussion</u>

One surprising aspect of the scan melt regions was their shallow depth and their broad width relative to the beam diameter. Simple calculations predicted a narrower, more dish-shaped fusion zone. In this case, the explanation related to the absorption coefficient data shown in Figure 2.

A two-dimensional thermal conduction computer code with a moving heat source (ANSYS) was applied to the problem employing laser absorption coefficients represented by the solid curve in Figure 2. The code provided for temperature-dependent thermal conductivity and specific heat, but not for laser absorption. Thus, the code was run until the surface temperature reached about 400°C at which time the absorbed flux was changed manually. Another change was made when surface melting was achieved. With this set of input parameters, the calculated results confirmed observations.

Figure 7 plots the calculated and measured melt and heat affected zone profiles for the sample shown in Figure 3. The computer-calculated data were interpolated from a rather coarse calculational mesh in the

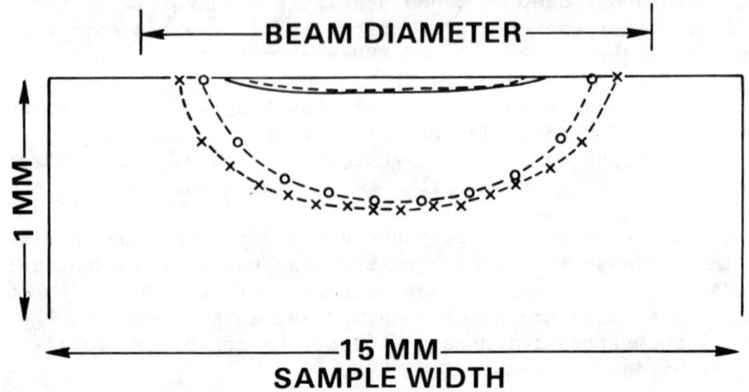

Figure 7.   Melt and heat affected zone profiles for the
homogenizing 10 kW laser melt pass of Figure 3.
X = measured HAZ depth; o = calculated 730°C
isotherm; solid line = measured fusion zone depth.

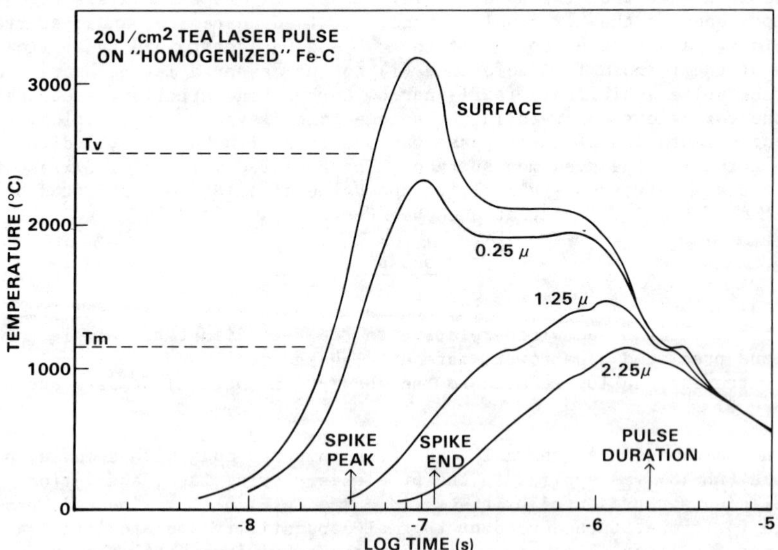

Figure 8.   Calculated time-temperature profiles for
different depths under the specimen surface
for a 1.5 μs pulsed laser melt on an homogenized
sample.

transverse direction (1 mm) and in time ($\sim$ 20 ms). Although the calculated profiles underestimated the data by about 10%, the agreement was within expectations when the resolution of the absorption curve and calculational elements were considered. Employing the same calculations with the assumption of a constant absorbed laser flux (25% of incident average flux) resulted in peak specimen temperatures below the melt point.

Reliable experimental data for the laser coupling coefficient of the liquid alloy do not exist. However, it is reasonable to assume that the molten absorptivity would lie closer to data for polished specimens than for roughened surfaces. On this basis, a rather liberal coupling estimate of 20% was chosen for molten surfaces. The calculated results indicate that it is (1) the increase in beam absorption with temperature which allowed the surface to reach melting at the given process condition and (2) the absorption coefficient reduction on melting which tended to reduce the melt superheat. These factors combined to produce a shallow, more uniform homogenized layer than is usually achieved.

A one-dimensional finite-difference thermal absorption and conduction program was created to model the pulsed laser melting and solidification conditions. The laser pulse time history was approximated by a triangular spike followed by a linearly decreasing tail flux. Temperature-dependent laser absorption and specimen specific heat were included along with heat of fusion effects. The input beam intensity was assumed to be spatially uniform and the surface to be planar. The temperatures calculated for the uppermost calculational cells were extrapolated to the front surface for the determination of the laser absorption.

Calculated time-temperature profiles for various depths of an homogenized Fe-4.2C sample exposed to a 20 J/cm^2 laser pulse are given in Figure 8. The initial 10.6 μm absorptivity measured for the lightly oxidized surface and input to the code was 17%. The calculated melt depth for this process condition is 2.25 μm. The overlapping pulsed melts of Figure 5 were produced with 50 J/cm^2 per pulse. This beam condition which produced 5 μm-deep melts is associated with a calculated melt depth of 4.25 μm.

Several results can be inferred from the curves of Figure 8. First, the peak surface temperature occurs near the end of the gain-switched spike. Most of the fusion zone remains molten for 3-4 μs, during which the temperature gradient in the liquid changes dramatically. Thirdly, the cooling rates for the various depths converge below the melt point to a range on the order of 10^8 °C/s. Finally, the speed of the solidification front can be estimated from these curves to be about 1.5 m/s.

Although the pulsed laser fusion zones appeared to be featureless in the optical and SEM micrographs, a weak cellular structure was revealed by thin-section transmission electron microscopy (TEM). The primary cell size ranged from 0.15-0.20 μm which corresponded to the grain size in the homogenized melt zones. A finer subcell structure was also observed. Additional TEM and selected area diffraction data will be required to fully determine the rapid solidification structure.

Conclusions

To summarize, continuous wave and pulsed CO_2 laser beams have been applied to binary eutectic iron-carbon samples to produce thin surface melt layers. On the basis of these results, the following conclusions may be drawn.

(1) Changes in laser absorption due to differing surface conditions
     can be used to help control the melt process. In particular,
     increasing the absorption coefficient of the solid by surface
     roughening to a value higher than the molten absorptivity, tends
     to minimize the melt superheat and produce a uniform melt pool.

(2) The fusion zone microstructure exhibits a strong epitaxial rela-
     tionship to the substructure.

(3) For CW laser melting at slow scan rates, the predominant harden-
     ing mechanism for Fe-4.2C is the production of high-carbon
     martensite.

(4) Using 1.5 μs laser pulses in vacuum, uniform fusion zones from
     2-10 μm in depth are produced. Calculations indicate quench
     rates on the order of 10^8 °C/s and an average solidification
     speed around 1.5 m/s.

(5) The pulsed laser melt layers contain microcrystalline phases
     comparable to splat-quenched specimens of the same nominal alloy
     composition.

Acknowledgements

The authors wish to express gratitude to B. Campbell and G. Long of
the Battelle laser lab facilities for assistance with the laser
processing experiments, to H. Dooley and A. Skidmore for extensive
metallographic preparation and analysis, and to P. Schumacher for the
X-ray diffractometer traces.

References

1. R. C. Ruhl and M. Cohen, "Splat Quenching of Iron-Carbon Alloys", Trans. AIME, 245 (1969) pp. 241-251.

2. P. H. Shingu, K. Kobayashi, K. Shimomura, and R. Ozaki, "An Amorphous Phase in a Splat-Cooled Fe-3.8 wt. % C Alloy", Scripta Met., 8 (1974), pp. 1317-1319.

3. M. C. Cadeville, G. Kaul, M. F. Lapierre, C. Lerner, and Ph. Maitrepierre, "On Metastable Dilute Binary NiBx and FeCx Alloys Obtained by Splat Quenching", pp. 143-155 in Rapidly Quenched Metals, Proceedings from the Second International Conference, Section I, N. J. Grant and B. C. Giessen, eds.; MIT, Cambridge, Mass., 1976.

4. B. G. Lewis, D. A. Gilbert, and P. R. Strutt, "Laser and Electron Beam Melting of Iron-Base Hard Materials", pp. 747-753 in Laser and Electron Beam Processing of Materials, C. W. White and P. S. Peercy, eds.; Academic Press, NY, 1980.

5. P. R. Strutt, D. A. Gilbert, and H. Nowotny, "Heat Treatment of Laser Melted M2 Tool Steel", pp. 232-238 in Laser - Solid Interactions and Laser Processing - 1978, S. D. Ferris et al., eds.; American Inst. of Physics, NY, 1979.

6. C. T. Walters, A. H. Clauer, and B. P. Fairand, "Pulsed Laser Surface Melting of Fe-Base Alloys", pp. 241-245 in Rapid Solidification Processing - Principles and Technologies, II, R. Mehrabian, B. H. Kear, and M. Cohen, eds.; Claitor's Publishing Div., Baton Rouge, LA, 1980.

7. C. T. Walters, T. R. Tucker, S. L. Ream, and A. H. Clauer, "Thermal Coupling of CO_2 Laser Radiation to Metals", this volume.

LASER SURFACE ALLOYING: THE STATE OF THE ART

Clifton W. Draper
Western Electric Engineering Research Center
Princeton, New Jersey 08540

In assembling this review article, it became apparent that the metal processing area known as laser surface alloying is far removed from maturity. In fact, it could hardly be placed in its adolescence. A reasonable comparison to emphasize our lack of progress can be made by examining the real advances made in the equivalent processing of semiconductor materials [1-3]. In essentially three years, several dozen research groups have taken a new field and developed a mature science and functional understanding in a large number of binary systems which are far removed from equilibrium both dynamically during processing and constitutionally in post-processing composition. The literature available in one journal (Applied Physics Letters) in one year alone (1980) on the formation of metal-silicon surface alloys out numbers all the literature in the surface alloying of metals.

Although numerous factors could be put forth to explain these vastly different rates of progress, only two are really relevant: 1. The potential for significant payback is perceived by the semiconductor industry, and 2. Semiconductor processing has historically pushed technological advances instead of being forced to change by them - the industry thrives on new ways of doing things.

What does this tell us about the prospects for surface alloying of metals in the near future? First, when the cost savings realizable through surface rather than bulk alloying become significant, we can expect greater interest and commensurate progress. For precious and strategic metals that time is nearly upon us. The second point may be more crucial when applied to metals processing. It probably means that domestic bulk alloys will be competing with cheaper foreign surface alloyed equivalents at some point in the not too distant future.

One may justifiably ask at this point, why then bother reviewing this field. There are three reasons, first, the published literature in this area is widely scattered. Much of it is hidden in published conference proceedings. Most of the Russian work is not translated. An attempt has been made here to reference all the available scientific work. Secondly, laser surface alloying dates back to 1964. A review is simply overdue. Third, a new direction in metal surface alloying research is just getting underway. Q-switched laser processing is extending the attainable quenching rates to the 10^{10} K sec^{-1} region and resolidification interface velocities to the 1-20 m sec^{-1} range. The use of very thin layers (hundreds or thousands of Angstroms) and well characterized host lattice metals will supplement the continuing interest in transition metal and carbide addition to ferrous alloy systems. Analytical techniques borrowed from thin film technology are finding wider use in metal surface alloy compositional analysis. It is, therefore, appropriate that we bring together in one place a hopefully coherent picture of where we presumably stand so as to appreciate perhaps more fully where we are going.

In what follows the popular laser systems employed for making surface alloys at the metal-environment interface will be briefly mentioned. The emphasis will be on the positive aspects and shortcomings of the particular laser system and associated parameters as they apply to the making of surface alloys. A chronologically tabulated reference list will be given and summarized. Finally, selected research efforts will be discussed in the hope that a meaningful status quo may be assessed.

The realm of laser surface alloying (LSA) covered in this review encompasses all metallic substrates in which a concentration gradient normal to the surface has been formed due to the incorporation of an alloying species during the self quenching of a surface liquid formed by laser irradiation. As will be seen in greater detail below, the depth over which this alloying takes place can vary from fractions of - to hundreds of - micrometers.

The methods used to predeposit and in some cases simultaneously deliver the alloying elements to the metallic substrate are far too numerous to discuss here. It is sufficient to say that all forms of coating techniques have been employed including vacuum evaporation, plating, powder coating, thin foil application, ion implantation and on

and on. The LSA patent literature is particularly over weighted with
lengthy descriptions of preirradiation coating methods. This author is
aware of only one [17] systematic study comparing various preirradiation
methods. Another point worth detailed investigation is the role of
interface cleanliness. How will the evaporation of organic contaminants
or outgassing of surface oxides effect the LSA processing?

Assume for the present that a spatially uniform coating of element
"A" has been applied to metallic substrate "B" as shown in part (a) of
Figure 1, with stepped concentration gradient as seen in part (b). The
purpose of LSA processing is to transform that film and substrate structure
to a surface alloy, as represented schematically in parts (c) and (d) of
Figure 1.

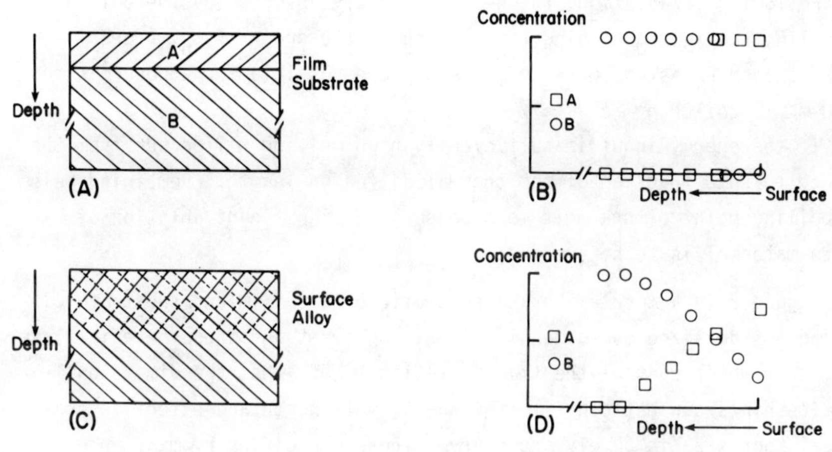

Figure 1 - Schematic representation of film +
substrate structure and concentration
depth profile, before (A, B) and after
(C, D) laser surface alloying.

In order to appreciate how the various laser systems and associated
processing parameters effect the manner of the transformation sketched
in Figure 1, some appreciation of the absorption of electromagnetic
radiation and subsequent heat diffusion is needed. This topic has
adequately been described [4-9] elsewhere and the sequence presented here
is simply meant to be illustrative.

The temperature rise induced in a material by laser irradiation depends
both on the thermophysical properties of the material and on the energy-
time characteristics of the focussed laser output. The optical properties

characterizing the metal surface are the absorptance, α, and reflectance, R_o. A portion of the incident laser intensity I, is absorbed by the metal within its electromagnetic skin depth, typically 10-100 nm, and is converted into heat. The heat rapidly diffuses away from this near surface region to depths given by the thermal diffusion length, $(2Dt_p)^{1/2}$, where t_p is the dwell time or pulse time of the laser output and the thermal diffusivity $D = K/C\rho$ where K, C and ρ are the thermal conducitivity, specific heat and density, respectively. For a square wave laser pulse, the average temperature rise within this thermal diffusion layer is:

$$\Delta T = \frac{(1 - R_o)It_p}{C\,\rho(2Dt_p)^{1/2}}$$

Following irradiation, the heat is rapidly absorbed by the bulk of the solid, the self quenching times being on the order of t_p. For ΔT = 1000K, quench rates of $10^5 - 10^{10}$ K sec^{-1} are possible for exposure times of 10 m sec to 100 n sec.

If the energy input is sufficiently high melting will occur from the surface to some fraction of the thermal diffusion length. Remaining below the boiling point of the melt is necessary if significant ablation of molten material is to be avoided.

Figure 2 is a more elaborate schematic of the film and substrate, including an idealized one-dimensional melting event and those material/laser parameters which control the transformation. Note that the dimensional relationship shown between film thickness, melt depth and effective or optical spot size is purely arbitrary. These can differ by many orders of magnitude. In general, only the reflectance of A is important, whereas both the thermal diffusivity of A and B control the thermal diffusion length and melt depth. By selecting A & B combinations where the melting points are not too far apart problems can be avoided. High vapor pressure elements should be avoided where possible.

Because the thermal diffusion length is a function of t_p, the dwell or pulse time, this laser parameter offers the greatest access for processing control. The significance of this parameter, as well as those alluded to above, will become clear in many of the examples given below.

The lasers which to date have been utilized in LSA processing include the continuous CO_2, electrically pulsed Ruby, Nd-YAG and Nd-Glass, and the Q-switched Nd-YAG and Ruby. Table 1 compares some of the laser characteristics of these systems. It is not the object of this

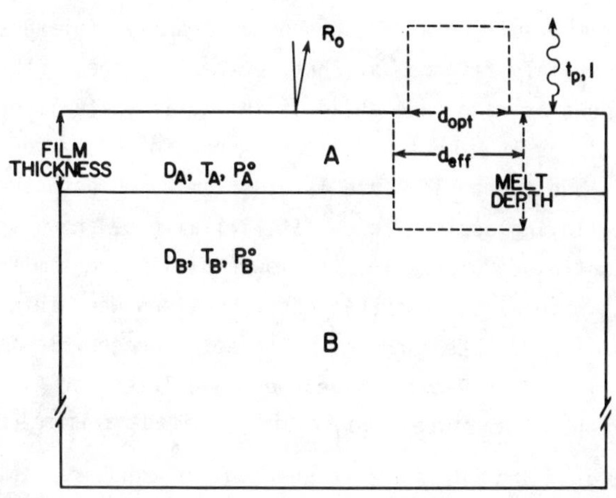

Figure 2 - Schematic of film, substrate and laser event. Important material and laser parameters are as follows: D_A, T_A, P_A^o D_B, T_B, P_B^o - the thermal diffusivity, melting point and vapor pressure of A (the film) and B (the substrate) respectively. R_o - the normal spectral reflectance of A. I, t_p - the incident laser intensity and pulse or dwell time. d_{opt}, d_{eff} - the optical and effective spot sizes.

TABLE I

COMPARISON OF LASER CHARACTERISTICS

FOR LASER SOURCES USED IN LSA PROCESSING

Laser Source	Laser Wavelength (μm)	Power Output (W)	Dwell or Pulse Time (s)	Repetition Rate (Hz)
Continuous CO_2	10.6	500-20,000	10^{-1}-10^{-5}	--
Pulsed Nd:YAG	1.06	200 (average)	10^{-2}-10^{-3}	100
Pulsed Nd:Glass	1.06	10^6 (peak)	10^{-2}-10^{-3}	1
Pulsed Ruby	0.69	10^6 (peak)	10^{-2}-10^{-3}	0.2
Q-Switched Nd:YAG	1.06, 0.53	1-10 (average)	1.5×10^{-7}-3×10^{-7}	10^3-5×10^4
Q-Switched Ruby	0.69	10^8 (peak)	5×10^{-9}-5×10^{-8}	2.0

paper to enumerate the selling advantages of one family of lasers over another. It is certainly fair to say that, presently, the continuous CO_2 is several orders of magnitude ahead of the solid state lasers in processing capabilities when measured as area transformed per unit time. This does not preclude the solid state lasers from being both superior and cheaper in point alloying applications. It will also be shown that for relative motion continuous processing the lower dwell time limit is still too long to effect high alloying specie concentrations when thin films (<5000 Å) are utilized. The range of melt depths accessible for the dwell/pulse times listed in Table 1 have been calculated for Al, Fe and Ni by Hsu, et. al. [5]. Surface vaporization limited-maximum Ni melt depths of 5000 Å, 5 μm and 50 μm are calculated for exposure times of 100 nsec, 10 μsec and 1 msec respectively. Judicious comparisons with experimentally determined melt depths are difficult owing to the uncertainties in the spectral reflectances. Calculated melt depths can provide guidance in choosing appropriate film thicknesses to be melted through in order to reach the substate. A 20,000 Å thick film of Ni on any metallic substrate will not melt through to that substrate if irradiated with 100 nsec Q-switched laser pulses.

In tabulating the references below, great lengths have been taken in attempting to include all scientific literature that falls within the scope of laser surface alloying. Search methods included computer keyword and author searches of Chemical Abstracts, Metals Abstracts and Laser Abstracts. Written correspondence and telephone conversations with other active researchers in this field have been very helpful. Much of the Russian literature was backtracked through citations, unfortunately limited access and funds have prohibited an exhaustive effort in searching the foreign literature.

Several related subject areas have specifically not been included. The formation of metastable, glassy or unusually microstructured surfaces via laser surface melting is an important, active research area. Many examples exist of modified surface behavior directly attributable to microstructural changes induced by laser melting/rapid quenching. This area has been amply covered by the literature on rapid solidification processing. Laser cladding is another closely related technology. The object of that processing, however, is to avoid alloying as much as possible. Discussions of laser cladding can be found in general reviews of material processing with lasers and in reviews of metallurgical coating techniques.

Table II is a chronological list of the laser surface alloying
literature. No attempts have been made to order the literature on
a finer scale then the year of publication. Table headings include
the reference year, type of laser source, alloying specie/host substrate,
post irradiation analysis of structure, composition or modification of
surface sensitive behavior, and finally the first author, nationality and
numerical reference. Abbreviations used are explained in Table III.
Taking as an example the 1978, Kovalenko work. Samples of Fe coated
with V were alloyed with a pulsed Nd-glass laser. Alloyed samples were
analyzed structurally using optical microscopy and X-ray diffraction.
Compositions are given based on lattice parameter changes in the X-ray
diffraction patterns. The microhardness of the irradiated samples is
reported.

TABLE II

Year	Laser	Alloying Specie/Substrate	Analysis	Reference
1964	P-Ruby	C,W,TiC,WC/St	Comp:EPM SSB:MH	Cunningham (Am.) [10]
1969	P-NdG	C/Fe	STRT:XRD SSB:MH	Mirkin (USSR) [11]
1969	P-Ruby	WC/St	STRT:OM Comp:EPM SSB:WR	Schmidt (Am.) [12]
1972	P-Ruby	C,WC/St	STRT:XRD SSB:MH	Betaneli (USSR) [13]
1974	P-NdG	C,W/Fe;Co/W;Al/Nb	STRT:OM, XRD	Gazuko (USSR) [14]
1975	$CW-CO_2$	Cr,Ni,W,V,Mn,C/St	Comp:ANG SSB:MH	Seaman (Am.) [15]
1976	P-NdG	Ni,Mo,Ti,Ta,Nb,V/Fe,St	SSB:MH	Kovalenko (USSR) [16]
1977	P-NdG	Mo,Nb/Fe	Comp:ANG SSB:MH	Kovalenko (USSR) [17]
1977	P-NdG	Mo/Fe	STRT:OM,XRD Comp:XRD SSB:MH	Belotskii (USSR) [18]
1978	$CW-CO_2$	Cu/Ag-7 Cu	STRT:OM,SEM	Copely (Am.) [19]
1978	P-NdG	V/Fe	STRT:OM,XRD Comp:XRD SSB:MH	Kovalenko (USSR) [20]
1978	Q-Ruby	Cu/Al	STRT:CHN Comp:RBS	Della Mea (It.) [21]

TABLE II (continued)

1978	CW-CO$_2$	Cr,Ni/St	STRT:OM Comp:EPM,EDX	Weinman (Am.) [22,23]
1978	CW-CO$_2$	Cr,C,Mn,Al/St	STRT:OM Comp:EPM SSB:MH	Gnanamuthu (Am.) [24]
1979	P-NdG	Fe/Nb	STRT:MOSB,XRD	Dekhtyar (USSR) [25]
1979	CW-CO$_2$	Mo-Ni-Si-Cr-C/St	STRT:OM,XRD Comp:EDX,ANG SSB:MH,HTS,WR	Belmondo (It.) [26]
1979	CW-CO$_2$	Cr/St	STRT:OM,SEM Comp:EPM,EDX,AES	Moore (Am.) [27]
1979	CW-CO$_2$,Q-NdY	Au,Ag,Pd,Sn,Ta/Ni	STRT:OM,SEM,CHN Comp:RBS,EDX,AES, ESCA SSB:MH,COR	Draper (Am.) [28]
1979	CW-CO$_2$	TiC,WC/Ti-6Al-4V, Inconel	STRT:OM,SEM	Schaefer (Am.) [29]
1980	Q-NdG	Sb/Al	STRT:OM,XRD Comp:RBS	Jain (Ind.) [30]
1980	CW-CO$_2$	TiC,WC/St,Ti-6Al-4V, Inconel,5052Al	STRT:OM SSB:MH	Ayers (Am.) [31]
1980	P-NdY	Re/Fe-50 Ni	STRT:OM,SEM Comp:EPM	Takei (Jap.) [32]
1980	Q-NdY	Au,Ag,Ta/Ni	STRT:OM,CHN Comp:RBS	Buene (Am.) [33]
1980	Q-dNdY	Pb/Cu	STRT:CHN Comp:RBS	Rimini (Am.) [34]
1980	CW-CO$_2$	TiC/304 SS	STRT:SEM SSB:MH	Ayers (Am.) [35]
1980	Q-NdY	Pd/Ti	Comp:RBS SSB:COR	Draper (Am.) [36]
1980	CW-CO$_2$,Q-NdY, Q-dNdY	Au/Ni	STRT:CHN Comp:RBS	Draper (Am.) [37]
1980	CW-CO$_2$,Q-NdY, Q-dNdY,Q-Ruby	Au/Ni	STRT:CHN Comp:RBS	Jacobson (Am.) [38]
1980	Q-Ruby	Ni/Au/Ni	STRT:TEM Comp:RBS	Pronko (Am.) [MRS-80][39]
1980	Q-NdY	Au,Ag,Pd,Ta,Sn/Ni	STRT:TEM,CHN Comp:RBS	Buene (Am.) [MRS-80][40]
1980	Q-NdY	Hf/Ni	STRT:CHN,PAC Comp:RBS,AES	Kaufmann (Am.) [MRS-80][41]
1980	Q-Ruby	Sb,Mo,Cd/Al	STRT:CHN Comp:RBS	Battaglin (It.) [MRS-80][42]
1980	Q-Ruby	Sb/Al	STRT:TEM,CHN Comp:RBS	Peercy (Am.) [MRS-80][43]

TABLE III

ABBREVIATIONS USED IN TABLE II

LASER

```
   P - Pulsed
  CW - Continuous
   Q - Q-Switched
 CO₂ - Carbon Dioxide
 NdY - Neodymium Yttrium Aluminum Garnet
 NdG - Neodymium Glass
dNdY - Frequency Doubled NdY
```

ALLOYING SPECIE/SUBSTRATE - Elemental Symbols

```
St               - Steel
Commas (,)       - Denote Individual Experiments
Dashes (-)       - Denote a Mixture of Elements
Semicolon (;)    - Denote Individual Experiments (Added Clarity)

For Example      - C,WC/St - Denotes carbon film and tungsten carbide film
                             on a steel substrate.

                   Re/Fe-50Ni - Denotes a rhenium film on an iron-50 atomic
                             % nickel alloy substrate.
```

ANALYSIS

```
STRT - Structure
TEM  - Transmission electron microscopy
OM   - Optical microscopy
XRD  - X-ray diffraction
CHN  - Channeling
MOSB - Mossbauer spectroscopy
PAC  - Perturbed angle correlation


Comp - Composition
RBS  - Rutherford backscattering spectroscopy
EPM  - Electron probe microanalysis
EDAX - Energy dispersive X-ray, elemental mapping
ESCA - Electron spectroscopy for chemical analysis
AES  - Auger electron spectroscopy


SSB  - Surface sensitive behavior
MH   - Microhardness
COR  - Corrosion
WR   - Wear
HTS  - High temperature stability


ANG  - Results quoted but analytical method not given
```

Figure 3 is a bar graph of the number of LSA references versus the year through 1980. It is evident that a period of rapid growth in LSA research is underway. It is, of course, not possible to give an extract of each article in the space allotted. Most of the referenced work falls into one of two categories. Either work on the incorporation of transition metals or transition metal carbides into ferrous alloy systems, or the study of solute distribution when thin films or ion implants are irradiated with Q-switched laser pulses. The papers to be discussed will be separated into two groups following this format.

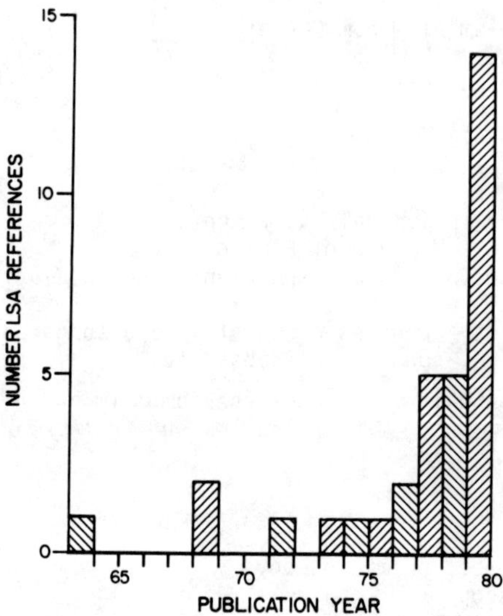

Figure 3 - Bar graph illustrating number of laser
surface alloying publications as
function of time.

Surface Alloying in Ferrous Systems

The technological importance of ferrous alloys is of course the main driving force behind the wide interest in this substrate material. From a purely historical standpoint, the Cunningham [10] pulsed ruby work on the surface alloying of steels is noteworthy. As far as this reviewer can determine, this Master's Thesis is the earliest reference-able work in this area. By and large, most of the results reported [10] were negative. All of Russian work [11, 13, 14, 16-18, 20, 25] is characterized by utilization of pulsed solid state lasers with pulse lengths of 1-10 m sec duration. The Belotskii, et. al. [18] work on

76

the surface alloying of high purity polycrystalline Fe with Mo is worth
noting as it clearly shows that the Russian scientists performing
surface alloying experiments appreciate the complex nature of the
mixing processes and liquid state diffusion occurring in the pulsed
laser regime of surface alloying. In this experiment thin foils of Mo
were rolled onto the Fe substrate, although other popular Russian
methods of deposition include powders and pastes of the alloying species
with a "water glass" binder. Postirradiation structural analysis
included optical microscopy of metallurgical cross sections and X-ray
diffraction work. Metastable α-Fe solid solutions of 28-36 wt % Mo
were formed over depths of 300-500 μm. The compositional analysis was
based on lattice spacing changes in the diffraction patterns. A
uniformly increased microhardness 1.5 times that of the unirradiated
bulk Fe was found as a function of depth within the alloyed melt zone.

American research [15, 22, 23, 24, 27] in surface alloying of
ferrous materials differs in two ways from that of the soviets. Most
of the American research has been done with multikilowatt cw CO_2
lasers in contrast to the low repetition rate solid state lasers. In
addition to an interest in producing hard, wear resistant surfaces,
there has been an active effort to produce stainless steel quality
surfaces on lesser corrosion resistant ferrous alloy substrates by the
addition of Cr, Mo and Ni through LSA processing.

One of the biggest advantages of relative motion cw CO_2 LSA processing is the inherent flexibility of the dwell time. Control over
the dwell time implies control over the melt depth and time in the
liquid phase. For a given predeposited film thickness this means
processing control over the resulting postirradiation concentrations.
Weinman, et. al [22, 23, 27] have reported results on the cw CO_2 LSA
processing of AISI 1018 steel with electroplated and sputtered films
of Cr or Cr/Ni. Film thicknesses ranged from 2 to 18 μm. A simple
rotating wheel configuration was used to produce line scans at various
transverse velocities. A portion of one of their tables is reproduced
in Table IV. In this case, the sputtered preirradiation Cr film thickness was 8 μm. Fairly homogeneous alloying was found throughout the
melt depth as seen in the Cr SEM X-ray map of a cross sectioned 50 cm
sec^{-1} melt stripe presented in Figure 4. Electron microprobe analysis
was used to obtain the Cr concentration at a depth of 10 μm from the
surface. Those values appear in the last column of Table IV. Although
there are some inconsistencies, the trends in linear velocity (dwell

TABLE IV

Melt #	Velocity cm/sec	Width μm	Depth μm	Chromium %
1	38	326	99	11
2	50	259	51	23
3	63	276	37	31
4	75	167	20	58
5	88	202	42	35
6	100	133	14	80
7	125	133	18	60

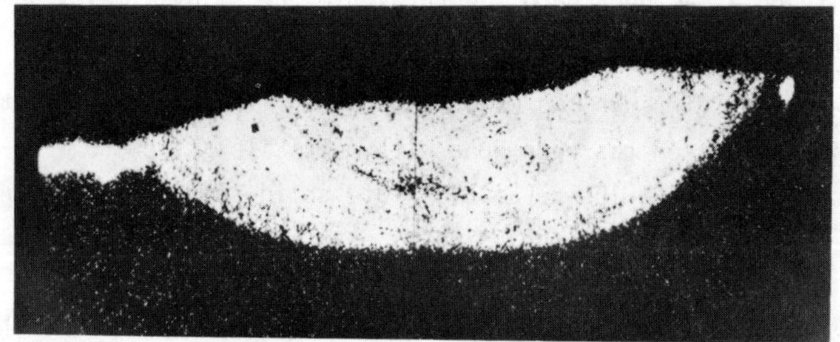

Figure 4 - SEM X-ray elemental map of Cr in cross sectioned 1018 steel [22, 23, 27]. Sputter deposited Cr was 8 μm thick. The laser dwell time was approximately 0.5 msec.

time), melt depths, and Cr concentrations are fairly clear. Longer dwell times result in greater penetration, longer melt times and less concentrated near surface alloying specie concentrations.

Gnanamuthu [24] has published a review of laser surface treatment using relative motion cw CO_2 lasers. The section on surface alloying describes the microstructures obtained when different alloy specie additives (in the form of metal powders suspended in liquid) were LSA processed into AISI 1018 steel. Addition of 3.5% Cr, 1.9% C and 1.3% Mn

transforms the austenite-cementite to a martensite microstructure. Depth
concentration profiles were determined using electron microprobe analysis.
The concentration profile, knoop microhardness impressions and microstructure are reproduced in Figure 5. The alloying depth is 125 μm.
The lower Mn concentration nearer the surface is attributed to selective
evaporation. Such high vapor pressure, selective, elemental loss behavior
has been observed [44, 45] in other laser irradiated metallic alloy
systems.

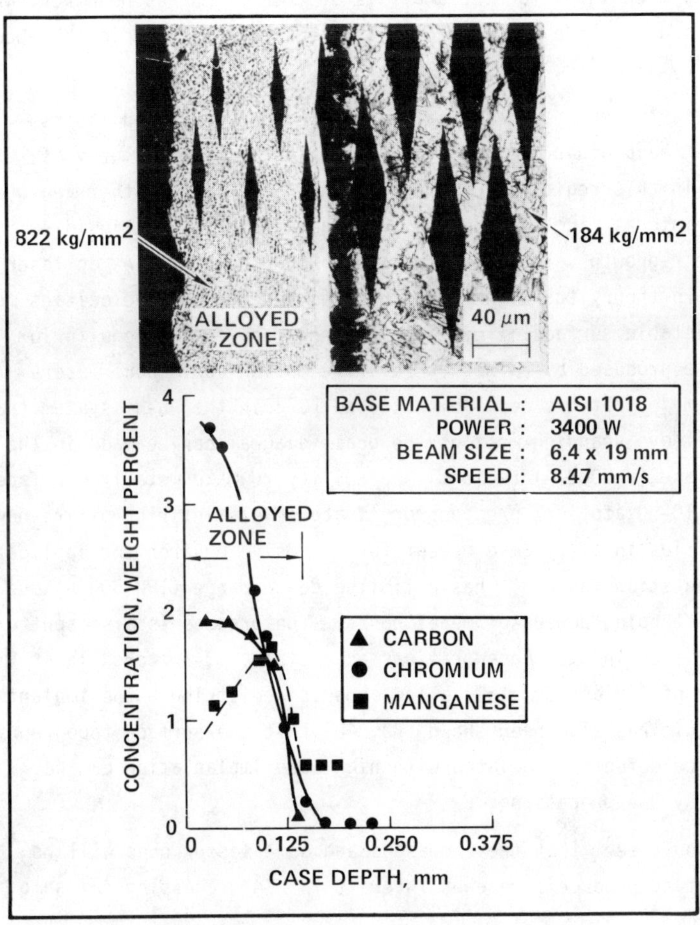

Figure 5 - Cross sectioned, laser surface alloyed
1018 steel [24]. Preirradiation films
were sprayed on (particles 10-50 μm in
size suspended in isopropyl alcohol).
Laser processing parameters, concentration profiles and microhardness results
are given in the figure.

Q-switched Laser Surface Alloying

References 21, 28, 30, 33, 34 and 36-43 all have two features in
common. The radiation source was a Q-switched laser and the analytical
technique utilized to determine the concentration profile was Rutherford
backscattering spectroscopy. The reason for this connection is that
the depth of melting associated with Q-switched laser irradiation of
most metals and the depths over which RBS can provide concentration
profiling are the same. That depth is $\leq 10{,}000$ $\overset{o}{A}$. The use of RBS and
channeling techniques for the analysis of thin films has been reviewed
[46]. Its use in the analysis of LSA processed Au-Ni samples has been
reported by Jacobson, et. al [38].

The fact that the analytical technique and Q-switched lasers
compliment each other does not fully explain the recent flurry of
interest in this regime of LSA processing. Some of the other reasons
for this interest include the very high quench rates and liquid-solid
interface regrowth velocities accessible with the 10-200 n s ec laser
pulses. In theory the higher quench rates and regrowth velocities mean
that metastable surface alloys even further removed from equilbrium
than those produced by pulsed or relative motion continuous lasers
could be made. It has been demonstrated [37] in the Au-Ni system that
surface alloys spanning most of the phase diagram can be made in the
near surface (< 5000 $\overset{o}{A}$) region. The ability to make metallic surfaces
that are 10-90 atomic % precious or strategic element rich offers new
possibilities in alloy development for corrosive environment applications.
Ion implantation in metals has a limited depth range (100-500 $\overset{o}{A}$) due to
the high stopping powers of metal hosts. The process is also sputter
damage limited in the concentrations accessible. Q-switched laser
annealing of implanted metals allows one to redistribute the implanted
species. It has also been shown [33, 40] that the surface topography
and lattice defects associated with high dose implantation can be
improved by LSA processing.

It would seem that these reasons and some lesser ones will be
sufficient to produce a renewed interest in LSA processing for some
time to come.

The Au-Ni system represents a binary alloy combination that has
received considerable attention [28, 33, 37-40]. From the laser processing point of view the combination offers some interesting problems, and
a great deal of information has been gained in the circumventing of

80

some of these constraints. The normal spectral reflectance of Au [47] is very high ($\geq$ 98%) at wavelengths longer than $\sim$0.7 μm. The thermal diffusivities of the two elements differ by a factor of 5 and this has interesting consequences in the melt depth as a function of Au film thickness. Au and Ni are sufficiently different in atomic number as to make them an excellent choice for RBS analysis.

Figures 6 and 7 contain before and after irradiation RBS spectra [38] for 3500 Å of Au on polycrystalline Ni LSA processed with cw CO_2 and Q-switched Nd-YAG lasers respectively. The relative heights of the Au and Ni backscattering yields are used to calculate concentrations. Figure 8 shows the calculated [37] depth profiles for the spectra in Figures 6 and 7 along with results on Au films irradiated with frequency doubled Nd-YAG, and Au films capped with 250 Å of Ni and irradiated with the fundamental Nd-YAG.

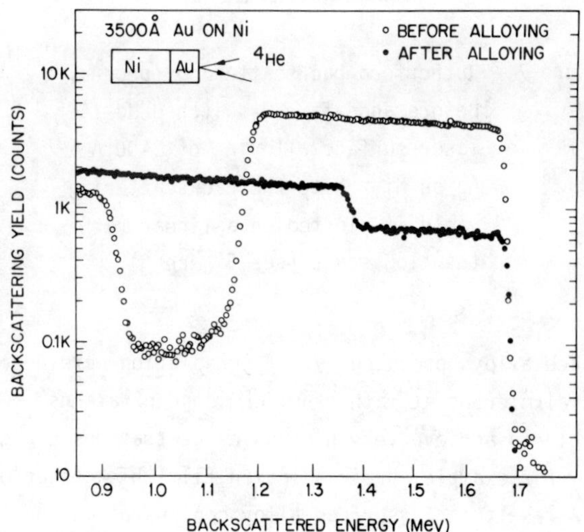

Figure 6 - Rutherford backscattering spectra.
Before and after cw CO_2 laser surface
alloying of 3500 Å Au on Ni. The width
of the Au peak is used to calculate
the preirradiation film thickness.
Following irradiation the ratio of the
peak heights gives the Au and Ni
concentrations.

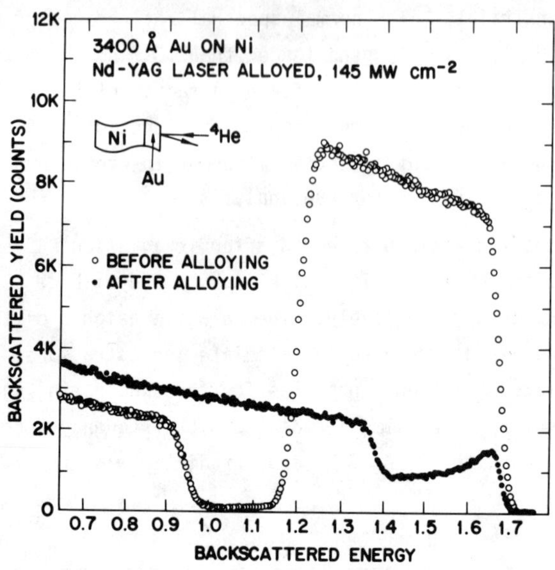

Figure 7 - Rutherford backscattering spectra.
Before and after Q-switched Nd-YAG
laser surface alloying of 3400 Å
Au on Ni. Note the backscattering
yield is plotted on a linear rather
than log scale (see Figure 6).

The Au surface alloys produced by CO_2 irradiation were determined to
extend over several microns at rather uniform concentrations. Therefore,
it was not possible to achieve very high Au concentrations using reasonably thin films. For example, preirradiation film thicknesses of 2,000,
3,500 and 6,100 Å result in Au surface alloy concentrations of 4, 11 and
20 at % respectively. Furthermore, with the higher power densities
required to alloy thick Au films, surface evaporation of the already
alloyed surface occurs since overlapping melt stripes are needed for area
coverage. The maximum surface concentration obtained for the Q-switched
Nd-YAG laser was about 15 Au at %, regardless of deposited film thickness.
The limitation for this processing proved to be the evaporation and/or
explosive removal [48] of the film. Both the high Au film reflectance
at 1.06 μm [47] and high peak power associated with the Q-switched pulse
contribute to the film loss problem. The normal spectral reflectance of
Au films thicker than about 500 Å at 10.6 μm is >98%. However, controlled

surface alloying can be achieved with CO_2 laser irradiation because of
the lower rate of energy deposition and concurrent lower thermal gradients.

The normal spectral reflectance at 0.53 μm (frequency doubled Nd-YAG)
is down by a factor of 2 relative to either 1.06 or 10.6 μm. For this
reason the incident power densities required for coupling to the film and
substrate are substantially lower. Very little Au is lost through
evaporation and concentrated surface alloys can be produced. Figure 8
shows that surface alloy concentrations of 50 at % Au can be produced by
frequency doubled Nd-YAG irradiation of a 1500 Å film. An alternative
approach to overcome the high reflectance at 1.06 μm is to cap the Au
with a Ni thin film. With this method an Au surface concentration of
> 50 at % is obtained by Nd-YAG irradiation of a 2000 Å Au film with
a 250 Å Ni cap.

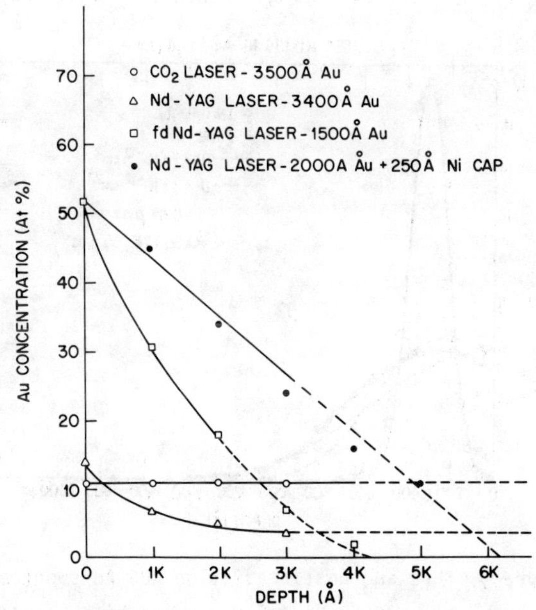

Figure 8 - Calculated depth profiles for laser
surface alloyed Au-Ni structures. Data
points come from RBS spectra like those
of Figures 6 and 7. Note uniform concentration for long melt time cw CO_2
and sharply sloped profiles for Q-
switched lasers.

Pronko, et. al have reported [39] the use of Q-switched ruby laser radiation of 35 n sec duration to surface alloy Ni-Au-Ni sandwich structures similar to those reported in references 37 and 38 except that the Au film thickness is 200 Å instead of 2000 Å. Figure 9 presents the backscattering yield for the Au before and after irradiation along with three solid curves. The Gaussion solid curve which fits the "as deposited" Au film is used in modeling the diffusion equation to a single adjustable parameter, Dt, the product of the diffusion coefficient and the melt time. Although the post-irradiation curves do not exactly fit the experimentally determined pro-files the agreement is impressive. It is reasonable to conclude that, in contrast to pulsed or relative motion cw CO_2 processing, convective mixing is not needed to explain postirradiation alloying specie distribu-tions in Q-switched LSA processing. Simple liquid state diffusion co-efficients and melt times on the order of the Q-switched pulse duration are sufficient to explain the resulting "frozen in" concentration gradients.

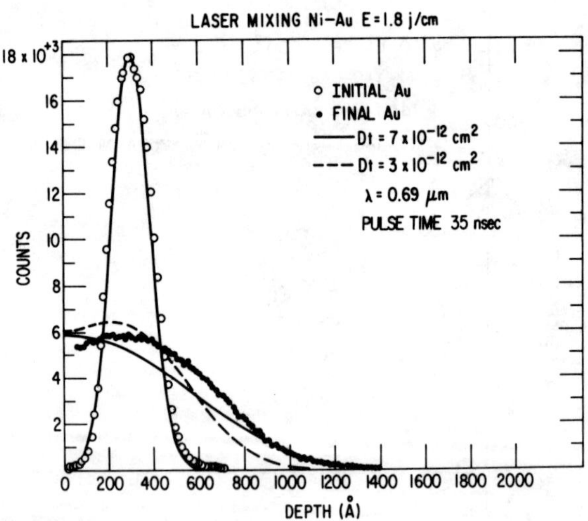

Figure 9 - Pre and postirradiation RBS Au spectra for 200 Å Ni + 200 Å Au on Ni-surface alloyed with a Q-switched Ruby laser. The dashed and solid lines represent theoretically calculated postirradiation spectra. Liquid state diffusion co-efficients and melt times on the order of the laser pulse length are sufficient to describe the postirradiation con-centration gradients.

As was mentioned above in introducing the Au-Ni system the differences in the thermal diffusivities of the two elements leads to an interesting aspect in Q-switched processing. It has been observed [49] in a series of Q-switched 1.06 μm irradiated 300 Å Ni - (z) Au-Ni <110> sandwich structures, where z = 500, 1000, 2000 and 5000 Å, that even though they were irradiated with the same incident power densities, the melt depth increased significantly with increasing Au layer thickness. Exactly the behavior one would expect if the apparent thermal diffusivity were sensitive to changes in relative, elemental film thickness over the thermal diffusion length of the 100 n sec pulse.

Figure 10 presents the final point to be made in examining the use of Q-switched lasers for surface alloying the Au-Ni system. In this case sandwich structures of 250 Å Ni - 2000 Å Au - Ni <110> were irradiated with both Q-switched Nd-YAG (t_p = 130 n sec) and Q-switched Ruby (t_p = 30 n sec). RBS analysis requires a sampling region of ∿ 1 mm diameter. The Ruby can produce an alloyed region of this size in a single shot,

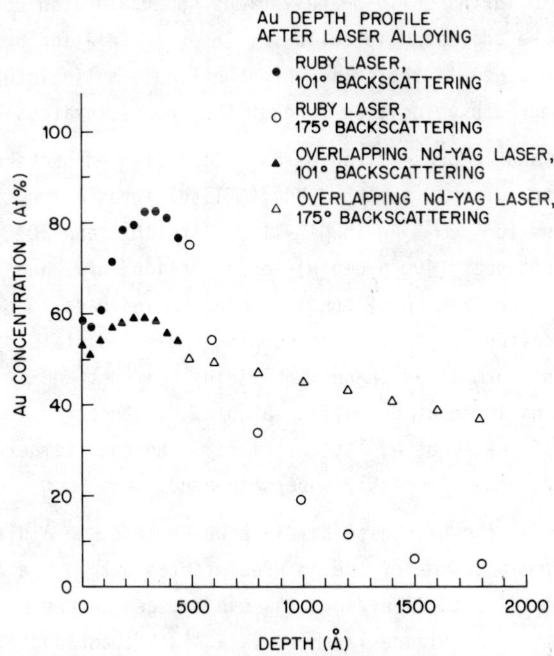

Figure 10 - Calculated depth profiles for Ni-Au-Ni
          structures surface alloyed with both
          Q-switched Nd-YAG (overlapped pulses)
          and Q-switched Ruby (single shot).

whereas overlapping melt spots, each ~ 30 μm in diameter, must be raster
scanned to produce sufficient area for analysis in the case of the Q-
switched Nd-YAG processing. Depending on the extent of overlap repetitive
melting and therefore increased homogenization through liquid state
diffusion will occur in the scanned case. The glancing angle data [38]
of Figure 10 provides improved depth resolution and shows significant
differences in the Au distributions in the range between the atomic
surface and a depth of 500 Å. The differences are due in part to both
the longer pulse time of the Q-switched Nd-YAG and the repetitive nature
of that LSA processing.

Since LSA processing relies upon liquid state diffusion in order to
transform the film + substrate to an alloyed region, one is confined by
thermodynamic constraints. The host and alloying specie must have liquid
state solubility. An interesting example of a system with a rather large
miscibility gap in the liquid is Ag-Ni [50]. This system has been examined
[33, 40] by means of ion implantation of Ag in single crystal Ni <110>
followed by Q-switched Nd-YAG surface melting. The implantation allows
one to distribute in the solid phase various supersaturated concentrations.
The question to be addressed is following laser irradiation how much of
the Ag can be retained in the solid when the liquid-solid interface moves
at these high regrowth velocities through the supersaturated liquid.

Figures 11 and 12 present random and <110> aligned backscattering
yields, before and after Q-switched Nd-YAG laser irradiation, for implant
doses of 10^{16} and 10^{17} cm^{-2} Ag in Ni <110>. The 10^{16} and 10^{17} cm^{-2} Ag
implants represent peak (400 Å depth) concentrations of 1 and 12 atomic
% respectively. Irradiation of the low dose sample leads to broadening
of the Ag backscattering yield corresponding to liquid state diffusion.
The fact that the normalized channeling yields for host and implant are
the same following irradiation may be taken as evidence for 100% sub-
stitutionality of the Ag on Ni lattice sites. The room temperature
metastable alloy formed is highly supersaturated.

Irradiation of the high dose sample however leads to a different
result. In this case, some of the Ag precipitates out in the liquid
ahead of the liquid-solid interface. As can be seen in the postirradiation spectrum of Figure 12 there is a significant redistribution
of Ag toward the surface. TEM work on thin sections [40] of the high
dose Ag in Ni <110> shows Ag dentrites and a high density of dislocations
in the host lattice following irradiation.

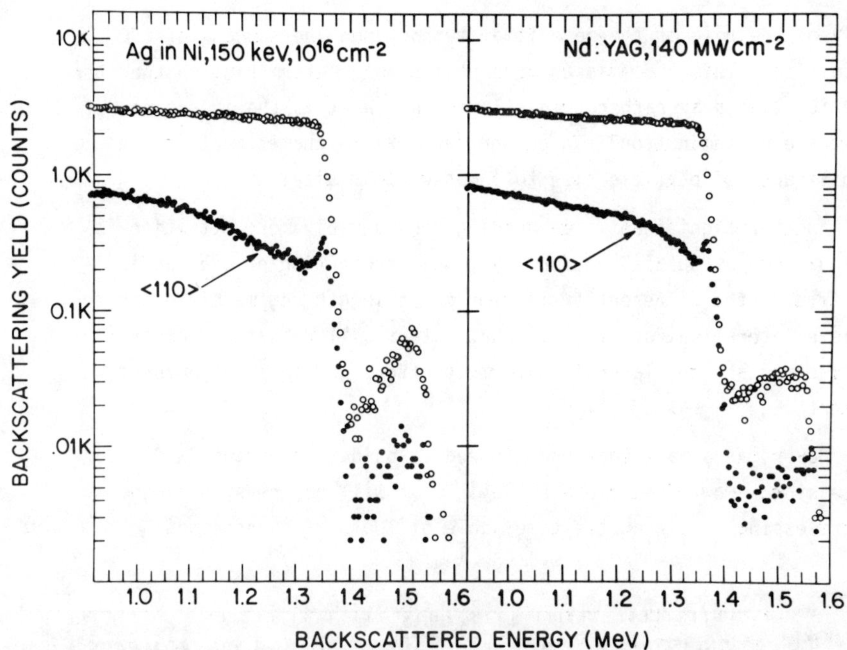

Figure 11 - Random and <110> aligned backscattering yields, before and after
Q-switched Nd-YAG laser irradiation.
Implanted dose of 10^{16} Ag cm^{-2}
represents a peak concentration of 1
atomic %. Following irradiation the
Ag is redistributed and 100% substitutional in the Ni lattice.

Summary

It should be clear at this point that there exists a rather broad
range of processing variables accessible through the different laser
sources one has available to utilize in LSA processing. Both deep and
shallow alloying have been demonstrated. Concentration gradients
ranging from flat to sharply sloped have been produced. A wide variety
of solute specie and host substrates have been examined. Solid state
surface alloys far removed from equilbrium in terms of solute concentrations have been produced.

There still remain some important experimental questions to be
addressed. Probably the most important is a systematic, reproducible

study of the role of the deposition method upon the success of LSA
processing. This encompasses more then simply determining whether the
film is blown away rather than alloyed in. What is the importance of
substrate contamination? In powder deposition schemes what is the
significance of particle size to focussed beam size?

Another significant area needing considerably more attention is
the question of spatial uniformity. What proportion of overlap is
required in the different dwell time-pulse length regimes in order to
achieve lateral uniformity? To what extent will repeated processing
in order to achieve lateral uniformity effect gradients as a function
of depth?

Cover gases have been used to reduce oxidation during cw CO_2 LSA
processing. How effective are they? What will be the significance
of processing in air on the time scale of 100's of nanoseconds.

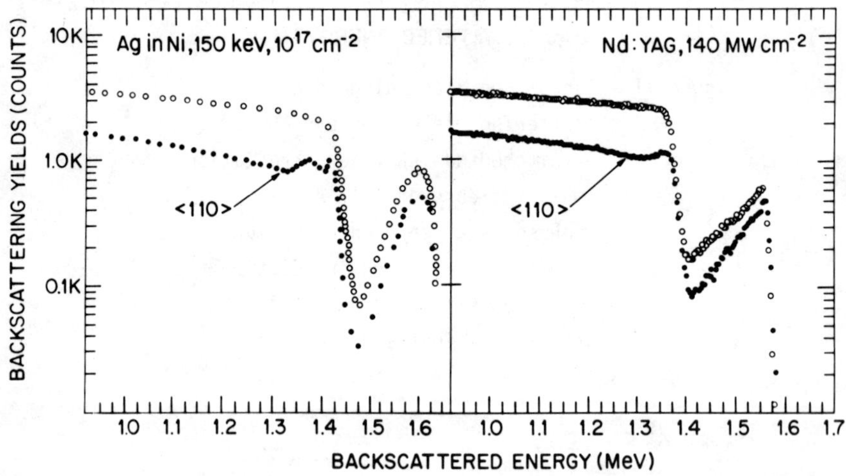

Figure 12 - Random and <110> aligned backscattering yields, before and after
Q-switched Nd-YAG laser irradiation.
Implanted dose of 10^{17} Ag cm^{-2}
represents a peak concentration of
12 atomic %. Following irradiation
much of the Ag precipitates in the
liquid ahead of the resolidification
interface.

It is interesting, and somewhat suprising, that the structural analyses performed on the surface alloys referenced in Table II is so lacking in TEM work.

To date, electron microscopy offers the only definitive means of gaining a fundamental understanding of the microstructural nature of the laser quenched surfaces. Preliminary results on the Au-Ni system [39, 51] indicate a complicated set of cell structures of two distinctly different sizes. A microstructural understanding is clearly a required step in fully characterizing the resulting surface alloys. What improvements in surface sensitive behavior may be expected? The only thoroughly addressed topic has been the increased microhardness achievable in ferrous alloy systems. The fact that the laser melted, self-quenched surfaces have a microscopic roughness associated with LSA processing has been a major drawback in those systems in which wear applications are envisioned. There is no shortage of examples where laser hardened-solid state transformations are finding real world production applications. There are only a sprinkling of references in Table II where a surface sensitive material property has been examined. These include tool life studies [12] of high-speed steels surface alloyed with WC, high temperature stability and sliding wear studies [26] of steels, surface alloyed with Mo-Ni-Si-Cr-C, electrochemical studies [28] of Au-Ni surface alloys, and reducing acid corrosion studies [36] of surface alloyed Pd-Ti.

It is perhaps in the area of tailoring surfaces for corrosion protection that surface alloying methods can make a significant impact. A 60% precious metal bulk alloy would never be considered for a general use container application, but now that surface alloys of high concentration can be made perhaps it is time to examine materials with just such compositions for just such corrosion environments.

From a forecasting vantage point those research areas which will probably grow most rapidly in the next decade are: 1. the use of the rapid quenching associated with LSA processing to make surface alloys of a radically different nature than made by conventional metallurgy - new materials and substitutes; 2. the use of LSA processing to make surface equivalents of established bulk alloys - material conservation (of strategic and noble metals); 3. the demonstration of improved material properties - a judicious choice of both substrate and laser source will be a prerequisite for commercial application.

Laser surface alloying may be far removed from maturity, and
even further from practical application in the immediate future, but
the methodology is established. It is up to those with imagination
to apply what we presently know as capabilities and limitations to
specific film/substrate systems, and to demonstrate that useful
improvements in surface sensitive behavior can result.

Acknowledgements

The author wishes to thank Peter G. Moore (NRL), Dan S. Gnanamuthu
(Rockwell International) and Peter P. Pronko (Universal Energy Systems)
for the materials used in producing Figures 4, 5 and 9 respectively.

References

1. Laser-Solid Interactions and Laser Processing, proceedings of the
    1978 Materials Research Society, edited by S. D. Ferris, H. J.
    Leamy and J. M. Poate. AIP Conf. Proceed. No. 50, New York, 1979.

2. Laser and Electron Beam Processing of Materials, proceedings of the
    1979 Materials Research Society, edited by C. W. White and P. S.
    Peercy. Academic Press, New York, 1980.

3. Laser and Electron-Beam Solid Interactions and Materials Processing,
    proceedings of the 1980 Materials Research Society, edited by J. F.
    Gibbons, L. D. Hess and T. W. Sigmon. Elsevier North-Holland, New
    York, 1981.

4. J. Mazumder and W. M. Steen, J. Appl. Phys. $\underline{51}$, 941 (1980).

5. S. C. Hsu, S. Chakravorty and R. Mehrabian, Met. Trans. $\underline{9B}$, 221
    (1978).

6. H. E. Cline and T. R. Anthony, J. Appl. Phys. $\underline{48}$, 3895 (1977).

7. J. A. McKay and J. T. Schriempf, Appl. Phys. Lett. $\underline{31}$, 369 (1977).

8. J. M. Bechtel, J. Appl. Phys. $\underline{46}$, 1585 (1975).

9. M. Sparks and E. Lok, J. Opt. Soc. Am. $\underline{69}$, 847 (1979).

10. F. E. Cunningham, M. S. Thesis, "The Use of Lasers for the Production
    of Surface Alloys," M.I.T., June, 1964.

11. L. I. Mirkin, Doklady Akademii Nauk SSSR $\underline{186}$ (2), 305 (1969).
    English Ref. Soviet Physics-Doklady $\underline{14}$ (5), 494 (1969).

12. A. O. Schmidt, J. Engineering for Industry $\underline{91}$, 549 (1969).

13. A. I. Betaneli, L. I. Danilenko, T. N. Loladze, E. F. Semiletova,
    B. M. Zhiryakov and A. K. Fannibo, Fiz. I. Khim. Obrab. Mater.
    (6), 22 (1972).

14. I. V. Gazuko, L. L. Krapivin and L. I. Mirkin, Poroshkovaya
    Metallurgiya $\underline{133}$ (1), 27 (1974).

15. F. D. Seaman and D. S. Gnanamuthu, Metal Progress 108 (3), 67 (1975).

16. V. S. Kovalenko and V. I. Volgin, Tekhnologiya I Organizatsiya Proizvodstva (7), 60 (1976).

17. V. S. Kovalenko, V. I. Volgin and V. V. Mikhailov, Tekhnologiya I Organizatsiya Proizvodstva (3), 50 (1977).

18. A. V. Belotskii, V. S. Kovalenko, V. I. Volgin and V. I. Pshenichnyi, Fiz. I. Khim. Obrab. Mater. (3), 24 (1977).

19. S. M. Copley, M. Bass, E. W. Van Stryland, D. G. Beck and O. Esquivel, in Rapidly Quenched Metals III. Vol. 1. edited by B. Cantor (The Metals Society, London, 1979) pp. 147-150.

20. V. S. Kovalenko and V. I. Volgin, Fiz. I. Khim. Obrab. Mater. (3), 28 (1978).

21. G. Della Mea and P. Mazzoldi, in text of ref. 1 above, pp. 212-213.

22. L. S. Weinman and J. N. Devault, in text of ref. 1 above, pp. 239-244.

23. L. S. Weinman, J. N. Devault and P. Moore, in Applications of Lasers in Materials Processing, edited by E. A. Metzbower. (American Society for Metals, Metals Park, 1979) pp. 245-258.

24. D. S. Gnanamuthu, ibid. pp. 177-211., Opt. Eng. 19, 783 (1980).

25. I. Ya. Dekhtyar, M. M. Nishchenko, V. V. Bukhalenko and S. Ya. Kharitonskii. Fiz. Metal. Metalloved. 47, 887 (1979).

26. A. Belmondo and M. Castagna, Thin Solid Films 64, 249 (1979).

27. P. M. Moore and L. S. Weinman, SPIE 198, 120 (1979).

28. C. W. Draper, C. M. Preece, L. Buene, D. C. Jacobson and J. M. Poate, in text of ref. 2 above, pp. 721-727.

29. R. J. Schaefer, T. R. Tucker and J. D. Ayers, in text of ref. 2 above, pp. 754-759.

30. A. K. Jain, V. N. Kulkarni, D. K. Sood, M. Sundararaman and R. D. S. Yadav, Nucl. Instrum. Methods 168, 275 (1980).

31. J. D. Ayers, T. R. Tucker and R. J. Schaefer in Proceed. 2nd International Conf. on Rapid Solidification Processing. (Claitors Publ. Div., Baton Rouge, 1980) pp. 212-220.

32. K. Takei, S. Fujimori and K. Nagai, J. Appl. Phys. 51, 2903 (1980).

33. L. Buene, J. M. Poate, D. C. Jacobson, C. W. Draper and J. K. Hirvonen, Appl. Phys. Lett. 37, 385 (1980).

34. E. Rimini and J. E. E. Baglin, Appl. Phys. Lett. 37, 481 (1980).

35. J. D. Ayers and T. R. Tucker, Thin Solid Films 73, 201 (1980).

36. C. W. Draper, L. S. Meyer, D. C. Jacobson, L. Buene and J. M. Poate, Thin Solid Films 75, 237 (1981).

37. C. W. Draper, L. S. Meyer, L. Buene, D. C. Jacobson and J. M. Poate, Appl. Surf. Sci. in press.

38. D. C. Jacobson, W. M. Augustyniak, L. Buene, J. M. Poate and C. W. Draper, proceedings of the I.E.E.E., Transactions on Nuclear Science, April, 1981.

39. P. P. Pronko, H. Wiedersich, K. Seshan, A. L. Helling, T. A. Lograsso and P. M. Baldo, in text of ref. 3 above, pp. 599-606.

40. L. Buene, D. C. Jacobson, S. Nakahara, J. M. Poate, C. W. Draper and J. K. Hiroven, in text of ref. 3 above, pp. 583-590.

41. L. Buene, E. N. Kaufmann, M. L. McDonald, J. Kothaus, R. Vianden, K. Freitag and C. W. Draper, in Nuclear and Electron Resonance Spectroscopy, edited by E. N. Kaufmann and G. Schenoy, Elsevier North-Holland, New York, 1981. pp. 391-396.

42. G. Battaglin, A. Carneara, G. Della Mea, P. Mazzoldi, A. K. Jain, V. N. Kulkarni and D. K. Sood, in text of ref. 3 above, pp. 615-622.

43. D. M. Follstaedt, W. R. Wampler and P. S. Peercy, in text of ref. 3 above, pp. 567-574.

44. J. M. Baldwin, Appl. Spect. 24, 429 (1970).

45. C. W. Draper, S. P. Sharma, J. L. Yeh and S. L. Bernasek, Surf. Interface Anal. 2, 179 (1980).

46. Thin Films - Interdiffusion and Reactions, edited by J. M. Poate, K. N. Tu and J. W. Mayer (John Wiley and Sons, New York, 1978).

47. Thermophysical Properties of Matter, Vol. 7, edited by Y. S. Touloukian and D. P. DeWitt (IFI/Plenum, New York, 1970) p. 258.

48. V. J. Zaleckas and J. C. Koo, Appl. Phys. Lett. 31 (1977) 615.

49. C. W. Draper, D. C. Jacobson, L. Buene and J. M. Poate, unpublished results.

50. M. Hansen and K. Anderko, "Constitution of Binary Alloys," 2nd Ed. (McGraw Hill, New York 1958), pp. 36-37.

51. P. P. Pronko, private communication.

VIBRO LASER CLADDING

J. Powell & W.M. Steen
Department of Metallurgy & Materials Science
Imperial College
London

Laser cladding by powder fusion has been practised in laboratories since around 1974 and in production since 1978.

As a cladding process, it exhibits many advantages over alternative methods such as plasma spraying and arc welding. These advantages include a reduction in dilution, a reduction in waste due to thermal distortion (very little energy is absorbed by the substrate in comparison with the two mentioned alternatives), a reduction in deposit porosity and a reduction in post cladding machining costs because the material can be more accurately placed.

One drawback which has not been overcome by laser cladding is
thermal stress induced cracking of the solidifying deposit. Also deposit porosity, although it is much reduced in comparison with other methods, is still considered a problem particularly when the deposit thickness exceeds 1mm. This paper analyses the two problems of cracking and porosity and discusses a series of experiments using in process ultrasonic vibration of the substrate whilst laser cladding which appears to reduce both the thermal stress and the porosity. These improvements are observed by standard metallographic techniques and by the analysis of acoustic emission signals recorded from the cooling clad deposits.

Introduction

Multi-kilowatt lasers have, over recent years, become accepted as engineering tools in the field of welding, cutting and surface treatment (2, 3).

Laser surface treatments currently include transformation hardening, surface homogenisation, superficial rapid solidification, or glazing and surface coating either cladding or alloying (3).

In surface cladding with a laser several different processes have been and are being explored. The most straightforward is to pass a defocussed laser beam ($\sim$2mm spot size) over a uniform layer of powder ($\sim$1-2mm thick) resting on the surface to be clad (4). An attractive alternative is to blow the powder into the laser melt zone. The attraction being that the process then becomes a non-contact method of cladding, which would be highly flexible and easily automated as well as allowing some form of gas shielding to the melt zone.

Some success has been achieved by placing the powder as a paste (5) often this results in increased porosity or surface roughness of the trace due to the ejection of vapour from the binder. Wire feed has also been tried as in standard plasma cladding though the shadowing effect of a solid wire does not seem to couple the laser energy as effectively as powder.

A totally different style of cladding with a laser is also possible by blowing thermally sensitive vapours onto a laser generated hot spot, and thereby cladding by chemical vapour deposition (6).

In all these methods residual thermal stresses are present. Some alloys are able to accommodate these stresses without cracking and so subsequent tempering can remove them altogether. Most alloys, however, are likely to crack and no amount of post cladding heat treatment can heal already formed cracks. Such alloys can be successfully clad by preheating the substrate (an interesting rule of thumb is 1°C of preheat for each unit of Vickers hardness). Preheating to 300-500°C necessitates special workpiece handling and some form of gas shielding for the whole sample during processing, both of which add greatly to the complexity and cost of the process.

This paper reports a preliminary exploration into the effect of laser cladding by blowing powder into a moving weld pool on an ultrasonically vibrated substrate. Vibration has been shown over many years to reduce or modify internal stress patterns in castings, forgings and welded structures (7, 8). Both ultrasonic and sonic vibrations have been shown to modify conventional macro and micro structures as well as reduce porosity in castings (8) when applied to metallic melts during solidification. It is shown here that if ultrasonic vibration is applied to a substrate which is undergoing laser cladding the residual stress levels and the porosity are both considerably reduced.

Experimental Work

The experimental arrangement is shown in Fig. 1. The laser used was

94

a Control Laser 2kW continuous wave CO_2 laser model 901. The ultrasonic vibrators were standard powerful cleaning transducers (Lead-Zirconate-Titanate piezo-electric crystals with a frequency of $25kH_z$) arranged as a set of four working in phase and supplied with a special 300W amplifier by C.J. Electronics Ltd. The powder supply system was an adapted Deloro Stellite Ltd spray gun. The powder was spheroidised S.F.40. from Deloro Stellite Ltd, a Nickel-based hardfacing alloy designed for plasma spray cladding. It was blown at various rates governed by the Argon flow rate. The substrate was a surface ground 35 x 7.5 x 1.3 cm^3 En3 steel plate spring clamped to the aluminium plate attached to the transducers. A grease film between the aluminium and the steel plates ensured good transmission of the ultrasonic vibrations. The amplitude of vibration of the substrate was of the order of $\sim$20 μm at a frequency of 25kHz.

The experiments reported here consisted of single track and multiple track deposits as shown in Fig. 2. The single track runs were examined by acoustic emission techniques and optical microscopy while the multiple track runs were analysed only by optical microscopy since the longer processing time made acoustic studies irrelevant.

Single Track Deposits

Acoustic Emission Studies(II) The monitoring of the acoustic signals emitted from a cooling clad track gives information which reflects the magnitude of the residual thermal stress in the deposit. To make the relevant measurements a piezo-electric 'Dunegan and Endevco' D140B acoustic emission transducer was spring clamped, using the usual inter-mediate grease layer, to the surface of the En3 plate to be clad. A track was then deposited under set process parameters and of a standard length (the width of the plate), on completion of this track all the cladding equipment was immediately turned off and the acoustic emission monitoring equipment (standard Dunegan and Endevco amplifiers etc shown schematically in Fig. 5) was simultaneously switched on.

The ring down count (N_{RT}) and the super threshold event count (N_{ET}) together with the amplitude analysis were plotted until the noise from the cooling track had died down to the background level. At this point the clad track was repeated using exactly the same process conditions except that the substrate was ultrasonically vibrated, the acoustic emission measurements were then carried out as before.

The acoustic emission information from both of the tracks in each such pair was then compared as in Fig. 2.

Macro and Microscopic Examination Cross sections and longitudinal sections of the clad deposits were photographed for comparison (see Fig. 3) of the vibrated and unvibrated specimens. Careful examination was made of each specimen using optical and scanning electron microscopy in a search for cracks and pores.

Multiple Track Runs

Multiple track deposits made with varying degrees of overlap were similarly examined by sectioning. Due to the total deposition time for a multiple track deposit (eg 10 tracks side by side) being of the same order of magnitude as the acoustic emission record time for a single track, acoustic emission studies of multiple track deposits were not considered to

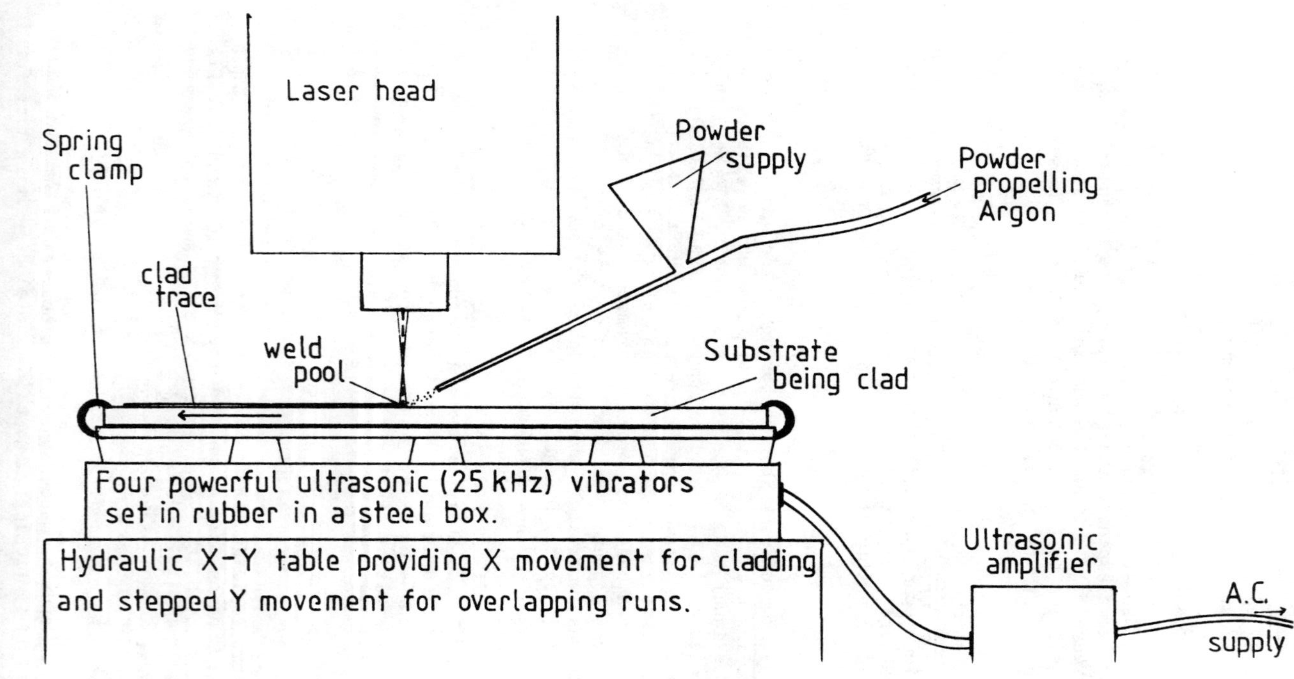

FIG. 1 - The experimental arrangement for Vibro Laser Cladding

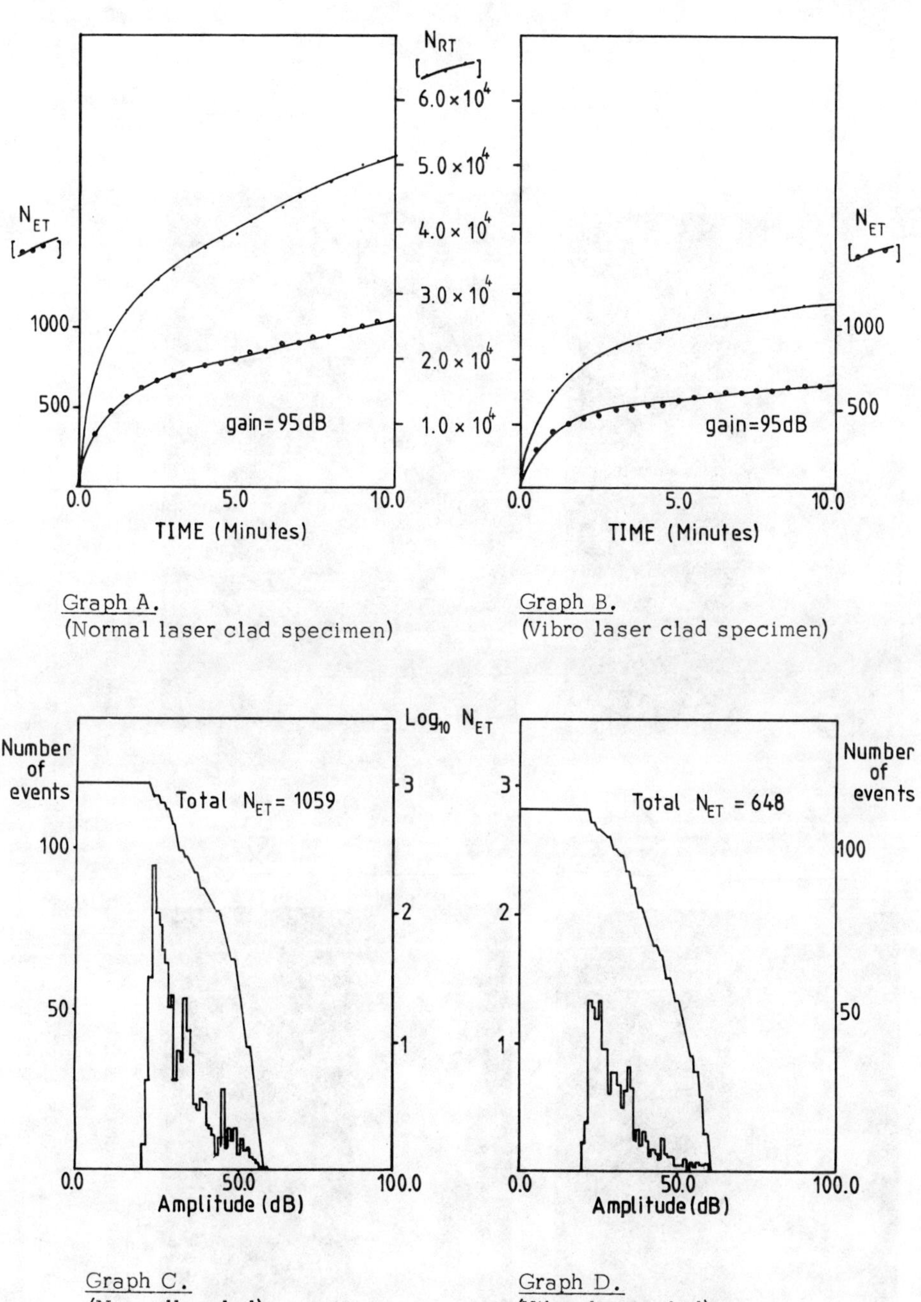

Graph A.
(Normal laser clad specimen)

Graph B.
(Vibro laser clad specimen)

Graph C.
(Normally clad)

Graph D.
(Vibro laser clad)

FIG. 2 - A sample of the results of the acoustic emission survey

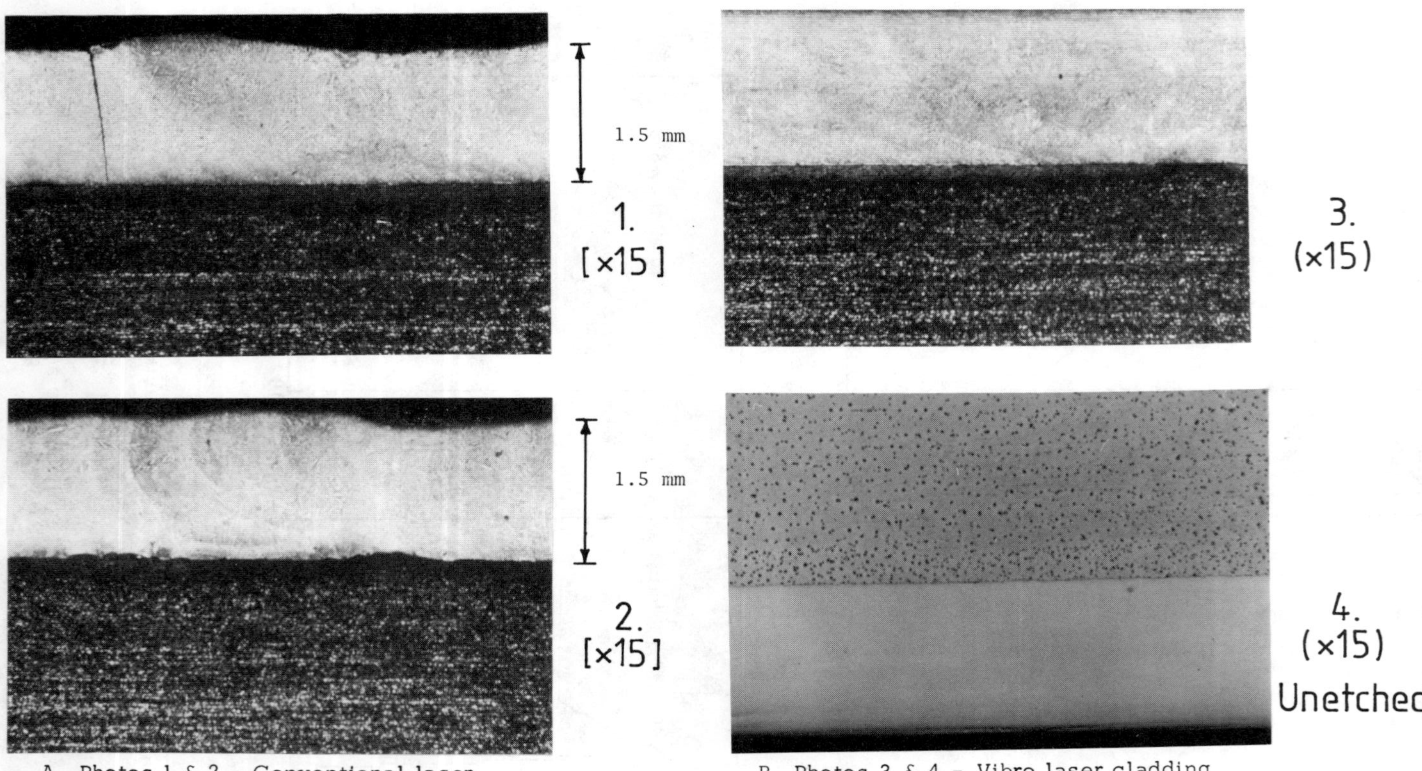

A. Photos 1 & 2 - Conventional laser cladding (longitudinal cross-sections)

B. Photos 3 & 4 - Vibro laser cladding (longitudinal cross-sections)

FIG. 3.- Cross-section macrographs comparing conventional and vibro laser clad specimens (single track deposits).

be useful and were therefore not carried out.

Results

Acoustic Emission Results   The mechanical movements within the cooling
track (ie dislocation movements, phase changes, external movement due to
shrinkage, and crack formation) result in mini 'seismographic' shocks in
the substrate which are picked up by the acoustic emission transducer and
converted into electrical signals. The number of signals or events N_{ET}
are recorded together with their amplitude by the acoustic emission
monitoring equipment. A signal 'scores' if it exceeds a certain set
threshold amplitude. These super threshold events can also reverberate
through the steel substrate and within the lightly damped electronics,
sometimes resulting in a multiple super threshold signal (or ringdown)
produced by a single event. The 'ringdown count' (N_{RT}) recorded by the
acoustic emission monitoring equipment is a measure of this effect. A full
explanation of the technique can be found in the more specialised papers
and textbooks on the subject (eg 10). In summary it can be stated that N_{ET}
and N_{RT} are both measures of the amount of stress induced noise emitted
from the cooling clad track and thus of the residual stress in the track,
but analysis of the type of mechanical movement indicated by a particular
signal trace is, as yet, virtually impossible.

Figure 2 shows a typical set of graphs comparing the acoustic signals
picked up from two otherwise identical cooling tracks, one ultrasonically
vibrated and one not.

Graphs 2a and 2b show that the vibrated specimen generated only 61% of
the number of super-threshold events generated by the unvibrated specimen
(N_{ET} dropped from 1059 to 648). In addition to this the value of N_{RT}
(ringdown count total) after ten minutes dropped from 5.1×10^4 to 2.9×10^4
when vibro laser cladding was employed rather than conventional laser
cladding.

This reduction in noise level is indicative of a reduction in stress
induced mechanical movement on cooling and is therefore evidence supporting
the argument that vibro laser cladding reduces the internal stress in clad
deposits. The reduction to 61% of the unvibrated value compares well with
the stress reduction measured by Yu (7) when ultrasonically vibrating cold
welds.

The amplitude analysis shown in Figs 2 (c and d) as a histogram of
number of events vs amplitude shows a fairly uniform reduction for all
amplitudes. Crack formation would be a loud event compared to a dislocation
movement and hence our main concern is with the larger amplitude events.
In Fig. 2c the small peak of events around 50dB is missing in Fig. 2d,
though events of this amplitude are still recorded.

Similar graphs were recorded for other specimens examined. A summary
of these results is given in Table 1 for seven pairs made under the
following conditions:

Constants   Laser power (P)   = 1500W

                 Laser spot diameter
                 at work piece surface (D) = 2mm

99

Clad track length = 7.5 cm.

Variables between Pairs

Traverse speed (V) = 20 → 50 mm/s

Argon gas flow rate = 1 → 2 arbitrary units.
(The gas flow rate controls the velocity and mass flow rate of the
powder feed and this was varied over a 2:1 range).

TABLE 1

TOTAL EVENT COUNTS FOR VIBRATED AND UNVIBRATED PAIRS

PAIR	NET TOTAL AFTER 10 MINS		
	NOT VIBRATED	VIBRATED	RATIO VIBRO/NONVIBRO
A	129	105	$81^\%$
B	139	75	54
C	171	117	68.5
D	314	300	95.5
E	588	455	77
F	651	256	39
G	1059	648	61
TOTAL	3051	1956	476
AVERAGE	436	280	64 ± 25%

The variation in noise between pairs is a larger subject than can
be discussed here, except to say that the less deposit the less noise,
which accounts for some of the variation.

Results from Macrographs   Fig 3(i) and (ii) show typical longitudinal sections through a single unvibrated clad trace in which cracking
is found (3(i)) and interface porosity in thick deposits (3(ii)), together
with very small pores in the bulk (3(ii)). This is the same sample
whose acoustic trace is shown in Fig. 2 (a & c).

Figs 3(iii) and (iv) are the longitudinal section through the
'identical' vibrated twin to that of Figs. 3(i) & (ii). Fig. 3(iv) was
specially etched to show the clean interface.

Note also the smoother surface profile seen in the vibrated specimen.
Although in this pair of specimens no cracks or pores were found in the
vibrated specimen while they were in the non-vibrated it would only be
true to claim a reduction in these events in other specimens rather than
their total elimination.

In Fig. 4 comparison is made between vibrated and non-vibrated
multiple runs. These runs are particularly prone to inter-run inclusions
and porosity as seen in Fig. 4(iii). This porosity is greatly reduced
in the vibrated twin to Fig. 4(iii) as shown in Fig. 4(iv). For this
reason there is a striking difference in the porosity shown in the

Photo 1 – Conventional laser cladding
(longitudinal cross-section x 15)

Photo 2 – Vibro laser cladding
(longitudinal cross-section x 15)

Photo 3 – Conventional laser cladding
(transvers cross-section x 30)

Photo 4 – Vibro laser cladding
(transvers cross-section x 30)

FIG. 4 – Cross-section macrographs comparing multiple overlapping track deposits.

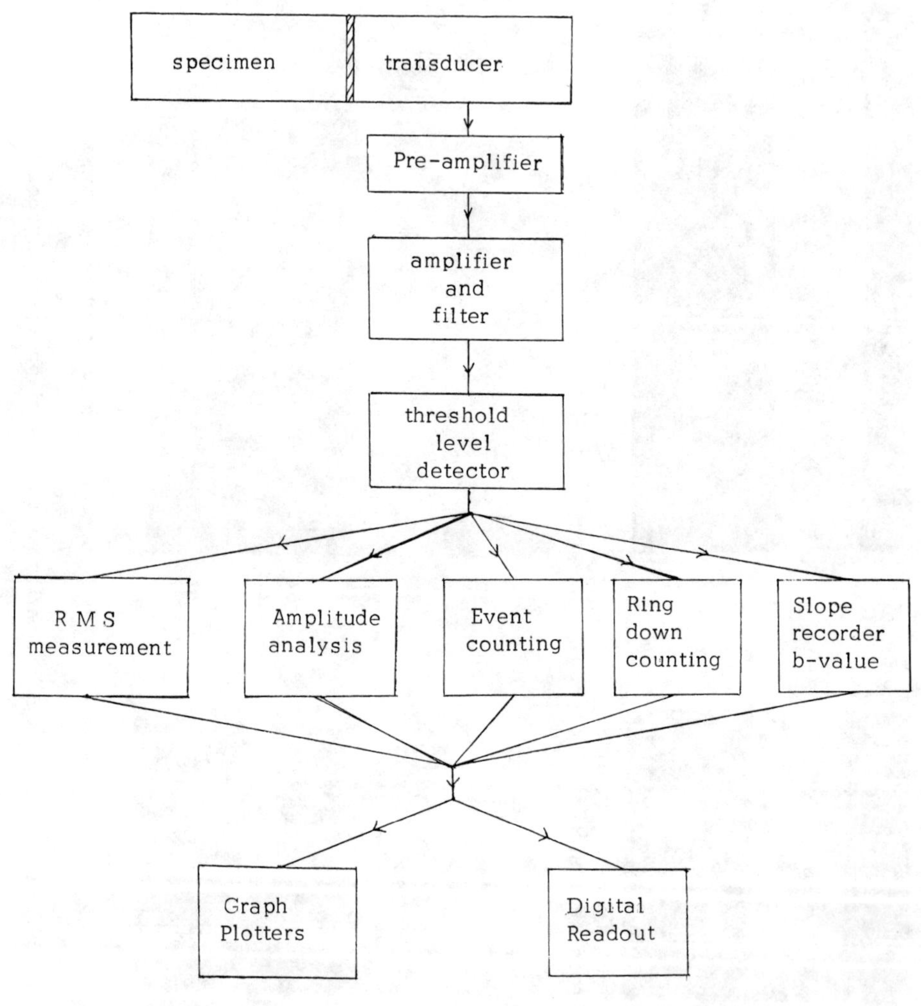

FIG. 5. - Schematic diagram of the acoustic emission system.

two longitudinal sections Fig. 4(i) and Fig. 4(ii).

Discussion

The macroscopic examination, acoustic emission signals and SEM examination (not discussed here) all point to a reduction in cracking, porosity and stress when laser cladding onto an ultrasonically vibrated substrate.

The reduction in internal stress is relatively easily understood. All stress results in structures where the atoms are in metastable positions (which is different from saying all metastable structures are stressed). To relieve the stress the atoms require sufficient energy to 'migrate' out of their metastable positions to more stable positions. Thus all stress relieving processes attempt to give atoms this energy in a controlled manner, eg by heating to a relatively low temperature, or by physically vibrating the structure. Ultrasonic vibration will impart a kinetic vibration to atoms which will allow some atoms to move to more stable sites, thus resulting in less internal stress.

The reduction in porosity is not so easily understood for there are several conflicting theories one or all of which could be correct.

A vibrated liquid tends to have greater fluidity and hence bubbles can escape faster; a vibrated liquid will enhance the liberation of bubbles from an interface (two examples are the effect of knocking the side of a glass of soda water and ultrasonic cleaning);and finally ultrasonically vibrated liquids are capable of generating bubble nuclei faster and hence allowing more time for the bubble to form and escape. Apart from vapour or gas generated within the melt or trapped from the propellant gas, there is a further possible source of porosity ie that due to solidification shrinkage. Studies on the porosity of castings (8) have indicated that vibration reduces shrinkage porosity because the increased fluidity allows the liquid to pass through the interdendritic spaces more easily.

This same increased fluidity appears to allow bubbles to float free faster and to be dislodged more easily from interface regions. The extent to which bubble nucleation plays a part is at present uncertain.

Conclusion

From the limited number of samples examined so far, it is apparent that cladding by blowing powder into a laser melted zone while vibrating the substrate ultrasonically reduces internal stress and hence cracking as well as reducing porosity in the clad trace. In the experiments recorded an ultrasonic vibration of 25 kHz and $\sim$20 µm amplitude reduced the acoustic signal which is related to stress to 64% of its original value on the system SF40/ mild steel.

Acknowledgements

The authors would like to thank Dr R.D. Rawlings and S. Chan of the Metallurgy Department, Imperial College for their help with the acoustic emission measurements. They would also like to thank Control Laser Ltd, British Steel Corp, Rolls Royce (1971) Ltd for funds and interest in this work. Mr J. Powell would like to express his gratitude to the Science Research Council and Rolls Royce (1971) Ltd for a student research grant.

References

1. F.D. Seaman and D.S. Gnanamuthu , "Using the Industrial Laser to
    Surface Harden and Alloy", Met. Prog. 1975, <u>108</u> (3), 67-74.

2. W.M. Steen and M.C. Sharp, "The Laser for Welding, Cutting and Heat
    Treatment", paper VA2, UIE 9 Conference, Cannes, Oct. 1980.

3. W.M. Steen and J. Powell, "Laser Surface Treatment", International
    Journal of Materials in Engineering Applications, to be published
    1981.

4. W.M. Steen and C.G.H. Courtney, "Hardfacing of Nimonic 75 Using
    2kW Continuous Wave CO_2 Laser", Metals Technology, pp232-237, June
    1980.

5. A.V. La Rocca, H. Walther and P.C. Cappelli, "Development Activities
    for the Industrial Application of High Power Lasers to Metal
    Working", International Seminar on Laser Processing of Engineering
    Materials, June 30 1977, pp13-14.

6. W.M. Steen, "Surface Coating Using a Laser", Paper 3 Conf. Advances
    in Surface Coating Technology, Welding Inst. London, Feb 1978.

7. Yu V. Kholopov and Uniieso, "Ultrasonic Treatment of Welded Joints
    in Metals to Relieve Residual Stresses", Svar. Proiz. 1973, No 12,
    pp20-21.

8. G.G. Saunders and R.A. Claxton, "VSR-(Vibratory Stress Relief) - A
    Current State of the Art Appraisal", Conf. Residual Stresses in Welded
    Construction and their Effects, The Welding Inst. 1977, paper 29,
    pp173-183.

9. A.A. Grudz et al, "Reducing the Welding Stresses in Plates by
    Vibration", Aut Svarka 1972, No 7, pp75-76.

10. T.A. Strivens and R.D. Rawlings, "The Application of Acoustic
    Emission to the Study of Paint Failure," J. of Oil, Cot., Chem. Assoc.,
    pp412-418, Vol. 63, 1980.

11. R.D. Rawlings and W.M. Steen, "Acoustic Emission - Monitoring of
    Surface Hardening by Laser", to be published.

SURFACE RIPPLING INDUCED IN THIN FILMS

BY A SCANNING LASER

H. E. Cline
General Electric Corporation
Research and Development Center
Schenectady, NY 12301

Introduction

A scanning laser beam has been used to melt the surface of materials either to manufacture articles or to modify the material properties. The resolidified surface in the path of the moving laser beam resembles that of a weld; some regions are raised and other regions are depressed below the original planar surface. This surface rippling was previously explained by the fluid flow induced in the molten metal by surface tension gradients that result from the local heating of the laser beam.(1) Flow occurs from the hot region in the center of the liquid toward the edge of the liquid pool.

Recently there has been increasing interest in using lasers to melt thin films to form a surface alloy(2, 3) or to recrystallize semiconductor films. (4, 5) Surface relief was also observed in thin films melted either by a pulsed(6, 7, 8) or scanning(9) laser beam. The thin film case is qualitatively different from the bulk case because the substrate constrains the flow of liquid, and capillary effects are relatively more important in thin films compared to the gravitational effects which occur in bulk material.

We extend the previous model(1) to the process of laser alloying and recrystallization of thin films. A model system consisting of Pb and Sn layers was melted with a scanning laser and then examined to compare the surface profiles with the calculations presented. These results are discussed and contrasted with other explanations of the surface relief in the literature.

Heat Flow

The temperature profile across the thickness of the film is described by the one-dimensional diffusion equation.

$$\partial T/\partial t = D_T \, (\partial^2 T/\partial Y^2) \tag{1}$$

where Y is the distance from the surface and D_T is the thermal diffusivity of the film. After a time t from the initial heating of the surface by the scanning laser, the solid-liquid interface moves a distance comparable to the thermal diffusion length given by

$$X_T = \sqrt{D_T t}\,. \qquad\qquad (2)$$

A thin film of thickness h melts in a time, h^2/D_T, and then the heat penetrates into the substrate. If the film is sufficiently thin, the thermal
properties of the substrate determine the temperature profile. The temperature distribution has been calculated(10, 11) for the case of scanning
Gaussian laser beam of radius R moving at a velocity V. With increasing
values of the dimensionless parameter RV/D_T, the thermal distribution
becomes asymmetric, the heating rate becomes greater than the cooling
rate, and the liquid pool becomes elongated behind the moving laser beam.
At sufficiently slow scan velocities ($RV/D_T < 1$), the temperature distribution is symmetric and the melt pool is circular in shape. In either case,
the temperature distribution at the surface across the path of the laser
beam is approximated by a Gaussian (Figure 1):

$$T = T_0 e^{-X^2/2R^2}\,. \qquad\qquad (3)$$

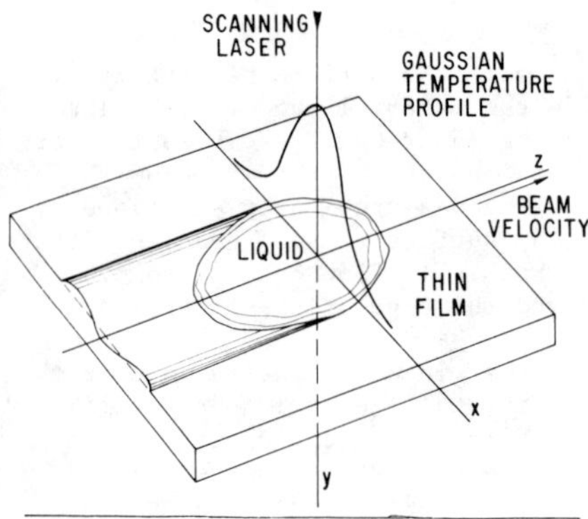

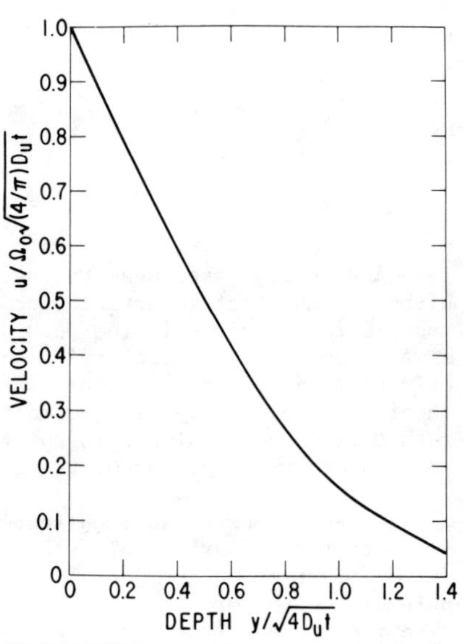

Figure 1. Schematic of a scanning
laser beam impinging on the surface
of a thin film, showing the surface
rippling and the temperature distribution across the molten region.

Figure 2. Velocity of the fluid as
a function of depth for a constant
surface shear Ω_0 after a time t,
normalized to give the shape of the
velocity profile.

Fluid Flow

During the melting of a thin film by a moving laser beam, a temperature
gradient extends radially away from the axis of the beam. At the center of
the molten pool of liquid, the temperature of the liquid is highest and the
surface tension is at its lowest value. The increase in surface tension away
from the beam axis results in a movement of liquid away from the center, which
depresses the surface of the liquid under the beam and raises the liquid
surface at the periphery of the liquid region. In the previous model(1) a
counterflow of liquid resulted from the pressure difference caused by gravity,
while in the case of thin films the gravitational effects are not significant in comparison with capillary effects.

At the surface of the liquid, there is a tangential force in the direction of increasing surface tension which is proportional to the surface tension gradient. Only the component of this force transverse to the motion of the beam, $\partial\gamma/\partial X$, is considered in this two-dimensional model that describes the surface profile across the molten region. This force at the surface of the liquid produces a shear or vorticity in the liquid that is inversely proportional to the viscosity of the liquid(1):

$$\Omega_o = (1/\eta)\,(\partial\gamma/\partial X).\tag{4}$$

Mathematically, it is easier to work with the vorticity field Ω which is the curl of the velocity distribution, $\partial u/\partial y$, than with the velocity field because the equation describing the vorticity is the familiar diffusion equation

$$\partial\Omega/\partial t = D_u(\partial^2\Omega)/\partial Y^2\tag{5}$$

where the kinematic viscosity, $D_u = \eta/\rho$, is the "diffusivity" of the vorticity penetrates into the liquid a distance comparable to the fluid diffusion distance:

$$X_u = \sqrt{D_u t}.\tag{6}$$

In the case of a metal, the thermal diffusion distance is greater than the fluid diffusion distance, which means that the film melts faster than the fluid flow profile, penetrates into the film thickness and, upon cooling, the film solidifies before the liquid flow has time to become redistributed. The surface rippling becomes frozen into the resolidified film. The distribution of vorticity changes with time according to the well known solution of the diffusion equation(12):

$$\Omega = \Omega_o\,\mathrm{erfc}\,(Y/2X_u).\tag{7}$$

With time the vorticity diffuses into the liquid. The velocity of the liquid u is obtained by integrating the vorticity using the boundary condition of vanishing velocity at the liquid substrate interface:

$$u = \int_y^h \Omega\,dY.\tag{8}$$

Fortunately the integral of the complementary error function has been calculated(13); the velocity profile is nearly linear (Figure 2). At the surface of the liquid, the velocity increases with time as the veolocity profile penetrates into the liquid at distance X_u

$$u_o = \Omega_o X_u \quad X_u < h\tag{9}$$

until the velocity profile reaches the substrate, after which the surface velocity remains constant

$$u_o = \Omega_o h \quad X_u > h.\tag{10}$$

The velocity profile may be approximated by a linear profile, especially for the case of full penetration at long times, which is relevant to thin films.

Thickness Profile

The total flux of liquid J increases with time until the diffusion distance equals the film thickness and then the flux becomes constant. Integrating the linear velocity profile from the surface to the diffusion depth, we obtain the flux

$$J = 1/2\,X_u^2\,\Omega_o \quad X_u < h\tag{11}$$

which is limited by the film thickness

$$J = 1/2\, h^2 \Omega_o \quad X_u > h. \tag{12}$$

A variation in the flux along the film results in the redistribution of the liquid and a variation in the film thickness. Conservation of the liquid volume is used to relate the rate of change of the film thickness to the flux gradient:

$$\partial h/\partial t = -\,\partial J/\partial X. \tag{13}$$

The thickness change Δh in the time t that the material is heated is obtained by integrating eq. (13) with respect to time using the expression for the flux [eq. (12)] to give

$$\Delta h = -\,h^2/2\,(t)\,(\partial \Omega_o/\partial X) + C \tag{14}$$

where the surface vorticity gradient is related to the second derivative of the temperature profile

$$\frac{\partial \Omega_o}{\partial X} = \frac{1}{\eta}\left(\frac{\partial \gamma}{\partial T}\right)\left(\frac{\partial^2 T}{\partial X^2}\right). \tag{15}$$

The magnitude of the surface rippling is determined by the rate of change of the liquid surface tension with increasing temperature.

Assume the molten pool is a circle of diameter W and that the liquid is at the melting point T_m at $X = W/2$; then the maximum temperature at the center of the liquid becomes

$$T_o = T_m e^{W^2/8R^2} \tag{16}$$

and the time t that the liquid is molten becomes approximately equal to W/V. Differentiating the temperature distribution twice with respect to X, we obtain the factor that determines the liquid thickness profile which is of the form

$$\Delta h = B\,\{e^{W^2/8R^2}\,[(X^2/R^2) - 1]\,e^{-X^2/2R^2} + 1\} \tag{17}$$

where

$$B = \frac{h^2 t}{2N}\,\frac{\partial \gamma}{\partial T}\,\frac{T_m}{R^2} \tag{18}$$

and the integration constant was determined by the conservation of liquid

$$\int_{-W/2}^{W/2} \Delta h\, dX = 0. \tag{19}$$

A plot of the calculated profile for $W = R$, $W = 2R$, and $W = 3R$ (Figure 3) shows increasing amounts of surface rippling. The discontinuous change in film thickness at the edge of the liquid is not realistic because capillary effects would prevent the surface from changing abruptly.

Capillary Effects

As the surface of the thin film becomes curved, a capillary pressure P induces a counterflow of liquid that limits the amount of surface depression. The flux of liquid in a film of thickness h is related to the pressure gradient(14):

$$J = (h^3/3\eta)\,(\partial P/\partial X). \tag{20}$$

We calculate the pressure gradient from the capillary pressure

$$P = \gamma\,(\partial^2 h/\partial X^2). \tag{21}$$

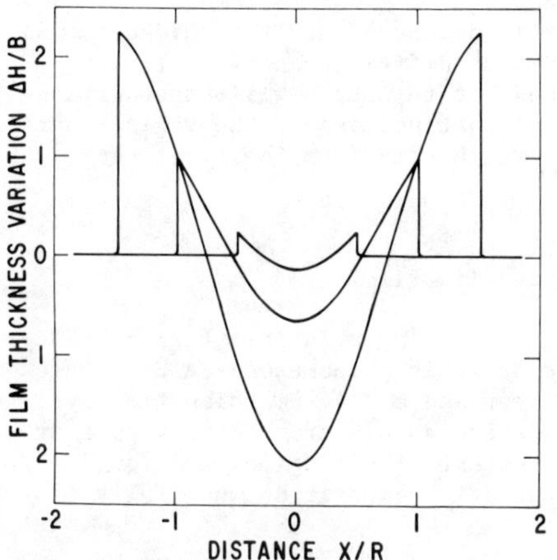

Figure 3. Film thickness variation calculated from surface tension gradients for different line widths: (a) W = R, (b) W = 2R, and (c) W = 3R.

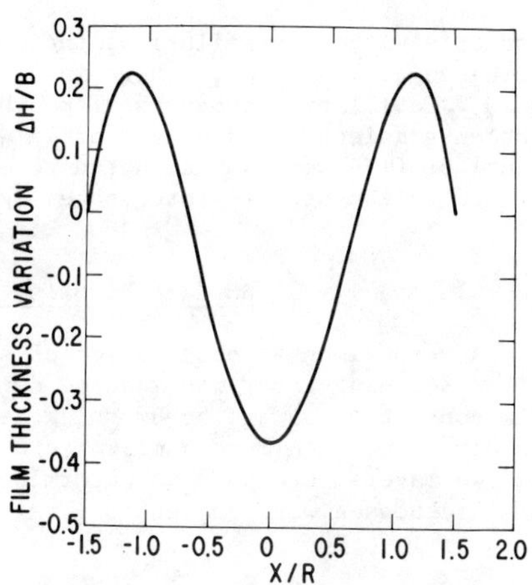

Figure 4. Film thickness variation calculated from capillarity effect in steady-state for W = 3R.

In thin films the capillary pressure is much larger than the gravitational pressure. Initially the surface is planar and the capillary pressure vanishes; however, with time the capillary pressure increases until the counterflow balances the flux induced by the surface tension gradients. Equating eq. (12) with eq. (20) gives a relationship between the pressure gradient and temperature gradient:

$$\partial P/\partial X = (3/2h)\,(\partial\gamma/\partial T)\,(\partial T/\partial X). \tag{22}$$

The steady-state thickness profile may be calculated by integrating with respect to X three times to give

$$\Delta h(x) = 3/(2\,h\gamma)\,(\partial\gamma/\partial T)\int_0^x\int_0^{x'} T(x')\,dx'\,dx$$

$$+\ \frac{CX^2}{2} \tag{23}$$

where the integration constant is obtained by the conservation of liquid. Substituting the Gaussian temperature profile [eq. (3)] into eq. (23) and performing the integrations gives an expression for the profile in terms of the tabulated error function integral

$$h(x/R) = B\,e^{W^2/8R^2}\int_0^{X/R} \mathrm{erf}\,(x/R)\,d(X/R)$$

$$+\ \frac{CX^2}{2} \tag{24}$$

where

$$B = 3/2\,\frac{T_m R'\,\dfrac{\partial\gamma}{\partial T}}{h\gamma}. \tag{25}$$

A calculated steady-state profile shape is shown in Figure 4 for W = 3R. The steady-state shape is similar to the shape induced by surface tension

gradients without capillarity; the center is depressed and the thickness
increased at the edges. There is an important difference between the two
cases. Capillarity produces a smooth steady-state shape, while the surface
tension gradients produce a profile with discontinuities at the edges. The
actual profile would be expected to evolve with time from the first case
treated to the steady-state shape.

Surface Rippling of Lead Tin Films

An experimental observation of surface rippling of a thin film melted
with a scanning laser was made to demonstrate this phenomenon. A 2-μm thick
film consisting of a 0.56-μm thick lead layer and a 1.44-μm thick tin layer
was deposited on a Pyrex* microscope slide with an electron beam evaporator.
The two layers were used to determine the extent of mixing in the liquid.
The thicknesses were chosen to give the eutectic composition when fully
mixed.

The film was melted with a Quantronix 116S, CW, Nd:YAG laser in the
TEM_{oo} mode scanned at 20 cm/sec. A Gaussian distribution of intensity was
observed as expected for the TEM_{oo} mode. The laser was configured with a
120-cm 12% front mirror and a 70-cm rear mirror without any beam expander;
the beam was focused with a 48-mm objective. A series of scans with
increasing power were made. Below a threshold level of 0.65 W, the tin
surface of the film remained solid, while above 1.8 W the film was removed,
leaving a scribed line on the glass slide that transmitted light (Figure 5).
An optical microscope was used first to measure the line width of the
resolidified material for each power level. The lines were quite uniform
in width and appearance; Normarski differential surface contrast revealed
the rippled surface (Figure 6). To make quantitative measurements of the
surface profile, a two-beam interferometer was used to give the thickness
profile (Figure 7). At the center of the line the thickness was reduced,
while at the edges the film thickness was increased to form shoulders. With
increasing power the depth of the depression in the center increased until,
at a critical power level, the liquid was removed from the substrate and
moved to the line edges. In this way a line of uniform width may be scribed
in a film without significant evaporation of material. The thickness profiles
of three lines of different widths show the increase in depth with increasing
power (Figure 8).

Measurement of mixing in the liquid was made with a microprobe set to
detect the K-alpha line of Pb. Only lead within about 0.2 μm of the surface
is detected because of the x-ray absorption of the Sn film. The resolidified
90-μm wide line shows Pb at the surface, indicating the Pb and Sn films
alloyed (Figure 9). At lower power levels, the Sn film melted over a 30-μm
wide region without Pb being detected at the surface. The tin flows away
from the center of the laser beam; with increasing power the lead liquid
moves to the surface in the center of the molten pool.

Calculated Profiles for Sn Thin Films

First the scanning laser melts the tin surface layer and then the
underlying lead layer. Initially the laser rippling is driven by the
surface tension gradient in pure tin. The measured profile has the same

---

*Trademark of Corning Glass Works

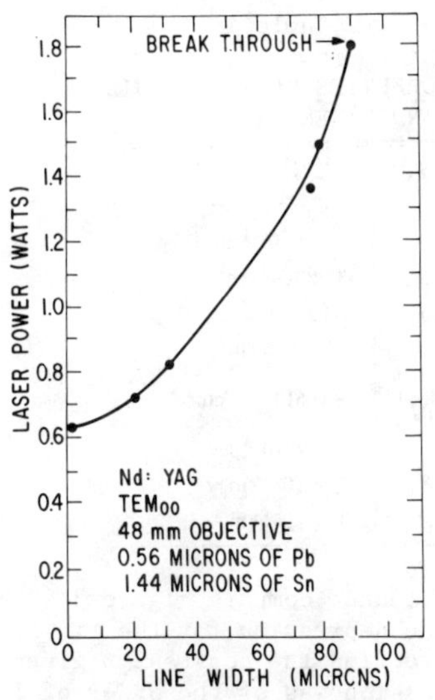

Figure 5. The laser power increases with line width from a threshold value needed to melt the Pb-Sn film to the power that breaks through the film.

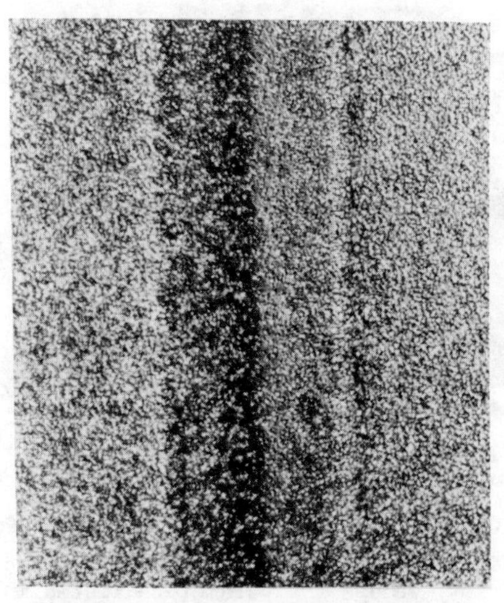

Figure 6. Normarski contrast reveals the surface relief of the 90-μm wide resolidified line in the path of the scanning laser beam.

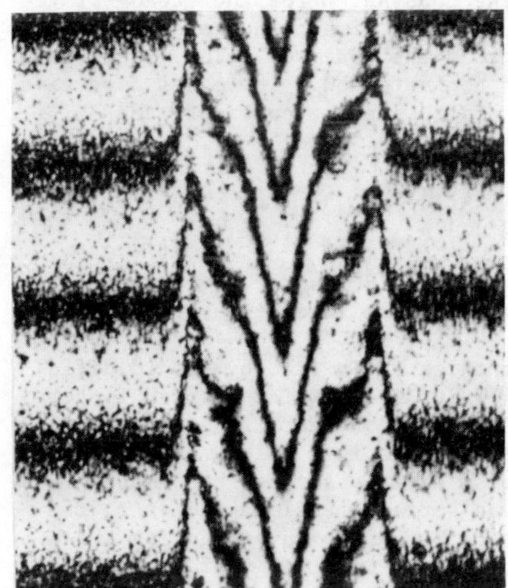

Figure 7. Two-beam inteferogram of the 90-μm side line used to measure the variation in film thickness. One fringe spacing equals 2700 Å.

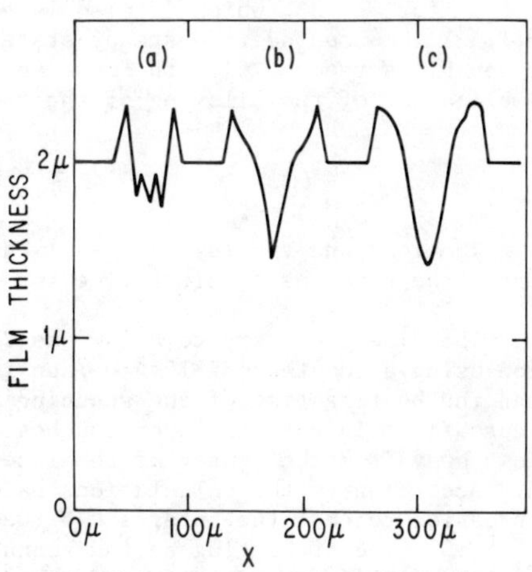

Figure 8. A series of film thickness profiles with increasing laser power: (a) 0.8 W, (b) 1.5 W, and (c) 1. 8 W.

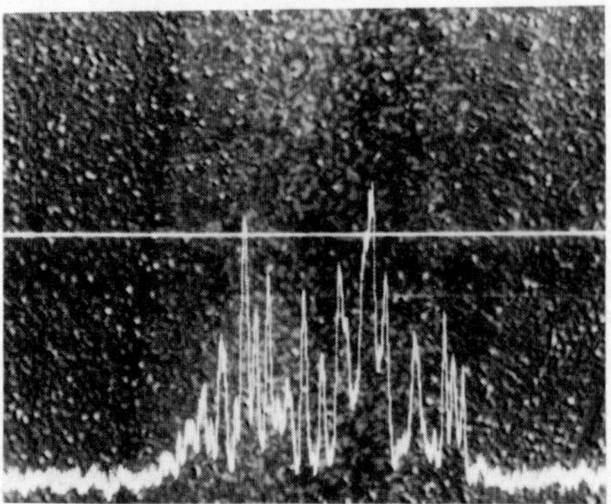

Figure 9. A microprobe x-ray scan
for the K-alpha line of Pb across
90-μm wide path of the scanning laser.

Table 1

PROPERTIES OF TIN LIQUID
AND OTHER PARAMETERS

T_m	232°C
ϱ	7.0 g/cm³
D_T	4.2×10^{-1} cm²/sec
D_u	3×10^{-3} cm²/sec
η	0.02 poise
γ	544 ergs/cm²
$\dfrac{\partial \gamma}{\partial T}$	-0.612 erg/cm²-°C
t	1.5×10^{-4} sec
h	2×10^{-4} cm
R	3×10^{-3} cm

shape as that calculated by surface tension gradients; compare Figure 3 with
Figure 8. In order to calculate the depth of the depression for the tin
layer, we use the parameters listed in Table 1 to find the constant B given
by eq. (18), which is 23 μm. The measured depression was of the order of 1
μm. The observed magnitude of the surface rippling is an order of magnitude
less than that calculated from the surface tension gradient in the time
available for heating the film to the liquid state.

Capillary effects retard the growth of surface rippling which could
explain the magnitude of the depression being less than calculated from the
surface tension gradient alone. However, the profile shape does not match
that of the steady-state which indicates that a steady-state was not reached.
Using the parameters listed in Table 1, the constant B for the steady-state
profile [eq. (25)] becomes equal to 170 μm, which gives a depression depth
of 6 μm for W = 3R, which is also larger than the observed value. However,
in 1.5 X 10⁻⁴ seconds the steady-state is not attained. The model does give
reasonable agreement with the observed profile when one considers the
complication of the alloying of the Pb and Sn film.

Alloying

The lead and tin layers are mixed by the fluid flow of the scanning
laser; there is insufficient time for the liquid to interdiffuse.

The time necessary to allow the 2-μm thick Pb-Sn film by liquid diffusion using a constant of 10^{-5}cm²/sec is 4 X 10⁻³ sec, which is much longer
than the heating time of the scanning laser. In contrast, fluid flow which
occurs first in the tin layer reaches the Pb layer in 6 X 10⁻⁵ sec. Tin
flows away from the center of the laser beam and the surface becomes depressed. According to the calculations based on surface tension gradients, the
depression reaches the interface between the Sn and Pb layers during the
1.5 X 10⁻⁴ sec of heating by the scanning laser. Once the surface contacts
the lead liquid at the center of the depression the surface tension changes
from that of liquid tin to that of liquid lead. The surface tension of lead
liquid is less than that of tin,(15) so the lead liquid would flow rapidly
over the surface. Flow of the lead is considerably slower than that of the
tin; the constant B for lead is only 1 μm, compared with 23 μm for tin.

112

Discussion

The observed thickness profiles were similar in shape to those calculated by surface tension gradients. Any differences could be explained by capillary effects. However, the magnitude of the observed surface rippling was less than calculated. Effects of alloying may be responsible for the small magnitude of the depression. It is the alloying case that is currently of interest in the semiconductor industry.

Previously, surface undulations after laser-alloying Pb-Si films with a pulsed laser were attributed to either shock waves(7) or acoustic waves.(6) Recently it was shown that the surface relief was related to the interference pattern generated by dust particles.(8) The Pb film on a silicon substrate is not uniformly heated by the pulsed laser because of this interference. Nonuniform heating induces fluid flow away from the hotter regions by surface tension gradients which may explain these effects. In forming silicides with a scanning laser,(9) the alloy formed preferentially in the center of the resolidified region. This effect may result from the fluid flow described here, which brings the liquid silicon to the surface at the center of the liquid pool.

The surface tension gradient-induced fluid flow observed previously in large welds(1) may be applied with some modification to the case of melting thin films to explain the surface relief that is observed.

Acknowledgments

The author wishes to acknowledge stimulating discussions with Dr. T. R. Anthony and Dr. G. Possin, preparation of thin films by Mrs. I. Mullins, interferometry by Mr. A. S. Holik, and microprobe analysis by Mr. N. Lewis.

References

1. T. R. Anthony and H. E. Cline, _J. Appl. Phys._ 48, 3888 (1977).

2. J. M. Poate, H. J. Leamy, T. T. Sheng, and G. K. Celler, _Appl. Phys. Lett._ 33, 918 (1978).

3. M. von Allmen, S. S. Lau, M. Maenpaa, and B. Y. Tsaur, _Appl. Phys. Lett._ 36, 205 (1980).

4. M. W. Geis, D. C. Flanders, and H. I. Smith, _Appl. Phys. Lett._ 35, 71 (1979).

5. J. C. C. Fan, H. J. Zeiger, R. P. Gale, and R. L. Chapman, _Appl. Phys. Lett._ 36, 158 (1980).

6. G. N. Maracas, G. L. Harris, C. A. Lee, and R. McFarlane, _Appl. Phys. Lett._ 33, 453 (1978).

7. A. G. Cullis, H. C. Weber, and N. G. Chew, _Appl. Phys. Lett._ 36, 547 (1980).

8. K. Affolter, W. Luthy, and M. Wittmer, _Appl. Phys. Lett._ 36, 559 (1980).

9. T. Shibata, J. F. Gibbons, and T. W. Sigmon, _Appl. Phys. Lett._ 36, 556 (1980).

10. H. E. Cline and T. R. Anthony, *J. Appl. Phys.* 48, 3895 (1977).

11. J. Mazumder and W. M. Steen, *J. Appl. Phys.* 51, 941 (1980).

12. H. S. Carslaw and J. C. Jaeger, *Conduction of Heat in Solids,* Oxford University Press, 2nd Ed. (1959).

13. M. Abramowitz and I. A. Stegun, *Handbook of Mathematical Functions,* Dover Pub., New York (1965).

14. G. K. Batchelor, F. R. S., *An Introduction to Fluid Dynamics,* Cambridge Univ. Press, 1970, p. 183.

15. *Metals Reference Book,* 5th Ed. (C. J. Smithells, ed.), Butterworths, London, 1976, p. 944-947.

PARTICULATE COMPOSITE SURFACES BY LASER PROCESSING

Jack D. Ayers

Naval Research Laboratory
Washington, D.C.   20375

In situ particulate composites have been formed in thin surface layers on a wide range of alloy types by forcibly injecting TiC, WC, or other carbide particles into a shallow melt pool produced on a moving sample by a high power CO_2 laser. Surface layers ranging in thickness from approximately 0.01 to 0.3cm are formed, the thickness being determined by the sample translation rate and by the laser power and beam size. Typical operating conditions employ a 2mm diameter, 6 kW beam and a sample translation rate of 5cm/sec. The surface layers so formed consist of the injected carbide particles in a matrix of the initial alloy, the composition of which is sometimes slightly altered by partial dissolution of the carbide.

This paper discusses the types of surface structures which have been produced by laser melt/particle injection processing and considers the effect upon these structures produced by varying processing parameters which alter the degree of dissolution of injected carbide particles. Special consideration is given to the degree of hardening produced in various metal matrices by carbide injection processing.

1. Introduction

The laser melt/particle injection process has, since its development in 1979 (1), proven to be capable of producing surface modifications of diverse types with a single experimental geometry. The most explored type of surface modification to date, and that which is the topic of this paper, is that of injecting carbide particles directly into the metal surface. This processing has produced hardened surfaces on iron, nickel, copper, titanium, and aluminum base alloys by injecting them with TiC, WC, or other carbide particles. As an example of the type of structure which can be produced in this way, Fig. 1(a) shows a single melt pass in a sample of Ti-6Al-4V which was injected with -70+140 mesh TiC. As the photo shows, the carbide particles are incorporated into the base metal which serves as a binder. With some metal/carbide pairs the injected carbide is partially dissolved in the molten metal matrix during processing, but most of this dissolved carbide resolidifies as finely dispersed carbide particles in the metal matrix. Figure 1(b) shows a deeply etched surface on Ti-6Al-4V where these dispersed carbide particles are evident. The large particle on the right is one of the injected particles which is covered with a rough layer of regrown TiC. The structures shown in Fig. 1 are discussed more fully in Reference 2. Because the partial dissolution and regrowth of the injected carbide is sometimes detrimental to the mechanical properties of the matrix (1,2,3), some factors affecting this dissolution are discussed in this paper.

A potentially important application of the laser melt/particle injection process which has only recently been explored (4) is in surface alloying. This processing differs from carbide particle injection only in that the injected particles are deliberately dissolved in the metal matrix, altering the composition of the surface to improve its corrosion or wear resistance. As a test of what could be achieved by this approach to surface alloying, -100 mesh silicon particles were injected into the surface of 5052 Al. After laser remelting to dissolve the injected particles the surface had the structure shown in Fig. 2. As Fig. 2(a) demonstrates the alloyed surface layer was relatively smooth and of uniform depth. The structure consisted of pro-eutectic silicon particles, evident in Fig. 2(b), in a silicon/aluminum alloy matrix. Reference 4 discusses the more complicated structures which occurred at the edge of laser/remelt passes.

In a third variation of this process, results of which have not been previously published, a cladding is formed on the surface by adjusting the operating conditions so as to fuse the injected powder with minimal substrate melting. Because this variation requires that the injected powder be capable of forming a self sustained melt pool on the surface, it must be of a material which can be fused in contact with the substrate. Figure 3 shows a sectional view of a cladding of nichrome made on 1018 steel by this technique. In another paper within this volume (5) Steen and Powell describe the formation of other claddings by an essentially identical process, modified by ultrasonic shaking of the substrate to lessen residual stress.

The versatility of this process recommends it for diverse surface modification applications, but full realization of its potential requires an understanding of how variations in materials and processing parameters influence the structure and properties of the modified surfaces. With this view in mind, this paper examines the role of carbide dissolution and regrowth in the hardening of metal surfaces injected with TiC.

116

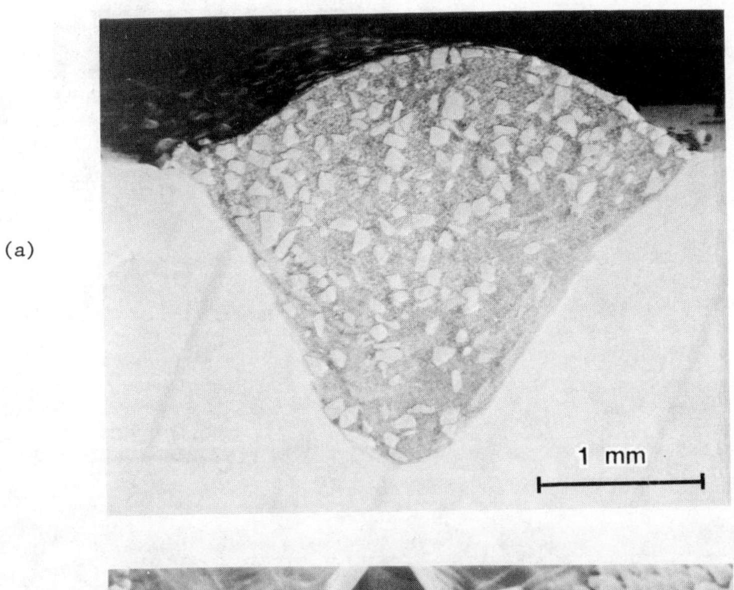

(a)

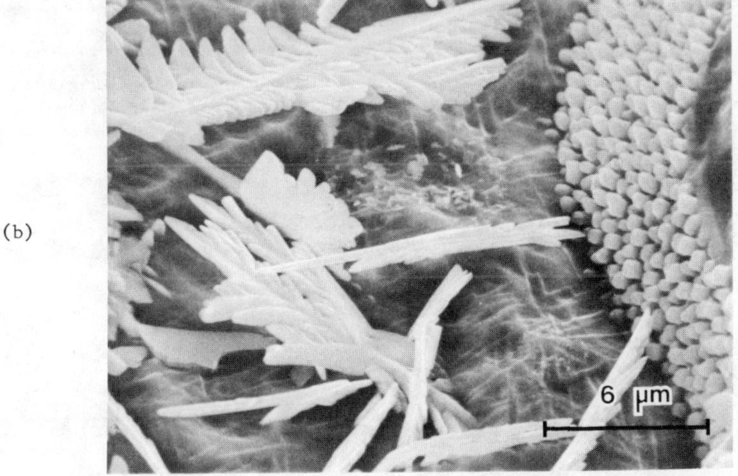

(b)

Fig. 1 - Microstructures of Ti-6Al-4V injected with TiC. The single
pass in 1(a) was processed with a laser power of 10 kW and
a sample translation speed of 5 cm/sec. The fine carbides
in 1(b) formed upon resolidification of a 6 kW melt pass in
the same sample.

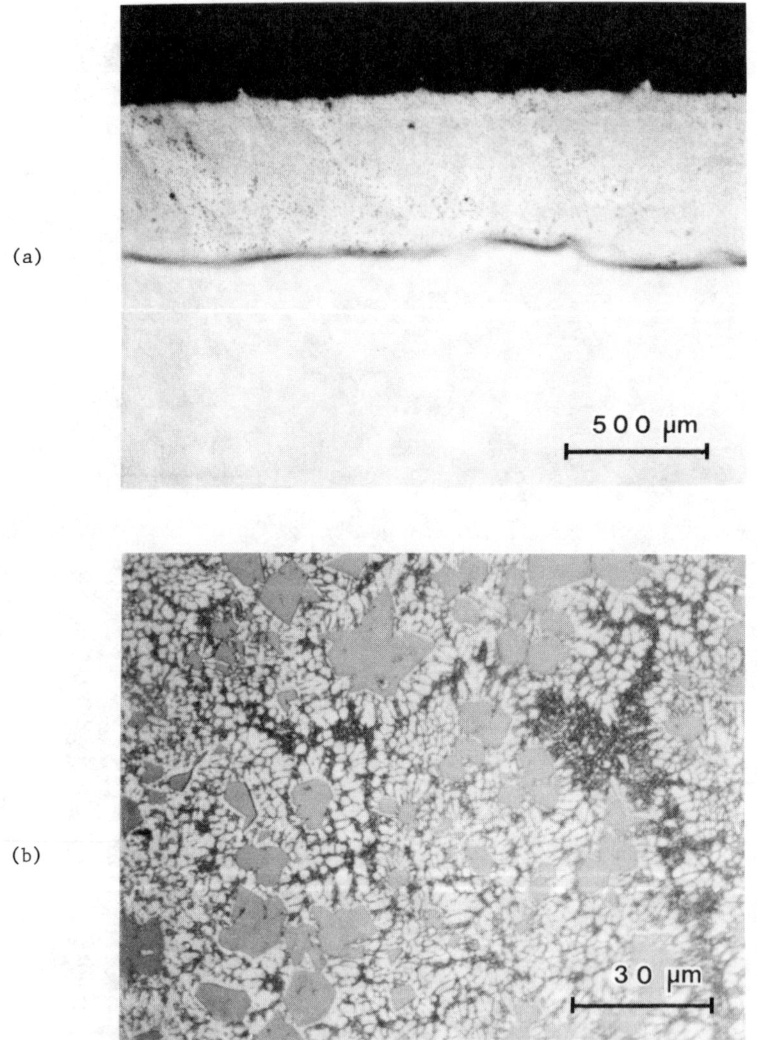

Fig. 2 - Surface alloying of 5052 Al accomplished by injecting silicon
powder at a laser power 5 kW followed by remelting at 4 kW.

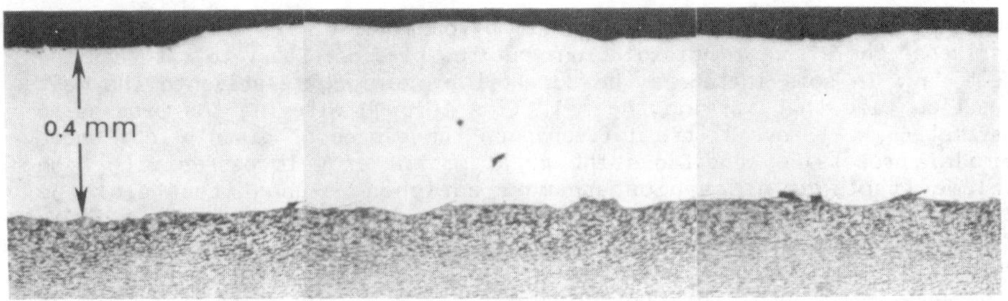

Fig. 3 - Surface of 1018 steel clad with nichrome by injection processing.
The surface was clad at 5 kW and smoothed by remelting at 4 kW.

2. Experimental

Because this laser processing technique has been described in earlier
papers (1-4), few details are included here. All of the carbide injected
samples considered in this paper were processed in a vacuum chamber which
was partially backfilled with helium, though recent work has demonstrated
that good results can be achieved when injecting TiC into aluminum at
ambient pressure, using helium gas shielding (4).

Preparation of the samples for injection processing normally consisted
of chemical cleaning only, though some of the rougher samples were ground
first. The carbide particles were screened to desired size fractions,
and were sometimes washed in alcohol to remove the fines (2). This practice was discontinued when microscopic examination of samples injected with
washed and unwashed carbides failed to reveal any microstructural differences. Samples were prepared for microscopic examination by mechanical
polishing sometimes followed by etching with standard reagents.

3. Factors Affecting Carbide Dissolution

The amount of carbide dissolution which occurs during injection
processing is dependent upon both materials parameters and upon processing
parameters. The relevant materials parameters are the alloy composition,
the carbide composition, and the carbide particle size. Important processing parameters include the temperature of the melt, the time available
for dissolution, the carbide volume fraction, and the amount of fluid flow
or "stirring" in the melt. Considering materials parameters first, the
solubility of any given carbide varies considerably from metal to metal.
Thus, no evidence has been found of TiC dissolving into 5052 Al during
injection processing (4), while extensive dissolution of TiC has been
noted in titanium (2) and iron (3) base alloys. The solubility of different carbides varies considerably in any given metal, partly because the
carbides differ substantially in their chemical stability. For example,
the free energies of formation at 1000°C of TiC and WC are -41,600 and
-8,000 calories per mole, respectively (6), and WC generally dissolves
more freely than TiC. The carbide size affects the kinetics of dissolution
because fine particles expose more surface to the melt and because (at a
given volume fraction of carbide) diffusion lengths in the melt are shorter
with finely dispersed particles.

The choice of processing parameters can alter the degree of carbide
dissolution in several ways. High melt temperatures favor dissolution both
by increasing the solubility of the carbide and by increasing the rate of
diffusion of the carbide constituents away from the dissolving particles.

The time available for dissolution at any particular melt temperature will influence the extent of dissolution, but the time necessary to saturate the melt may, in some instances, be so short or so long relative to the melt duration time that varying the melt time through altering the processing parameters will have little influence on the degree of dissolution. Under similar processing conditions the metal matrix of melt passes with high volume fractions of carbide are more enriched in constituents of the carbide than are passes with low carbide volume fractions (2). Fluid flow within the melt increases the rate of dissolution by hastening the removal of carbide constituents from the carbide/metal interface. Fluid flow can arise both from convection effects and from stirring caused by impingement of the carbide particles and the carrier gas on the metal surface. The enhanced stirring at high carrier gas pressures is quite evident in high speed motion pictures of the process. In Ti-6Al-4V injected with TiC under reduced pressure, passes injected with a carrier gas pressure of 5.3 X 10^4 Pa exhibited surface cracking while those injected at 6.7 X 10^3 Pa, with all other conditions constant, were crack-free. Cracking in the melt passes with enhanced stirring was apparently associated with the presence of more resolidified TiC which was observed microscopically.

With the many possible materials and processing variables it is difficult to draw general conclusions regarding the relative importance of various factors which influence the structure and properties of carbide injected surfaces. In order to estimate the importance of the contribution of carbide dissolution and regrowth upon the hardening of metal surfaces, the tests of the following section were conducted.

4. Hardness of Metals Injected with TiC

Carbide injection processing can harden the surfaces of metals by several different mechanisms. These include solid solution strengthening, dispersion hardening both by the large injected particles and by fine ones formed from the melt when partial dissolution of the injected carbide occurs, transformation hardening of the metal matrix, and by matrix hardening through microstructural refinement. In an effort to assess the relative importance of these different mechanisms, a group of five different alloys were injected with TiC and subjected to Vickers and Rockwell hardness testing. Both tests were applied to all metals in the as received condition and to the processed surfaces, but in the processed surfaces care was taken to obtain the Vickers hardness values from the metal matrix regions between injected TiC particles. The Vickers microhardness was also determined in the heat affected zone (HAZ) near the laser melt passes. Results of this testing are shown in Table 1. The samples included in the table were injected with 30 to 50 volume percent TiC which was first screened to

Table 1
Hardness of Titanium Carbide Injected Surfaces

ALLOY	As Received		HAZ	Carbide Injected	
	H_V	Rockwell	H_V	H_V	Rockwell
Ti-6Al-4V	345	C 35	405	460	C 45
304 Stainless	152	B 78	196	219	C 37
4340 Tool Steel	430	C 27	950	1061	C 62
5052 Al	78	B 50	70	71	B 61
Aluminum Bronze	225	C 25	239	287	C 27

-70+140 mesh (i.e. 110-210μm) or -230+325 mesh (i.e. 45-60μm). These differences in carbide size and volume fraction are believed to be of less significance than the other factors discussed here.

The metals included in Table 1 were selected to represent a wide range of behavior during the laser processing. The first three metals in the table all dissolve the injected TiC to some extent and could hence be expected to exhibit hardening by all of the mechanisms mentioned, except by transformation hardening* which should be limited to the 4340 tool steel. The 5052 Al and the aluminum bronze (Cu-9Al-5Ni-4Fe-3.5Mn) should not exhibit additional hardening by solid solution strengthening or through dispersion hardening from fine resolidified carbides because high magnification scanning electron microscopy has failed to yield any evidence of carbide dissolution in these metals.

The first point to be noted from examination of the data in Table 1 is that transformation hardening of the 4340 steel contributes more to the hardening of that alloy than does any of the other possible mechanisms. This point is partially evidenced by the fact that the microhardness increase in the HAZ over that in the as received alloy is substantially larger than the additional hardness increase noted in the carbide injected metal matrix. This additional increase (from 950 to 1061 H$_v$) represents the combined effects of solid solution strengthening, dispersion hardening by fine carbides, and microstructural refinement. This increase of about 100 vickers units is of the order of the total microhardness increase found for Ti-6Al-4V and 304 stainless, two alloys which exhibit comparable degrees of carbide dissolution during injection processing. The aluminum bronze exhibits a microhardness increase nearly as great as that of the Ti-6Al-4V and the 304 stainless but the hardening derives from the laser melting alone, because the TiC is not dissolved during injection processing. The contrast in melt behavior between 304 stainless and the aluminum bronze is demonstrated by Figures 4 and 5. The two samples were processed under very similar conditions, but the microstructures are dramatically different. In both metals thermal stress during cooling cracked the angular injected carbides, but the bronze matrix was better able to accomodate the imposed strain without propogating cracks through the metal matrix.[+]

The hardening produced by laser melting of the bronze probably results from a combination of microstructural refinement and of the formation of a hard martensitic product during rapid cooling of β phase produced during melting. The modest hardening in the HAZ may result from a solid state phase transformation, for optical microscopy showed some evidence of structural change in the unmelted HAZ. The microhardness increase within the HAZ of the 304 stainless is more puzzling, for optical microscopy did not show any evidence of microstructural modification in this alloy. Softening within the HAZ of the 5052 Al resulted from overaging of the precipitation hardened matrix during the short excursion to high tempera-

---

* By convention transformation hardening implies the hardening of steels though formation of martensite during quenching. As discussed in the text, however, hardening or softening of metals can occur by other solid-state phase transformations.

+ In either alloy injection of spheroidized, crack-free carbide particles probably would have eliminated the cracking (3). Unfortunately spheroidized TiC does not appear to be commercially available.

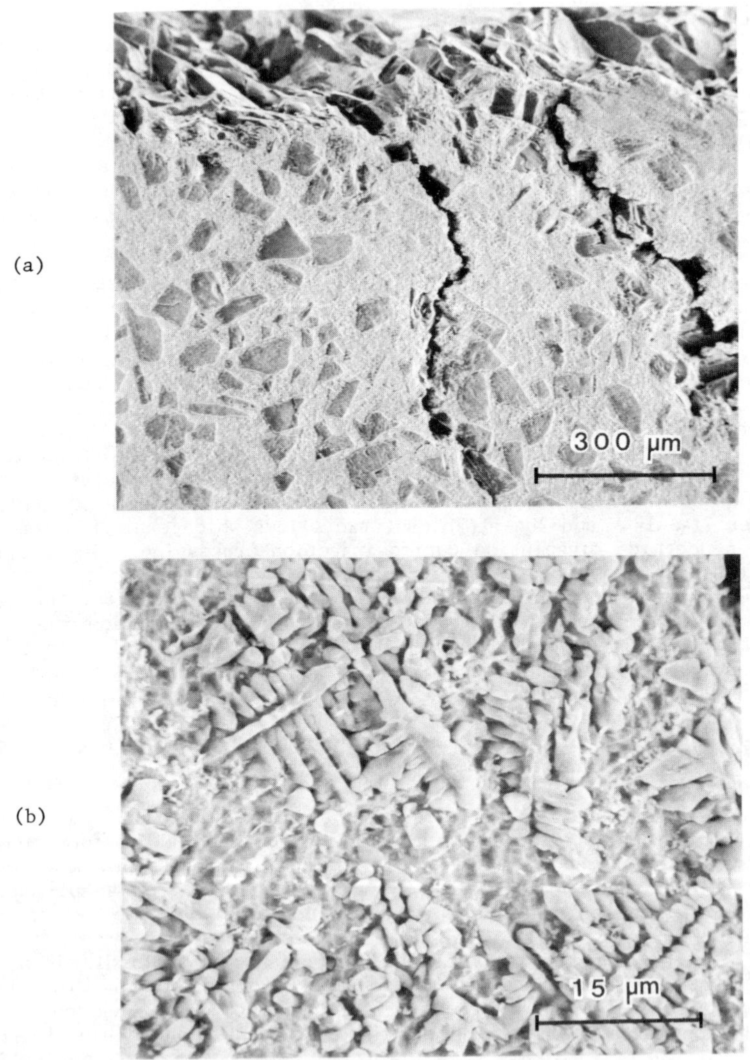

Fig. 4 - Carbide injected surface layer on 304 stainless showing TiC
dendrites formed during resolidification. Cracks in (a) formed
because embrittled matrix could not accomodate strain imposed
during rapid cooling.

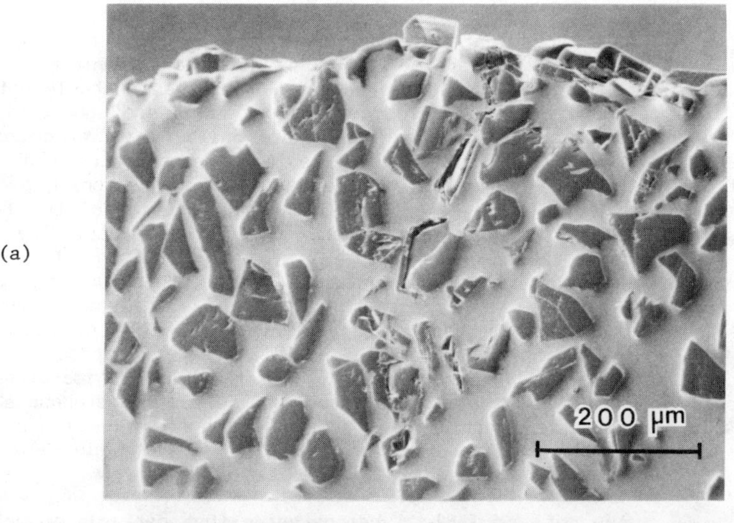

(a)

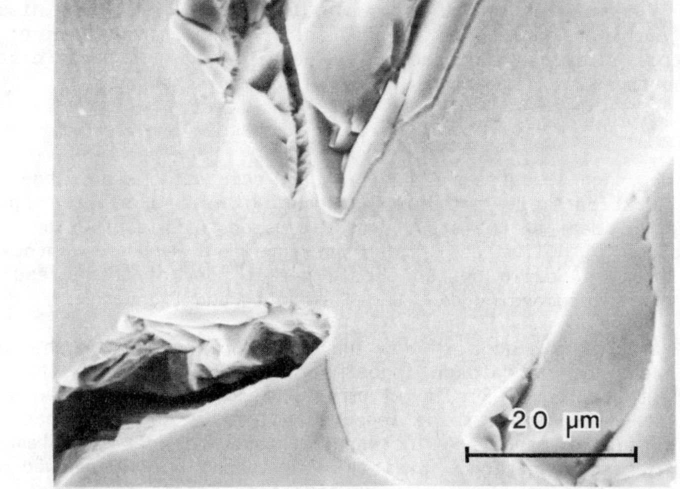

(b)

Fig. 5 - Aluminum bronze injected with TiC under conditions similar
to those used to produce the sample in Fig. 4. Individual
carbide particles were cracked by thermal stresses, but carbide
free metal matrix did not.

tures. The resolidified 5052 Al showed no evidence of age hardening at
room temperature or during artificial aging, so the laser processing
evidently did not trap alloying elements within the fcc matrix to a sufficient degree to permit precipitation hardening.

The microhardness results for these five alloys suggest that partial
dissolution of the injected carbide can contribute to hardening of the
metal matrix, but not to a pronounced degree. This result should not be
too surprising since the resolidified carbide particles are substantially
larger than are the fine precipitates which are deliberately produced to
strengthen many alloys. Examination of Table 1 demonstrates that the
contribution of the large injected carbide particles to hardening of the
alloys is also rather modest. For 5052 Al the contribution of the large
particles can be seen in the absence of other hardening effects. The data
shows the effect to be real because the laser processing softened the metal
matrix while producing a higher macrohardness. In the other four alloys of
Table 1, the macrohardness is increased to approximately the same degree as
the microhardness except in the 304 stainless for which the relative
increase in the macrohardness is greater. This greater increase in the
macrohardness probably derives from the very high level of carbide dissolution noted in this alloy at the upper surface (3). The microhardness shown
in the table for the metal matrix is an average value from near the center
of the surface layer. Values as high as 420 H_v were found near the
sample surface.

In summary, the data of Table 1 demonstrates that particle injection
processing can harden metal surfaces in several ways, but that no single
mechanism predominates in the various alloys. In alloys which dissolve the
injected carbide, hardnesses substantially higher than those of Table 1
can be achieved by choosing processing conditions which favor dissolution,
but the surfaces so formed generally contain many cracks.

5. Concluding Remarks

A wide range of surface structures and compositions have been produced
by laser melt/particle injection processing with a single experimental
configuration. Many of the structures discussed in this paper and elsewhere
were produced in helium at reduced pressure, but further improvements in
the gas shielded nozzle design should permit most of these results to be
duplicated with processing done under ambient conditions.

Only a limited amount of work has been done to date on characterizing
the wear behavior of carbide injected metal surfaces, but the available
data is encouraging. Results from dry sand/rubber wheel wear testing
(7) of as received and carbide injected alloys of several types showed at
least an order of magnitude increase in abrasive wear resistance in the
carbide injected condition. Even the TiC loaded 5052 Al showed wear test
results comparable to standard wear resisting materials. The only other
wear related test data so far available is from coefficient of friction
measurements on TiC injected Ti-6Al-4V and 304 stainless steel (8). Tests
using an unlubricated steel ball slider showed that the kinetic coefficient
of friction for both alloys was reduced from as received values of about
0.4 down to 0.2 or lower when the samples were injected with 40% or more of
TiC.

The available data suggest that carbide injection processing may be
suited for reducing wear rates with many materials. Given the tendency
toward cracking in some carbide injected metals, however, the process may
not be suitable for applications involving high level cyclic stress.

Studies of surface alloying by the particle injection process are in a very early stage, but initial results demonstrate that the technique can produce uniformly alloyed surfaces. This approach could well prove to be of commercial interest because it does not require precoating the surface with the alloying elements.

Acknowledgment

The able technical assistance of L.E. Richards is greatly appreciated.

References

1.   R. J. Schaefer, T. R. Tucker and J. D. Ayers, "Laser Surface Melting with Carbide Particle Injection," in Laser and Electron Beam Processing of Materials, C. A. White and P. S. Peercy, eds., Academic Press, N. Y., N. Y., (1980), p. 749.

2.   J. D. Ayers, T. R. Tucker, and R. J. Schaefer, "Wear Resisting Surfaces by Carbide Particle Injection," in Rapid Solidification Processing - Principles and Technologies, II, R. Mehrabian, B. H. Kear, and M. Cohen, eds., Claiter's Publishing Division, Baton Rouge, Louisiana, (1980), p. 212.

3.   J. D. Ayers and T. R. Tucker, "Particulate-TiC-Hardened Steel Surfaces by Laser Melt Injection," Thin Solid Films, 73, (1980), p. 201.

4.   J. D. Ayers, "Modification of Metal Surfaces by the Laser Melt/Particle Injection Process," Thin Solid Films, to be published.

5.   W. M. Steen and J. Powell, "Vibro-Laser Cladding," this volume.

6.   C. E. Wicks and F. E. Block, Thermodynamic Properties of 65 Elements - their Oxides, Halides, Carbides, and Nitrides, U. S. Government Printing Office, Washington, D. C., (1963).

7.   L. K. Ives, F. Matanzo and J. D. Ayers, unpublished research.

8.   J. D. Ayers, T. R. Tucker and R. C. Bowers, "A Reduction in the Coefficient of Friction for Ti-6Al-4V," Scripta Metallurgica, 14, (1980) p. 549, plus unpublished research.

LASER PROCESSING OF TITANIUM-6AL-4V*

Robert A. Bayles, Dale A. Meyn, Peter G. Moore
U. S. Naval Research Laboratory
Washington, DC 20375

Laser surface melting of Ti-6Al-4V produces a martensite microstructure which has a lower fatigue strength than the unprocessed material. Fractographic examination of unprocessed specimens indicates shear induced cleavage to be the characteristic fatigue initiation mechanism, while that of the processed specimens is shown to be simple tensile cracking.

* This work was supported by the U. S. Naval Air Systems Command.

Research Objectives

Laser surface melting and alloying of titanium alloys has the potential of producing surface finish and near surface microstructures which are superior in resisting fatigue crack initiation when compared to the bulk microstructures. Changing the processing conditions and concentration of additional alloying elements allows the production of a great variety of unique microstructures. The objective of the work described in this paper is to relate the mechanism of crack initiation in Ti-6Al-4V to the microstructure and topography of the surface layer, and to identify microstructures which have good fatigue crack initiation resistance.

Materials

Two heats of Ti-6Al-4V were used in this study. Heat A is a 2.6 mm thick rolled sheet which was $\alpha + \beta$-solution treated and aged. The microstructure is shown in Figure 1a. Heat B is a 9.6 mm thick rolled plate which was β-processed. The microstructure is shown in Figure 1b. Details of the composition and heat treatment of these materials are not yet available. The yield stress of the Heat A material is 924 MPa and the yield stress of the Heat B material is 852 MPa.

Laser Processing

The laser used to melt the metal surface is a continuous wave, CO_2 laser which has a wavelength of 10.6 μm. In the case of the processing described here, the typical beam power is 2 to 3 kW.

The specimens are mounted on the periphery of a rotating wheel which provides translational speeds of up to 2 m/s. A helium gas shield is used to protect the metal from oxidation and to control the plasma which couples the laser to the workpiece. Each pass of the specimen through the laser beam results in a ribbon of material being rapidly melted and then quenched from the melt by the cool substrate. The processed surface is built up of parallel, partially overlapping passes as illustrated in Figure 2. This figure shows the characteristic chevron markings on the surface of the processed area, and the depth of the previously melted and the heat affected zones. The heating of the workpiece was limited to 80C, and in most cases, the piece had time to cool between passes.

Laser surface melting increases the texture of the microstructure due to preferred solidification in the <100> direction of the high temperature β-phase, with the <100> direction nearly normal to the surface of the workpiece. The preferred orientation persists after the transformation to martensite occurs during cooling. Bowing of the 2.6 mm thick sheet indicates that residual tensile stresses are present in the 100 μm thick processed layer.

Fatigue Experiments

The fatigue machine consists of a fixed grip which holds the bottom end of the specimen and a constant-displacement crank-type drive with a load cell, operated at 10 Hz or 16 Hz. A cantilever bending fatigue configuration was used with the surface of interest in tension. A narrow waist was machined in the specimens to control the location of maximum stress. A load ratio of 0.2 was used and the load was measured with ± 1% accuracy. The specimens of Heat A had a rectangular cross section and the specimens of Heat B had a circular cross section.

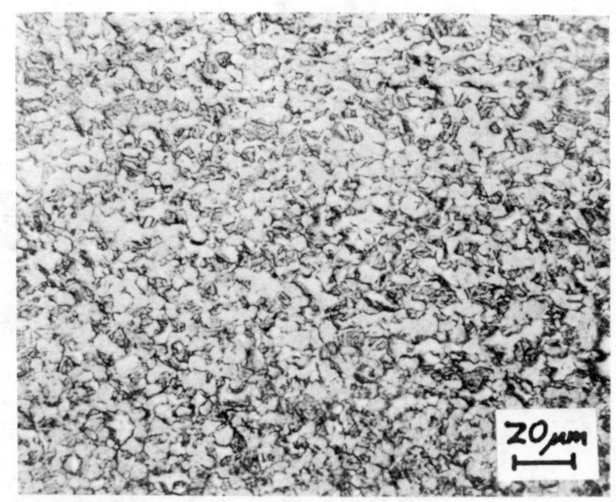

Fig. 1a. -   Heat A microstructure. Equiaxed primary α grains (white) in a matrix
of transformed ß (mottled).

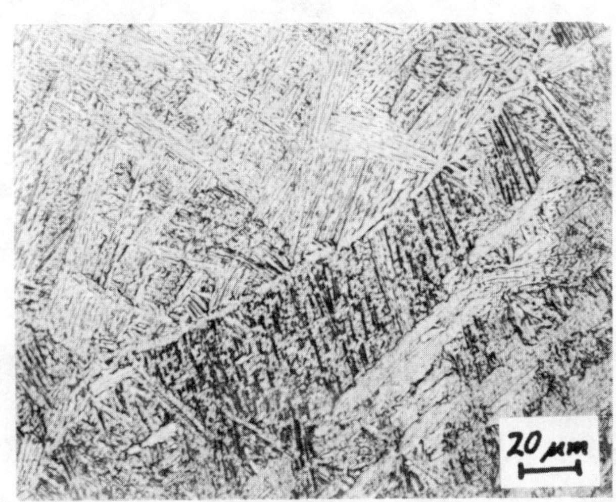

Fig. 1b. -   Heat B microstructure. Widmanstatten structure of alternating α and
ß plates and prior ß grain boundaries.

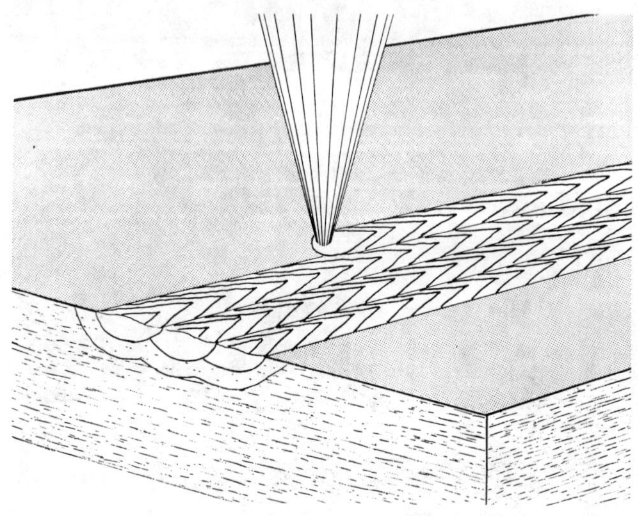

Fig. 2 –   Laser surface melting configuration illustrating laser beam, chevron markings, melt depth and heat affected zone.

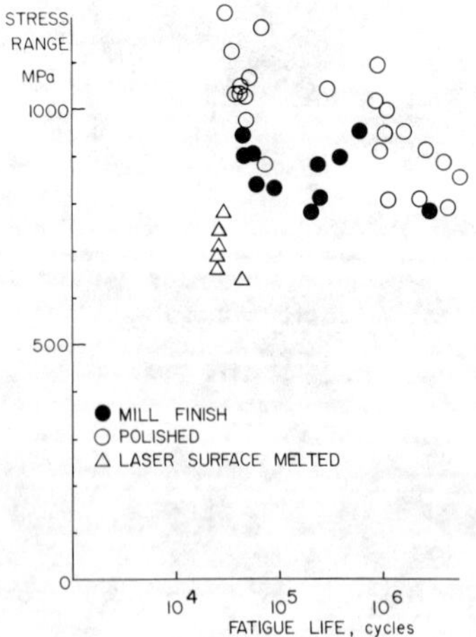

Fig. 3 –   Fatigue data from Heat A specimens.

130

Fatigue Results

The results of fatigue tests of the two types of specimens are shown in Figures 3 and 4. Cyclic stress range is plotted versus the number of cycles at which the value of the mean load dropped 5% from its original value. This number of cycles is about the same as the fatigue life to complete failure in unprocessed specimens, and long life processed specimens. In the case of short life processed specimens earlier crack initiation and easy propagation through the processed layer causes propagation through the bulk to be a significant fraction of the fatigue life.

The data from the Heat A specimens show much scatter. This scatter is believed to be due to microstructural variations, perhaps specimen-to-specimen texture variations (1) although the effect of the rectangular shape of the cross section is a possible source of scatter.

The data from the Heat B specimens show very little scatter. The fatigue strength of the laser surface melted specimens was more than 50% below that of the polished specimens for long lives, and for a given fatigue strength the fatigue life was more than 100 times shorter.

Fractography

In order to explain the degradation in fatigue strength, a fractographic examination was performed using a high-resolution scanning electron microscope. The fracture origin of an unprocessed polished fatigue specimen is shown in Figure 5. This origin is characterized by steeply inclined, large cleavage facets. This fracture surface indicates that crack initiation resulted from cleavage on slip planes weakened by to and fro shearing (2). Figure 6 shows the fracture origin of a laser surface melted fatigue specimen. This fracture surface is very flat and perpendicular to the tensile stress axis even on a microscopic scale. The martensite microstructure is apparent on the fracture surface. A close-up view of this origin area (Figure 7) shows some cleavage and some ductile tearing. Figure 8 shows a close-up view of a laser surface melted specimen tested at a higher stress range. The fracture surface is again flat, but in this case cleavage of the martensite plates is modified by dimples formed by microvoid coalescence. Figure 9 shows a point near the fracture origin on the face of a laser surface melted specimen. This shows the multitude of cracks indicating multiple crack origins which is typical of the laser surface melted material. Almost no multiple origins were observed in the case of the polished specimens. This micrograph also shows the microscopically smooth topography of the laser melted surface.

The flat, approximately planar fatigue origins in the laser treated specimens indicate a very easy microcrack initiation and growth process through several "grains" or martensite colonies of similar orientation. Also, the martensite is a single phase microstructure, affording overall easy slip penetration and tensile crack growth through the microstructure. By contrast, microcrack initiation in the as-received, two phase, $\alpha - \beta$ microstructure may be delayed by limitation of suitable slip processes to a single or a few large, isolated, favorably oriented grains, and propagation of such microcracks into surrounding grains or from α phase grains into β phase is probably made difficult by orientation and phase mismatch.

Conclusions

1. Laser surface melting and self-quenching (without subsequent heat treating) produces an HCP martensite surface layer. No glassy phases are observed by x-ray diffraction. The microstructure is highly textured.

131

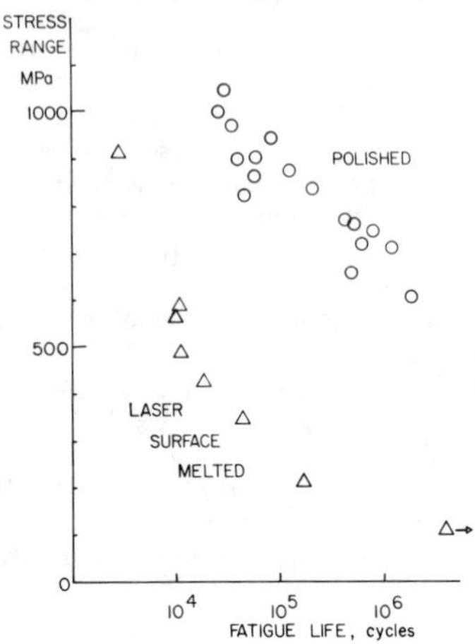

Fig. 4 -   Fatigue data from Heat B specimens.

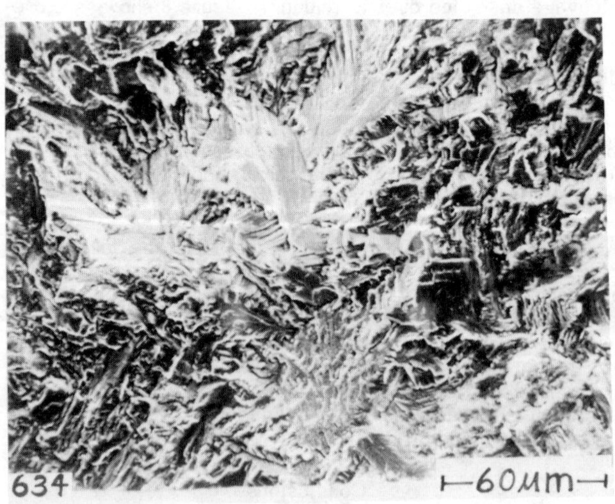

Fig. 5 -   Fracture origin from unprocessed material showing inclined cleavage
facets. Heat B specimen.

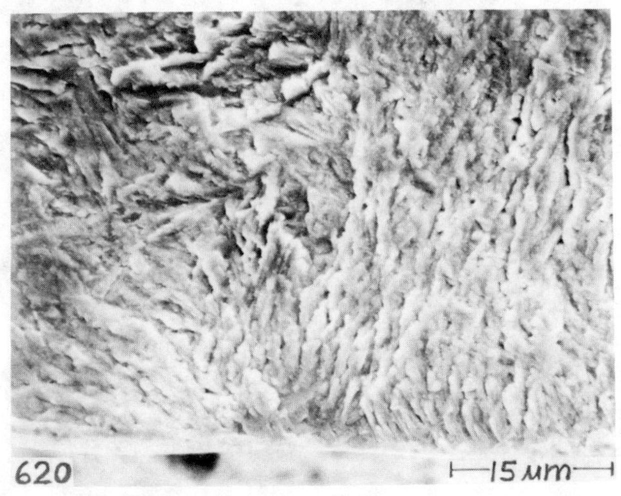

Fig. 6 –   Fracture surface from laser surface melted material showing flat tensile cracking. Heat A specimen.

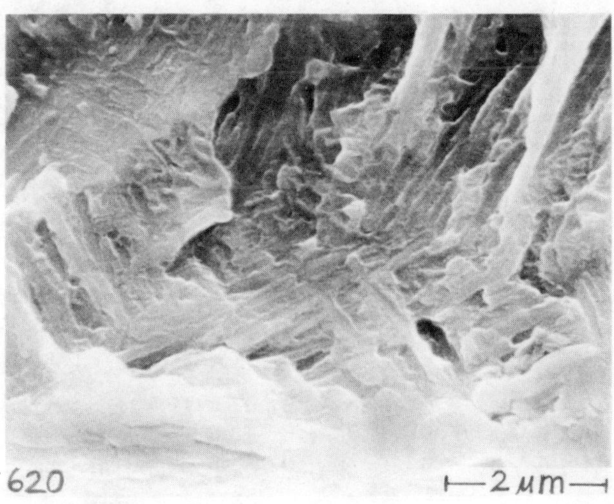

Fig. 7 –   Close-up view of Fig. 6 showing cleavage of martensite plates. Heat A specimen.

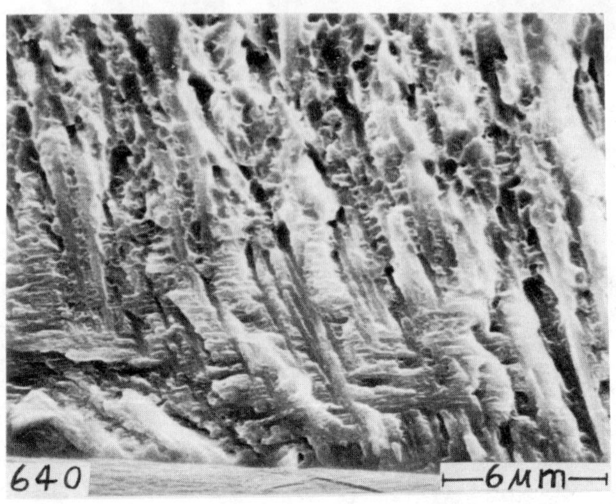

Fig. 8 –   Laser surface melted specimen tested at higher stress showing dimples and tensile cleavage. Heat B specimen.

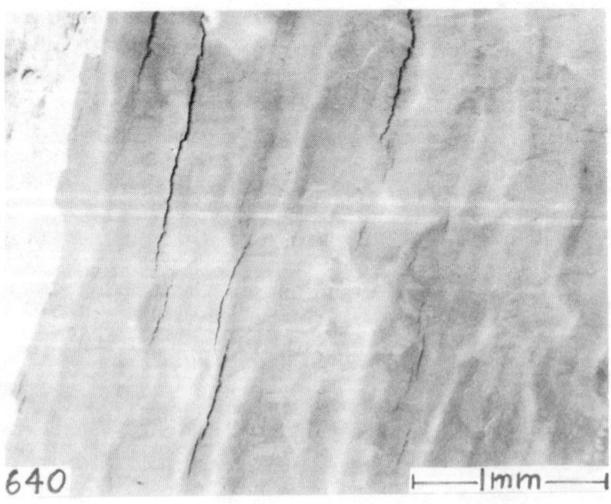

Fig. 9 –   Face of laser surface melted specimen showing multiple cracks and microscopically smooth surface. Heat B specimen.

2. Laser surface melting produces residual tensile stresses in the processed layer, and very smooth surface topography.

3. The processed surface has a microstructure which is sensitive to crack initiation, but crack propagation through the bulk is unaffected.

4. This sensitivity to crack initiation greatly reduces fatigue resistance of laser processed material.

5. Single crack origins are observed in unprocessed surfaces while multiple crack origins and multiple parallel cracks are observed in processed specimens.

6. In the unprocessed microstructures, the mechanism of crack initiation is associated with shear band cleavage inclined relative to the stress axis, while in laser surface melted material the mechanism is simple tensile cracking on planes perpendicular to the tensile stress axis.

Future Work

The present work will be augmented by studies of processes designed to alter the mechanism of crack initiation presently observed in laser melted surfaces. Various heat treatments to provide stress relief and strengthening will be used to produce microstructures which will be less sensitive to tensile cracking. Additional work will be done to confirm the source of scatter in the fatigue data for the α-β microstructure. Quantitative analysis of the residual stresses will be attempted.

References

1. C. A. Stubbington, "Metallurgical Aspects of Fatigue and Fracture in Titanium Alloys," Alloy Design For Fatigue and Fracture Resistance, AGARD Conf. Proceedings #185, January 1976, pp. 3-1 to 3-19.

2. C. H. Wells and C. P. Sullivan, "Low Cycle Fatigue Crack Initiation in Ti-6Al-4V," Trans. ASM Vol. 62 (1969), pp. 263-270.

SHOCK DEFORMATION AND MICROSTRUCTURAL EFFECTS

ASSOCIATED WITH PULSE LASER-INDUCED DAMAGE IN METALS*

K. Mukherjee
Michigan State University
East Lansing, MI 48824

T. H. Kim
&
W. T. Walter
Polytechnic Institute of New York
New York, NY 11201

Shock deformation and microstructural effects associated with pulse laser interaction with metallic surfaces, are reviewed. Some recent results of laser damage in low melting point metals are reported. A wavy deformation pattern in these metals is associated with a decaying elastic wave propagation through an elastically inhomogeneous medium. Defect structures and void formation in vapor deposited pure Cu are reported and discussed.

* This work was partially supported by the U. S. Office of Naval Research.

Introduction

It is now well known that stress waves and associated shock deformation of solid targets can be produced with Q-switched laser pulses (1-19). Highly localized thermal expansion, in the wake of a rapid laser pulse at the surface of an elastic solid, promotes a stress wave which then propagates into the interior of the solid. The magnitude of the peak stress becomes appreciable as the pulse duration becomes shorter. If the duration of the pulse is such that the stored elastic distortional energy is greater than or equal to that necessary for yielding, then plastic deformation will occur.

The initial impact of the laser energy pulse is analogous to a high strain rate explosive shock wave deformation of the material (12). In fact, a striking similarily between the defect structure and deformation, produced by explosive shock loading and Q-switched laser pulse damage has been reported (19). In unconfined thin metal foils (250Å to 1 μm thick) a peak stress of the order of 1 k. bar has been reported for 25 to 50 n. sec duration Q-switched pulses (11). An order of magnitude higher peak stress for the same materials was observed with a plasma confining transparent overlay (11). In a similar experiment with vapor deposited confined thin foils (1 μm to 10 μm) of various pure elements (B, C, Mg, Al, Si, Ti, Cr, Mn, Fe, Co, Ni, Cu, Zr, Mo, Ag, In, Sb, Pt, Au, Pb, and Bi), it was found that the peak stress is approximately a linear function of the laser energy fluence (12). Peak pressures as high as 250 k. bars have been estimated for W and Mo wires by comparing vacancy concentration produced by a Q-switched ruby laser (35 J/cm^2 fluence) and that produced by explosive shock loading (19).

The stress wave generated by a laser pulse can give rise to elastic and plastic deformation, formation and rearrangement of dislocations, formation of vacancies and formation of vacancy clusters leading to microporosity etc. The above mentioned effects are important from the metallurgical point of view since these can alter microstructure and mechanical properties of metals and alloys. Beyond a threshold energy density, for a given material, melting and crater formation, sputtering and plasmaformation are possible (17,18,20-28). If melting is involved, then subsequent ultra rapid solidification can produce metallic glasses, ultrafine grain structure, metastable phases and a very large concentration of vacancies (19,29-34).

Coupling of the incident laser light with the metallic target, on the other hand, can be influenced by inherent structural discontinuities such as vacancies, dislocations, grain boundaries and inclusions. A better understanding of the laser-materials interaction and the role of various structural singularities in this interaction, is essential for a more efficient application of lasers in materials processing. In this paper, shock wave phenomenon and microstructural aspects associated with pulse laser interaction with metals are briefly reviewed and some new results related to these effects are presented.

Mechanism(s) of Shock Wave Formation

Several mechanisms of laser induced shock wave formation can be postulated. The laser photon energy is almost instantaneously absorbed on the metal surface as thermal energy, producing a highly localized temperature excursion. If the local temperature reaches or exceeds the boiling point of the metal then a rapid vaporization occurs. The recoil pressure of the

escaping (vapor) atoms (without ionization) could possibly set up the shock
deformation in that case. If the laser energy fluence is high enough to
produce a plasma, then the expanding plasma can also provide the shock wave.
But shock phenomenon and/or plastic deformation have been reported for laser
energy fluences less than that for melting (18). In such a case, the thermal stress must be the predominant cause. Of these, the thermal stress and
plasma effects appear to be the two important modes as far as current experimental results are concerned.

Thermal Effects

Heating by Q-switch laser pulses is nearly approximated by a constant
volume process. As the laser energy is absorbed in a thin surface layer,
the internal energy of the irradiated volume of material increases. If the
rate of heating of this volume is very rapid, as in the case of a Q-switched
pulse, then the normal thermal expansion of this volume can not take place
because of the inability of the adjacent matrix to relax at a rate commensurate with the temperature rise. Thus a pressure wave in the form of a com-
pressive shock wave is produced, which travels through the material.

The amplitude of the elastic wave, produced for a given absorbed laser
energy fluence, depends on the elastic constraints applied at the heated
surface. If the heated surface is free (stress-free), the wave amplitude
could be relatively small compared with that for a constrained surface (such
as a transparent overlay). The nature of this thermo elastic stress can be
examined by considering a homogeneous isotropic solid as follows (2).

Let us consider a semi-infinite surface with a localized transient heat
source as shown in Figure 1.

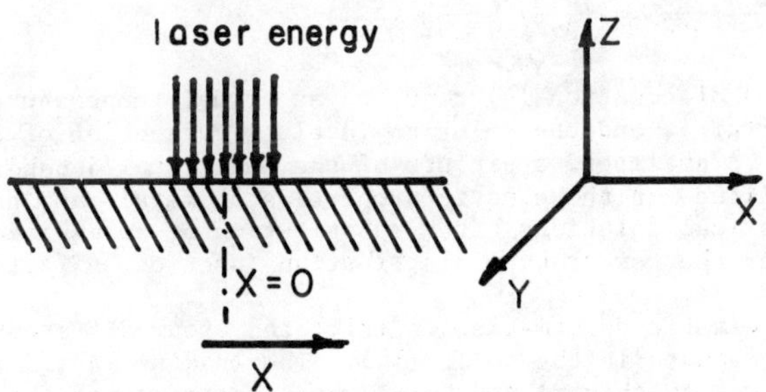

Fig. 1 - Transient, symmetric and uniform heat source.

For simplicity, it is assumed that heat is absorbed at the surface
Z = 0 and the laser energy is uniform across the spot diameter. In reality,
however, the energy distribution within the spot is ~ gausian and the heat
energy will be deposited in a layer of thickness δ, given by the skin depth
for the electromagnetic radiation:

$$\delta = C/2\omega \, |\varepsilon|^{\frac{1}{2}}$$

(1)

where c and ω are the velocity and angular frequency of the laser light and
ε is the real part of the dielectric constant of the metal.

As a result of this surface heat absorption there is a surface temperature rise and an associated elastic strain. At this point we consider an

one-dimensional problem, i.e., $e_{xx} = e_{yy} = 0$, since we have seen that the rigidity of the surrounding matrix during a Q-switched laser heating does not allow the lateral volume expansion. The only non zero strain e_{zz} is then:

$$e_{zz} = \frac{\partial u(z,t)}{\partial z} = \alpha \, \Delta T(z,t) \qquad (2)$$

where $u(z,t)$ is the z-component of particle displacement, α is the coefficient of linear thermal expansion and $\Delta T(z,t)$ is the temperature rise. If the strain, $e_{zz} = \alpha \Delta T$, was produced by a stress σ_{zz} without the temperature rise, then $\sigma_{zz} = -Be_{zz} = -B\alpha\Delta T$, where B is the bulk modulus of the metal. In the presence of both heating and actual stress, then (35)

$$\sigma_{zz} = (\lambda + 2\mu) \, e_{zz} - B\alpha\Delta T \qquad (3)$$

where λ and μ are the first Lamé constant and the modulus of rigidity respectively. The equation of motion for the particle displacement is thus given by:

$$\rho \left(\frac{\partial^2 u}{\partial t^2}\right) = \left(\frac{\partial \sigma_{zz}}{\partial z}\right) = (\lambda + 2\mu) \left(\frac{\partial^2 u}{\partial t^2}\right) - B\alpha \left(\frac{\partial \Delta T}{\partial z}\right) \qquad (4)$$

where ρ is the density of the metal. Utilizing the relationship

$$(\lambda + 2\mu) = \rho v^2 \qquad (5)$$

where v is the velocity of elastic wave propagation in the metal, we can write

$$\frac{1}{v^2} \left(\frac{\partial^2 u}{\partial t^2}\right) = \left(\frac{\partial^2 u}{\partial z^2}\right) - \frac{B\alpha}{\rho v^2} \left(\frac{\partial \Delta T}{\partial z}\right) \qquad (6)$$

A formal solution of equation (6) requires an actual temperature distribution function $\Delta T(z,t)$, and the solution involves evaluation of rather cumbersome integrals. A qualitative picture of the functional dependence of the stress-wave amplitude on the elastic properties of metal can, however, be obtained by numerical solutions after substitution of an empirical (12) exponential form of the temperature distribution function, $\Delta T(z,t)$.

Once, this elastic deformation is initiated, four different types of elastic waves propogate in the solid (11). The leading wave is an elastic dilatational wave. As this dilatational wave sweeps across the surface, a series of waves is generated. Using Huygens principle, an envelop of these waves can be defined and this envelop is called the Von Schmidt wave. The dilatational wave is followed by a shear wave with a velocity $v_s < v_e$ where v_e is the elastic wave velocity. A Rayleigh wave is also produced on the surface with a velocity $v_R < v_s$. Plastic deformation will occur if the distortional elastic energy in these waves exceeds that necessary for yielding. The qualitative nature of various elastic waves are shown in Figure 2.

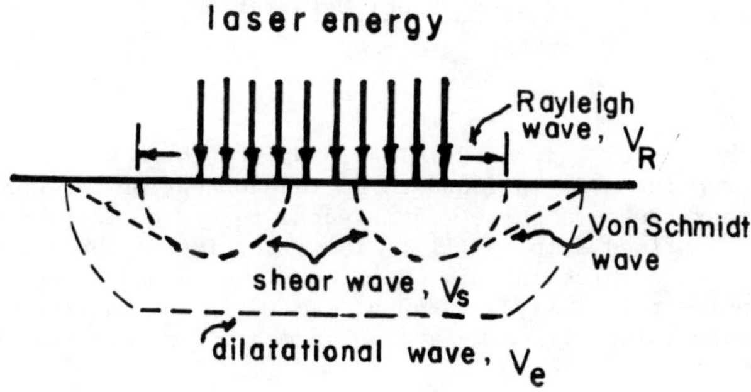

Fig. 2 - Elastic wave system induced by an uniform laser pulse source.

Plasma Effects

At a laser energy density above the threshold for melting and evaporation a hot plasma can form. The expanding plasma can induce a shock wave in the material. We are, however, considering the laser interaction with metals in vacuum. Thus the possibility of a LSD-wave proproduced in air, which can also induce a shock wave, is not discussed.

The essential difference between the thermoelastic shock wave as discussed earlier and the plasma induced shock wave is that the latter is more nearly a surface stress source than the former. Thus the boundary constraint in the plasma induced shock deformation is different from those in the thermo-elastic case. Further more, confinement of plasma, with an overlay, will have a much more pronounce effect on the shock intensity. In fact, experimental results indicate that the peak pressure can be at least an order of magnitude larger if the plasma is confined.

Experiments which measured momentum transfer to various target metals as a function of laser intensity, show an optimum intensity for each metal (36) as shown schematically in Figure 3.

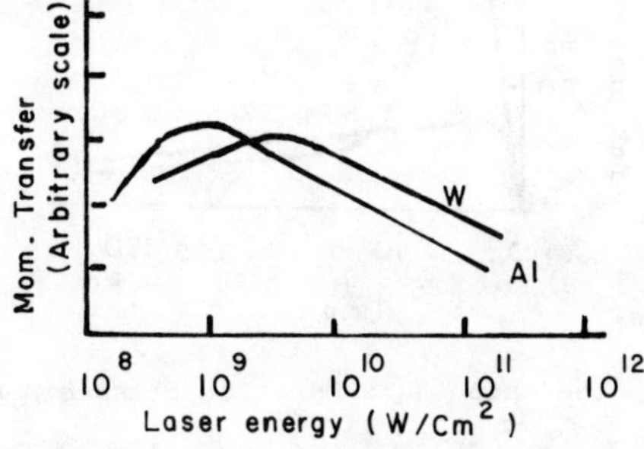

Fig. 3 - Momentum transfer as a function of laser energy density.

An explanation of the behavior, as shown in Figure 3, is that below the optimum energy, some energy is lost by thermal conduction and above the maximum, some energy goes in increasing ionization and temperature of the plasma. Thus it will appear that for a given material, the plasma induced shock pressure will decrease with increasing laser energy density beyond the optimum value.

Nature of Shock Deformation

Thin Films

In stainless steel and in 6061-T6 Al foils (0.4 to 0.6 mm thick), the permanent deformation mode is found to be independent of the confinement, although the peak pressure pulse was measured to be an order of magnitude higher for the confined samples (11). The rise time of the pressure pulse, for the unconfined samples in this experiment, was found to be comparable to the laser pulse rise time; ~ 25-50 n. sec. In the confined samples, both the pressure pulse rise time and pulse decay were considerably longer.

Shock wave peak-pressure pulse measurements have been reported for a large number of pure element foils, 1 μm to 10 μm thick at energy fluences up to $12J/cm^2$. These samples were confined between glass plates and a 15 n. sec. duration Q-switched ruby laser pulse was used. It was found that the peak shock pressure is approximately a linear function of laser fluence (12).

Bulk Samples

It has been reported (13) that in confined pure Fe samples, the peak shock pressure increases with the reduction in sample thickness at an energy density of ~ 10^9 w/cm^2 and a pulse duration of 20-30 n. sec. For example a peak pressure of 59.6 k. bar was observed for a 14 μm thick compared with a peak pressure of 9.2 k. bars for a 3 mm thick sample. Most importantly, the rise-time of the pressure pulse is comparable to the rise-time of the laser pulse in both cases. However, the decay time of the pressure pulse is much longer than the laser pulse in the thicker samples. A qualitative nature of the pressure pulse shape as a function of sample thickness is shown in Figure 4.

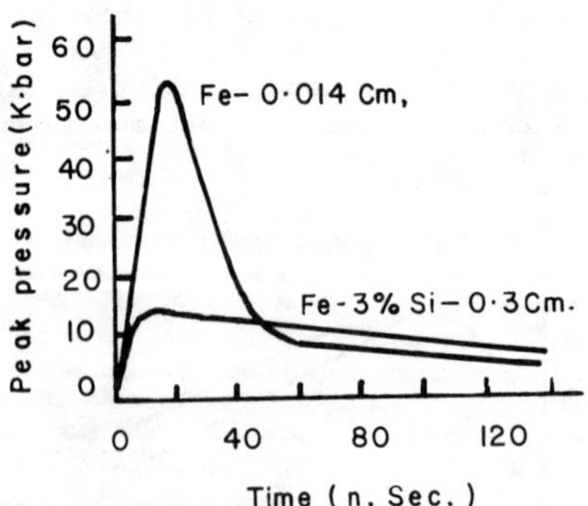

Fig. 4 - Pressure-pulse shape as a function of specimen thickness (13).

Slip, deformation and surface markings produced by the shock wave have been studied by several investigators (10,15-19,37-41). Both polycrystalline and single crystal metallic surface show crystallographic slip (10, 16-19) as well as slip and periodic ripple patterns (16,37) induced by high energy pulse laser. In semiconductor materials, a highly periodic ripple pattern is also reported. In a single crystal Si sample, a 50 n. sec. Q-switched ruby pulse, ~ 10^8 w/cm^2, produced twinned structures on (111) planes.

142

In single crystals of Cu and Al, multi threshold damage studies have
been reported (17,18). Intragranular crystallographic slip and grain boundary sliding in polycrystalline Cu and Al are observed at laser energies
well below the threshold for melting (17,18). These authors have also reported that the existing imperfections such as dislocations, grain boundries etc., lower the threshold for melting and other catastropic damage,
but increases the threshold for plastic deformation. These results are indeed important in terms of understanding the mechanism(s) of coupling of
pulse laser energy on metal surfaces and for the consideration of pulse
laser applications in materials and metals processing. One interesting observation (18) is that a symmetric six-fold slip pattern on the (111) surface of Cu single crystals is observed at laser fluences up to the threshold for plasma formation. The six-fold pattern is replaced by a 3-fold
pattern at intensities which produce a hot plasma. The six-fold slip pattern therefore, is associated with the thermoelastic shock wave and the 3-
fold pattern, perhaps is related to a different boundary constraint situation associated with the plasma shock wave as discussed earlier. The nature
of crystallographic slip on various orientations of Al single crystals are
also reported in the literature for ~ 100 n.s. duration pulse envelope from
a CO$_2$-TEA laser at energies $\geq$ 37J/cm^2.

Spallation and Fracture

Photographic evidence of back face spallation of 1.0 mm thick 6061-T6
Al alloy has been reported (14). These samples were irradiated in vacuum
with 1.5 n. sec., 1.06 μ laser pulses at fluences as high as 8600 J/cm^2.
It is also reported by this author that low-energy pre-pulsing lowered the
subsequent spallation threshold. It is believed that the prepulsing improves the coupling of the laser energy with the metal surface.

Non Equilibrium Vacancy Concentration

From field-ion microscopy studies, Murr et al. (42,43) have demonstrated that vacancy and vacancy clusters are produced in explosively
shock-loaded Mo. In a later experiment, field-ion microscopy was performed (19) on pulse laser irradiated Mo and W. The results of increasing
vacancy concentration with increasing laser fluence can be compared with
similar observations in the explosively shock loaded Mo and W. At a laser
fluence of 35J/cm^2 a vacancy concentration as high at 6 at .% was recorded
for Mo and a vacancy concentration of 0.9 at .percent was observed for W at
a laser fluence of 29J/cm^2. A direct comparison of vacancy concentration
produced by an explosive shock and a laser pulse, led these authors to estimate a peak pressure pulse ~ 250 k. bars at the maximum laser fluence.

We have observed (44) vacancy clusters produced in vapor deposited Cu
foils irradiated with Q-switched ruby pulse, ~ 10^9 w/cm^2 and 20 n. sec.
FWHM. These results will be presented later.

Dislocation Mechanisms

Recently (10), dislocation etch pit studies have been performed on
(111) planes of high purity single crystal Cu. This study shows a symmetric
six-fold etch-pit pattern around the pulse laser damage spot as shown
schematically in Figure 5.

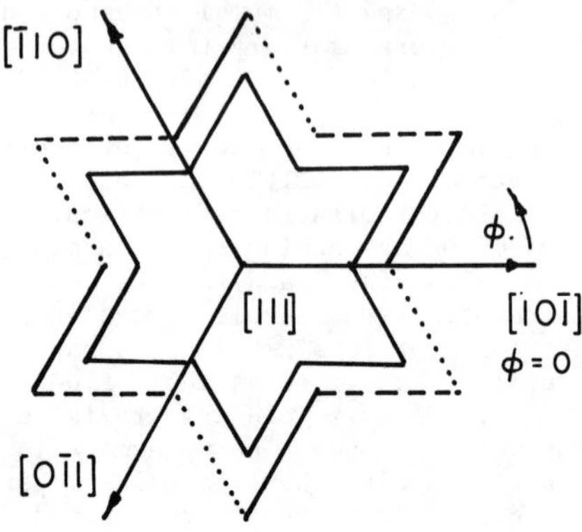

Fig. 5 – Schematic presentation of etch-pits on the (111) surface
of a Cu single crystal.

There are nine different orientations of edge dislocations and three
screw dislocations which intersect the (111) surface and therefore, may
lead to etch pit formation. The force F on a dislocation line L and Burgers
vector b is given by F = σ.b x L. In the case of a thermoelastic stress,
generated by a circular surface source, σ is a tensor. These authors
assumed a stress field, in a cylindrical co-ordinate system (origin of the
co-ordinate system is in the center of the laser spot of radius R) as
follows:

$$\sigma_{rr} = -\frac{1}{2} E.\beta.\Delta T \left(\frac{R}{r}\right)^2 \qquad \text{for } r > R$$

$$\sigma_{\phi\phi} = +\frac{1}{2} E.\beta.\Delta T \left(\frac{R}{r}\right)^2 \qquad \text{for } r > R \qquad (7)$$

$$\sigma_{rz} = \sigma_{zz} = 0$$

A calculation of forces on the twelve orientations of the dislocations using
the stress field shown in equation (7), shows that the largest forces act
on the three edge dislocations shown in Table I.

Table I.   The Nature of Dislocations on Which Maximum Force Acts

Dislocation	Direction of b	Direction of L	Slip Plane	Slip Trace $\parallel$ to
D_1	1 0 $\bar{1}$	$\bar{1}$ 2 $\bar{1}$	$\bar{1}$ 1 $\bar{1}$	1 0 $\bar{1}$
D_2	0 $\bar{1}$ 1	$\bar{2}$ $\bar{1}$ $\bar{1}$	1 $\bar{1}$ $\bar{1}$	0 $\bar{1}$ 1
D_3	$\bar{1}$ 1 0	$\bar{1}$ $\bar{1}$ $\bar{2}$	$\bar{1}$ $\bar{1}$ 1	$\bar{1}$ 1 0

The magnitudes of the forces on these dislocations D_1, D_2 and D_3 respectively have maximum for angles (note ϕ = 0 for [10$\bar{1}$] direction and Z = 0
on the (111) plane) as follows:

$$\phi_1 = 45^{\circ} + n \cdot 90^{\circ}$$

$$\phi_2 = 15^{\circ} + n \cdot 90^{\circ} \qquad n = 0,1,2,3 \qquad (8)$$

$$\phi_3 = 75^{\circ} + n \cdot 90^{\circ}$$

If we draw the traces of only those slip systems for which the acting forces are largest, then a parallel pattern at an angular interval $\Delta\phi = 30^{\circ}$, as shown in Figure 5 is obtained and it matches exactly with the experimental etch pit pattern.

Deformation Modes in Low Melting Point Metals

We have investigated crater and plasma formation and shock deformation in several high purity metals including low melting point Sn, Zn and Bi polycrystals. A Q-switched ruby laser with a pulse duration $\sim$ 20 n. sec. and energy densities up to $\sim 1.5 \times 10^9$ w./cm^2 was used for these experiments and samples were irradiated in vacuum. Figure 6, shows a scanning electron microstructure of the laser damaged crater as well as the adjacent surface in Sn at a fluence $\sim 1.5 \times 10^9$ w/cm^2.

Fig. 6 - Wavy deformation pattern in Sn. Scanning electron micrograph. Fluence $\sim 1.5 \times 10^9$ watts/cm^2.

A wavy surface deformation pattern is observed in this figure. Several points to be noted with reference to this observation: (a) the amplitude of the wavy deformation is maximum near the crater edge, (b) the amplitude of the wavy deformation decays with increasing radial distance r from the center of the crater, (c) the spacing of the ripples (wave length) decreases as r increases and (d) distortion and pinning of the wave front at grain boundries. The nature of this wavy deformation is schematically presented in Figure 7.

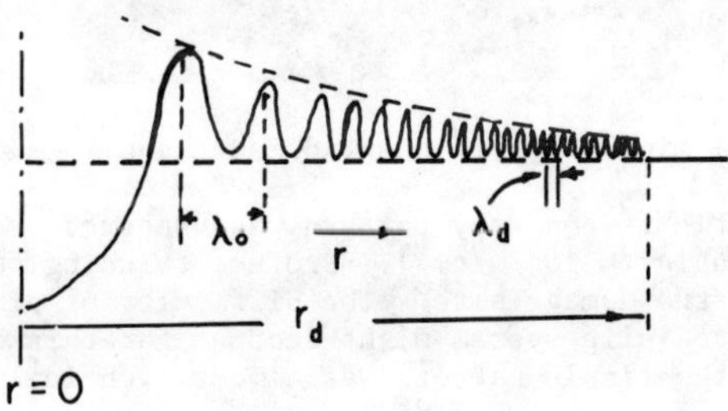

Fig. 7 - Schematic presentation of a crossection through the crater in Sn.

The measured values of the wave length near the crater λ_o, the value of λd and decay distance r_d (as defined in Figure 7) are as follows:

$$\lambda_o \approx 20 \times 10^{-4} \text{ cm.}$$

$$\lambda_d \approx 1.5 \times 10^{-4} \text{ cm.}$$

$$r_d \approx 400 \times 10^{-4} \text{ cm.}$$

The sound wave velocity in Sn at room temperature is 2.5×10^5 cm/sec. Assuming that the elastic wave propagates at this velocity, when the shock wave is produced, we can obtain a time t for the propagation of this wave to a distance λ_o. This time t is calculated to be $\sim 8 \times 10^{-9}$ sec. This time compares favorably with the laser pulse rise time of the order of $\sim 10 \times 10^{-9}$ sec. Similarly, the time t_d corresponding to the decay of the stress pulse amplitude is calculated to be $\sim 160 \times 10^{-9}$ sec.

It is realized that the material near the edge of the crater is heated to a temperature $\lesssim T_m$, where T_m is the melting point. Although, the temperature distribution $T(r,t)$ is not known, it can be assumed that an exponential decay type radial temperature distribution exists during the shock wave propagation through the material. Since the sound wave velocity and elastic moduli are inverse functions of the temperature, the wave propagates through a material where the modulus varies continuously. This could then account for the decay of λ. Two effects could account for the amplitude decay on the other hand. One is the dissipation of elastic energy in plastic work as the wave front moves out and the other is that because of the higher temperature near the edge of the creater the yield stress of the metal is low and thus a given amount of distortional elastic energy can produce a greater plastic yielding. It must be pointed out that non-crystallographic nature of the plastic yielding implies a non-conservative nature of dislocation motion during a very high strain rate associated with the shock pulse. Figure 8, shows an optical micrograph of laser damaged Sn at a laser energy density of $\sim 10^7$ watts/cm^2. At this relatively lower energy density more twinning and less ripple patter is observed.

Fig. 8 - Optical micrograph of Laser damage in Sn. Fluence $\sim 10^7$ watts/cm^2

In Zn, very little if any wavy pattern was observed even at the highest energy density available ($\sim 10^9$ w/cm^2). Profuse twinning, however, was observed adjacent to the damage site. The difficulty of plastic deformation due to the limited slip system might account for this difference. In polycrystalline Bi, the ripple pattern was almost identical to that for Sn.

Figure 9 shows a scanning electron micrograph for a laser damaged ($\sim 10^9$ w/cm^2
Bi. This micrograph shows the fine spacing of the wave pattern at $r \approx 200$ μ.

Fig. 9 - Scanning electron micrograph of wavy pattern in Bi.
Fluence $\sim 10^9$ watts/cm^2.

Microstructural Effects Associated Crater Formation

At laser energy densities $\sim 10^9$ w/cm^2, secondary molten craters at grain
boundrites and tripple points, at distances up to ~ 150 μm from the edge of the
laser spot was observed in bulk Cu, electro-chemically polished and etched
before laser irradiation. Figure 10 shows such an observation. The nature
of such localized temperature spikes is not clearly understood at this time.

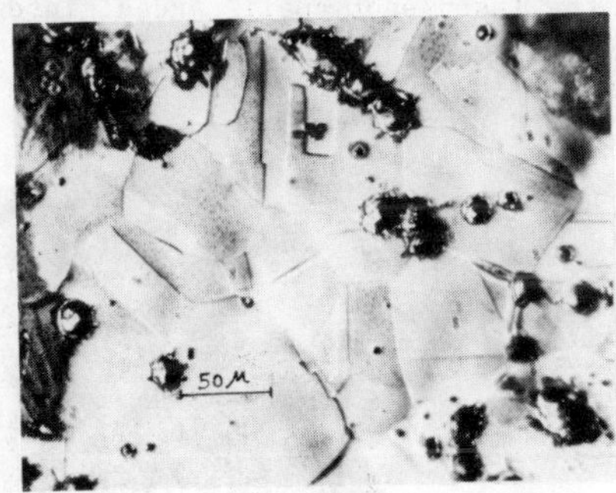

Fig. 10 - Secondary molten craters along grain boundries in a
polycrystalline Cu. Fluence $\sim 10^9$ watts/cm^2.

In a fully martensitic, medium carbon (1040) steel, reversion to austenite was achieved within a concentric circular area around the laser spot.
Within this reverted structure, profuse grain boundry melting and void formation was observed as shown in Figure 11.

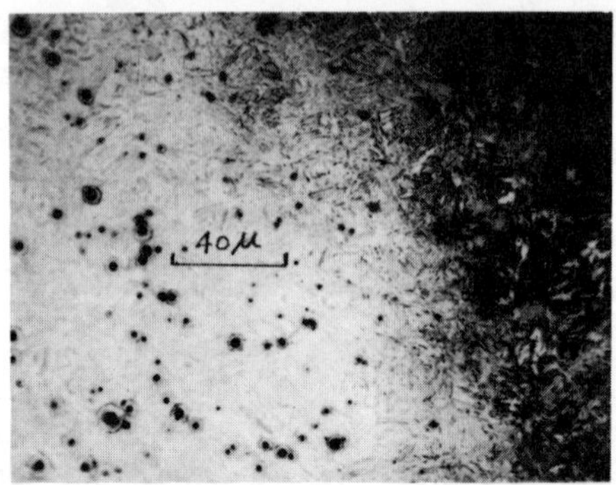

Fig. 11 - Void formation in reverted austenite in a 1040 steel. Fluence $\sim 10^9$ watts/cm^2.

Nature of Pulse Damage in Vapor Deposited Copper

Damage studies were carried out with ~ 2000 Å vapor deposited cupper foils which were backed up by an electro deposited Cu layer ~ 2mm thick. At identical laser energy fluences, the nature of the damage spot in the vapor deposited samples was quite different from that for the bulk Cu. The crater depth was almost uniform across the laser spot and no detectible slip, ripple pattern or liquid ejection were observed. It appears that the vapor deposited layer within the laser spot spalls and disintegrates rather than melts. Liquid metal ejection and bulk metal like behavior was, however, obtained after vacuum annealing which promoted grain growth in the vapor deposited layer.

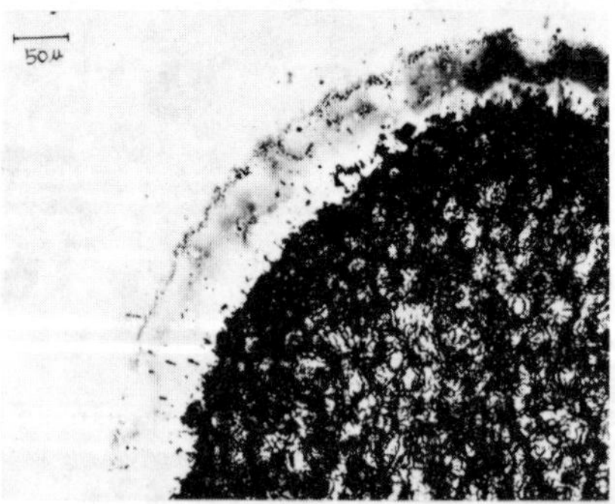

Fig. 12 - Micro-void formation around the laser damaged spot in a vapor deposited Cu-foil.

One very interesting effect observed in the as received vapor deposited Cu foils is, that a halo of bluish-green color appeared within a few hours to a day in the laboratory atmosphere, around the laser damage craters produced at energy densities $\sim 10^9$ w/cm^2. Areas away from the laser damage spot remained shiney. Microscopic observation revealed that a band of micro porosity deleniates this region as shown in Figure 12. It is concluded that there is a very high vacancy concentration and vacancy clusters in this region produced by the shock wave. High densities of ledge atoms and surface vacancies could provide activated sites for enhanced oxidation and corrosion in this region.

References

1. J. E. Michaels, *Planetary and Space Science*, Pergamon Press, NY, vol. 7 (1961).

2. R. M. White, J. Appl. Phys. **34**, 2123 (1963).

3. N. M. Kroll, J. Appl. Phys. **36**, 34 (1965).

4. S. S. Penner and O. P. Sharma, J. Appl. Phys., **37**, 2304 (1966).

5. C. E. Bell and J. A. Landt, Appl. Phys. Lett **10**, 46 (1967).

6. C. Mark Perceval, J. Appl. Phys., **38**, 5313 (1967).

7. E. F. Carome, E. M. Carrcira and C. J. Prochaska, Appl. Phys. Lett, **11**, 64 (1967).

8. L. S. Gournay, J. Acoust. Soc. Am., **40**, 1322 (1966).

9. J. C. Bushnell and D. J. McCloskey, J. App. Phys., **39**, 12, 5541 (1968).

10. F. Haessner and W. Seitz, J. Mat. Science, **6**, 16 (1971).

11. J. D. O'Keefe, C. H. Skeen and C. M. York, J. Appl. Phys., **44**, 4622 (1973).

12. L. C. Yang, J. Appl. Phys., **45**, 2601 (1974).

13. B. P. Fairand, A. H. Clauer, R. G. Jung and B. A. Wilcox, J. Appl. Phys., **25**, 431 (1974).

14. Jay A. Fox, Appl. Phys. Lett., **24**, 340 (1974).

15. B. Steverding and H. P. Dudel, J. Appl, Phys., **47**, 1940 (1976).

16. J. D. Porteous, M. J. Soileau and C. W. Fountain, Appl. Phys. Lett., **29**, 156 (1976).

17. J. D. Porteous, M. J. Soileau and C. W. Fountain, *Laser Induced Damage in Optical Materials*, NBS Special Publication, 462, p. 165 (1976).

18. J. D. Porteons, C. W. Fountain, J. L. Jernigan, W. N. Faith and H. E. Bennett, *Laser Induced Damage in Optical Materials*, NBS Special Publication, 509, p. 204 (1977).

19. O. T. Inal and L. E. Murr, J. Appl. Phys. **49**, 2427 (1978).

20. R. A. Olstad and D. R. Olander, J. Appl. Phys., **46**, 1499 (1975).

21. R. A. Olstad and D. R. Olander, J. Appl. Phys., **46**, 1509 (1975).

22. H. Schwarz, *Laser Interactions*, Plenum Press, NY, (1971).

23. J. M. Baldwin, J. Appl. Phys., **44**, 3362 (1973).

24. G. V. Plyatsko, N. I. Moisa and V. M. Khirovetskii, Sov. Mat. Science, (Eng. Trans.) **7**, No. 3, 300 (1971).

25. V. G. Dneprovskii and B. A. Osadin, Sov. Phys. Tech. Phys., $\underline{19}$, No. 2, 275, (1974).

26. A. M. Bonch et. al., Sov. Phys. Tech. Phys., $\underline{19}$, No. 11, 1474, (1975).

27. S. I. Anisimov et al., Sov. Phys. Tech. Phys., $\underline{11}$, No. 7, 945 (1967).

28. I. G. Karasev et al., Sov. Phys. Tech. Phys., $\underline{15}$, No. 9, 1523 (1971).

29. W. A. Elliott, F. P. Gagliano and G. Krauss, Appl. Phys. Lett., $\underline{21}$, 23, (1972).

30. W. A. Elliott, F. P. Gagliano and G. Krauss, Met. Trans. 4, 2031 (1973).

31. B. H. Kear, E. M. Breinan and L. E. Greenwood, Materials Technology, April 1979, p. 121.

32. L. S. Weinman, C. Kim, T. R. Tuckey and E. A. Metzbower, Appl. Optics, $\underline{17}$, 906 (1978).

33. P. G. Moore & L. S. Weinman, SPIE, Vol. 198, Laser Applications in Materials Processing, p. 120 (1979).

34. J. Narayan, J. Metals, June 1980, p. 15.

35. A. E. H. Love, The Mathematical Theory of Elasticity, Dover press 1944.

36. D. W. Gregg and S. J. Thomas, J. Appl. Phys., $\underline{37}$, 2787 (1966).

37. R. J. Murphy & G. J. Ritter, Nature, April 9, $\underline{210}$, 191, (1966).

38. G. Vitali, M. Bestolotti, G. Foti and E. Rimini, Phys. Lett., $\underline{63A}$, 351 (1977).

39. J. C. Koo and R. E. Slusher, Appl. Phys. Lett., $\underline{28}$, No. 10, 614, (1976).

40. H. J. Leamy, G. A. Razgonyi, T. T. Sheng and G. K. Celler, Appl. Phys. Lett., $\underline{32}$ (9), 535, (1978).

41. D. C. Emmony, R. P. Howson and L. J. Willis, Appl. Phys. Lett., $\underline{23}$, No. 11, 598, (1973).

42. L. E. Murr, O. T. Inal and A. A. Morales, Appl. Phys. Lett., $\underline{28}$, 432, (1976).

43. L. E. Murr, O. T. Inal and A. A. Morales, Acta. Met., $\underline{24}$, 261 (1976).

44. K. Mukherjee, T. H. Kim and W. T. Walter, to be published.

LASER PROCESSING OF SEMICONDUCTORS*

J. Narayan

Solid State Division, Oak Ridge National Laboratory
Oak Ridge, Tennessee 37830

High-power laser pulses provide an ideal source of heat by which thin layers of semiconductors can be rapidly melted and subsequently regrown with solidification rates of the order of several meters per second. Crystallized layers, under these rapid solidification conditions, were found to be remarkably defect-free. The details of rapid solidification are presented for binary silicon alloys, the alloying agent being one of the common dopants. Some of the applications of laser processing in device technology are illustrated.

*Research sponsored by the Division of Materials Sciences, U. S. Department of Energy under contract W-7405-eng-26 with Union Carbide Corporation.

<h2 style="text-align:center">Introduction</h2>

Laser light with wavelength in the visible part of the solar spectrum is heavily absorbed in the surface regions of semiconductors such as Si, Ge, and GaAs. As a result of this, Q-switched ruby or Nd-YAG lasers, with pulse duration (τ) 15-100 x 10^{-9} s, provide an ideal source of heat by which thin layers ($<$ 1μm) can be rapidly melted and resolidified with heating and cooling rates exceeding 10^{8}°C s^{-1} (1). Crystal growth velocities, under these conditions, have been estimated to be of the order of several meters per second compared to rates of about 10^{-5} ms^{-1} encountered under normal crystal growth (2). In this paper, we discuss the details of microstructural modifications and the dopant solubility limits under laser-melt quenching; and the correlations of the observed microstructures with the electrical properties of these materials are presented. Pulsed lasers can be used to atomically clean and to reorder surface layers with the crystal structure of bulk materials (3). A new technique of forming p-n junctions based upon dopant deposition and laser irradiation (laser-induced diffusion or surface alloying) (4), and improvements of thermally diffused junctions by removing the process-induced defects in the emitter regions (1), are discussed. Finally, results on the segregation of dopants, such as Sb, In, Ga, Fe, and Bi with distribution coefficients much less than unity ($k_o \ll$ 1), and on the interface instability leading to the formation of constitutional supercooling cells (5-7) are presented, which provide detailed information on the nature of rapid solidification and the physics of nonequilibrium (metastable) alloying.

<h2 style="text-align:center">Laser-Solid Interactions and Melting</h2>

The photon energy contained in laser beams can be very rapidly ($<$ 10^{-13} s) transferred to the electronic system of the semiconducting material. The energy from the electronic system can, in less than 10^{-9} s, be transferred to the lattice, and this energy is then utilized to heat and melt thin ($<$ 1 μm) surface layers of materials (1,2). Assuming unidirectional heat diffusion by thermal conduction, the problem of heating and melting has been solved taking into account reflectivity changes associated with phase transitions from solid to liquid, changes in specific heat and thermal conductivities with temperature, spatial variation (as a function of depth) of the absorption coefficient, and the latent heat of fusion, depending upon the damage state of the near-surface layer. An example from such calculations (2), showing the penetration of the melt-front as a function of time for laser pulses of different energy densities, is given in Fig. 1. The results for 25 ns pulses show that the surface starts melting after 15 to 25 ns of the laser pulse, and then the melt-front penetrates to several thousand Angstroms before receding back to the surface. The solidification velocity, which is given by the negative slopes of the curves, varies slightly with depth, its magnitude being lower near the surface. The solidification velocity also varies with pulse energy density. In the top ~1000 Å thick region, the average velocity was found to increase from 2.5 ms^{-1} (for a 1.75 Jcm^{-2} pulse) to 3.7 ms^{-1} (for a 1.0 Jcm^{-2} pulse). These calculations also show that the depth of melting can be controlled simply by varying the pulse energy density while keeping the other laser parameters fixed. One of the unique features of pulsed laser melting is that the total amount of heat input is small, such that the temperature of the substrate is not raised beyond a few tens of degrees above the ambient temperature. This localized nature of laser heating lends itself to many semiconductor applications.

152

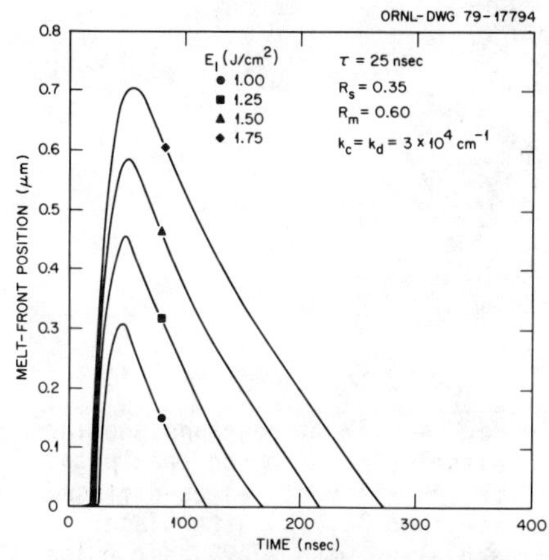

Fig. 1   Melt-front position as a function of time for ruby-laser pulses (λ = 0.694 μm).

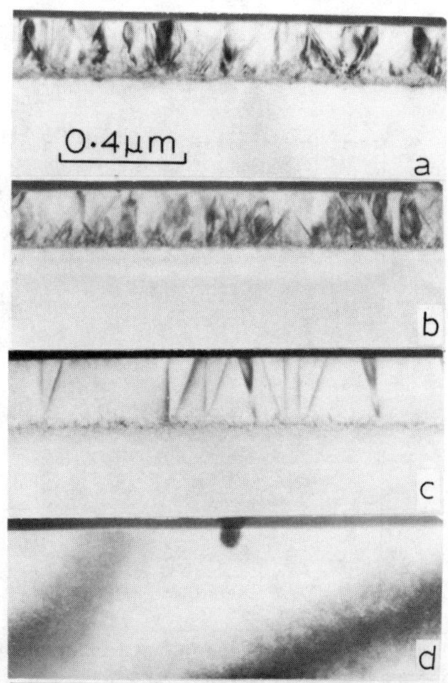

Fig. 2   TEM images of x-sections of (100) silicon implanted with 1.0 x 10^{15} 200 keV - As^+/cm^2: (a) E $\approx$ 0.5 Jcm^{-2}, (b) E $\approx$ 0.9 Jcm^{-2}, (c) E = 1.2 Jcm^{-2}, (d) 1.3 Jcm^{-2}.

Laser Crystallization

Figure 2 shows the results of laser crystallization of amorphous layers produced by ion implantation in silicon single crystal specimens. The amorphous layers were produced by 200 keV As^+ ions to a dose of 1.0 x 10^{15} cm^{-2}. This dose of As^+ ions produced ~1800 Å thick amorphous layer followed by a band (~700 Å thick) of dislocation loops. The cross-section micrographs in Fig. 2 show changes in microstructures (amorphous and polycrystalline regions, dislocations and loops) as a function of pulse energy density. The melt-front in Fig. 2(a) did not penetrate the amorphous layer, and the resulting microstructure is polycrystalline with an amorphous layer between the polycrystalline regions and the single crystal substrate. In Fig. 2(b) the melt-front penetrated the amorphous layer but the underlying single crystal substrate did not melt and could not act as a seed for crystal growth. In Fig. 2(c) the melt-front reached the dislocation band, and the loops intersected by the melt-front grew back as two segments of V-shaped dislocations. The dislocations of a given Burgers vector were found to grow in a specific direction that was characteristic of crystal growth from the melt. Thus, defect-free layers having the same orientation as the substrate are formed when the melt-front penetrates entirely through the damage layer [as shown in Fig. 2(d)]. The defect-free growth of layers at solidification velocities exceeding one meter per second is an unexpected result and it may be related to the unidirectional heat-flow conditions present during pulsed laser melting. Another interesting result was a much higher energy density that was required to melt the underlying crystalline layer [corresponding to the step in pulse energy density between Figs. 2(b) and 2(c)]; this is believed to be due to the differences in melting points of

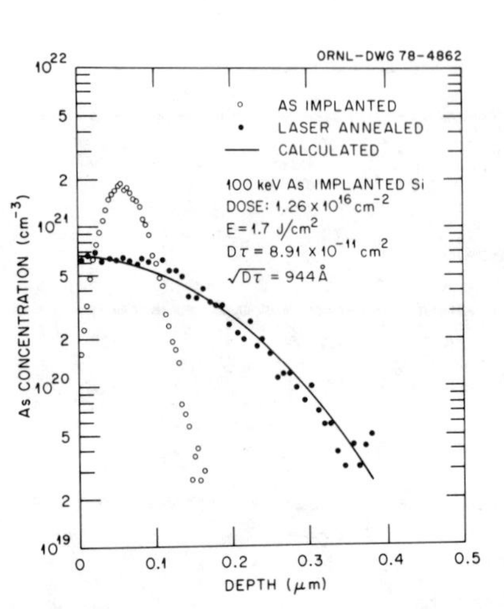

Fig. 3 Arsenic concentration profiles before and after laser annealing with a ruby laser pulse (τ=50 ns, E=1.7Jcm^{-2}).

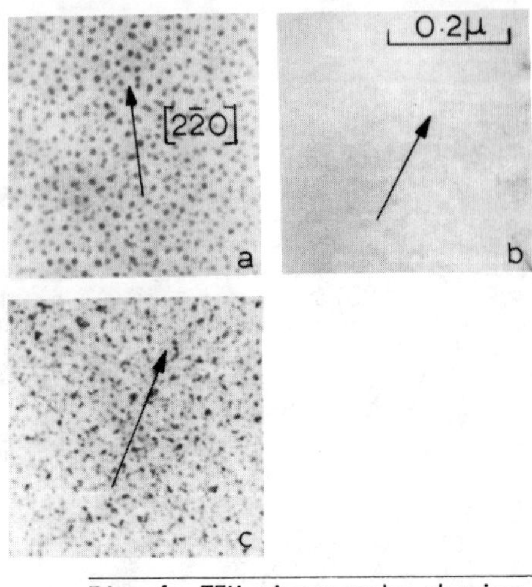

Fig. 4 TEM micrographs showing dissolution of boron precipitates in silicon: (a) as-diffused specimen D$_2$; (b) after laser annealing with ruby laser pulse E=1.7 Jcm^{-2}, τ=50 ns; (c) after heating the laser-annealed specimen at 950°C for 30 min.

amorphous and crystalline silicon. The melting point of amorphous Ge has been shown to be 20% lower than that of single crystal Ge and the situation in silicon is expected to be similar to that of Ge.

Figure 3 shows changes in dopant distributions after laser annealing of a 100 keV As$^+$ implanted, silicon specimen. The as-implanted profile is Gaussian as expected. After laser annealing, considerable redistribution occurs both toward the surface and deep into the crystal (2). Assuming unidirectional diffusion and taking into account melt-life time as a function of depth, the calculated profile is fitted to the observed profile and from the best fit, the value of D is extracted. A value of 3.0 x 10^{-4} cm^{2}s^{-1} for the D obtained from these calculations is in good agreement with the literature value of the arsenic diffusion coefficient in liquid silicon (2).

Dissolution of Boron Precipitates

The ultrarapid melting induced by pulsed lasers and subsequent recrystallization of melted layers may find wide application in semiconductor processing that does not involve ion implantation. Figure 4 shows the effect of laser irradiation on a silicon sample doped with boron at 950°C (D$_2$ sample in Table I) by conventional diffusion techniques (1,8). The conventional thermal diffusion results in the formation of boron precipitates in surface layers ~300 Å thick [as shown in Fig. 4(a)]. The boron contained in these precipitates is electrically inactive, besides, the precipitates act as recombination centers and traps for electrically active carriers. After irradiation of this sample with a single laser pulse, the precipitates dissolve resulting in defect-free doped layers [Fig. 4(b)].

Table I. Electrical Parameters of Boron Diffused Silicon

Sample	Diffusion Temperature and Time	Before Laser Treatment			After Laser Treatment		
		N_S (cm^{-2})	ρ_S $(\Omega/\square)$	μ $(cm^2/V\text{-}s)$	N_S (cm^{-2})	ρ_S $(\Omega/\square)$	μ $(cm^2/V\text{-}s)$
D1	900°C/50 min	2.1×10^{15}	77	39	1.4×10^{16}	13.0	31
D2	950°C/30 min	3.8×10^{15}	44	38	2.5×10^{16}	8.6	29
D3	1000°C/10 min	5.8×10^{15}	28	38	3.4×10^{16}	6.3	30

Electrical measurements (given in Table I) indicate about a six fold
increase in carrier concentration after the laser treatment. This increase
in carrier concentration was found to be consistent if all the boron contained in the precipitates dissolved and became electrically active.

Since this concentration of boron is 4-5 times the solubility limit at
950°C, it was possible to precipitate out excess boron on subsequent
heating as shown in Fig. 4(c). Secondary ion mass spectrometry (SIMS) profiles before and after the laser treatment of the diffused layers are shown
in Fig. 5, indicating the redistribution of boron contained in the precipitates near the surface. From the solubility limit indicated in the figure,
it was estimated that after the laser treatment, the concentration in the
top several thousand angstroms region is above the equilibrium concentration
at 950°C. The laser treatment of diffused crystals was found to improve the
p-n junction characteristics, as determined from electrical measurements on
simple diodes.

Doping By Laser Induced Diffusion (Surface Alloying)

Recently, laser radiation has been used to form large-area p-n junctions in silicon by the process of laser-induced diffusion or surface
alloying (4). In this process, the dopants are deposited by evaporation
onto the sample surface as a thin film approximately 100 A thick, and the
sample is then irradiated with pulsed laser light. The dopant diffusion
occurs in the liquid state and after recrystallization the surface region
is alloyed with dopants which are electrically active. A spatial profile
measured after laser irradiation of boron-deposited silicon is given in
Fig. 6; it shows redistribution of boron to a depth of several thousand
angstroms in the crystal. One of the desirable features of the dopant profiles produced by laser-induced diffusion is the continuous decrease of
dopant concentration without any plateau. Due to this feature, photo-
generated carriers in solar cells diffuse to the junction instead of to the
surface for recombination.

This technique has been applied to the fabrication of solar cells, and
efficiences comparable to those of ion implanted, laser annealed solar
cells (to be described in the next section) have been obtained. This type
of processing may become especially important for the formation of large-
area p-n junction devices, such as solar cells, because it bypasses the ion
implantation step and it does not require heating of the entire sample to
elevated temperatures. Thus, it is an attractive method for fabricating a
variety of devices which require shallow p-n junctions.

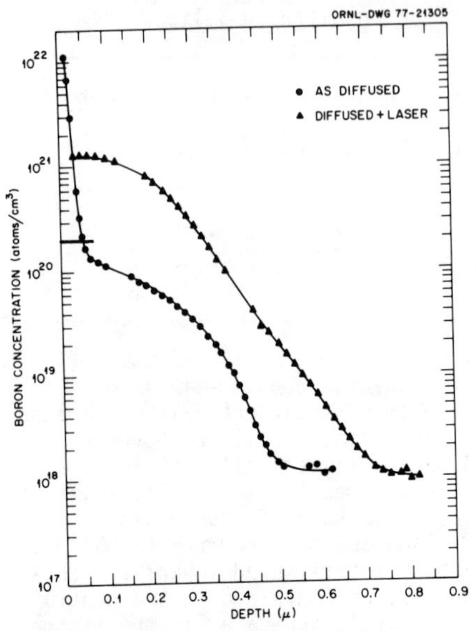

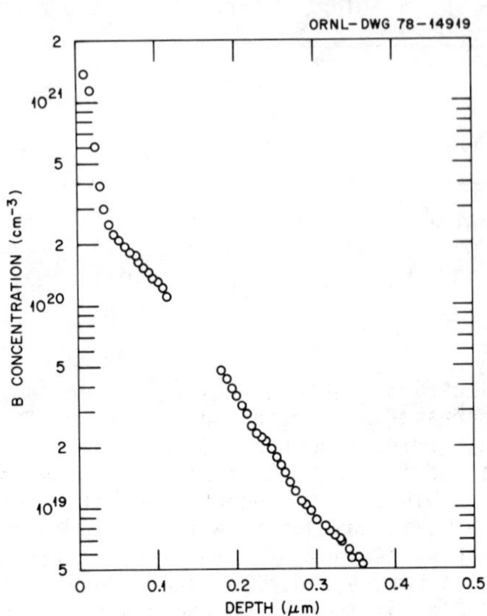

Fig. 5 Boron concentration profiles before and after laser annealing the as-diffused specimens.

Fig. 6 Boron concentration profile after laser annealing a boron-deposited specimen (τ=50 ns, E=1.5 Jcm^{-2}).

Solar Cells From Single and Polycrystalline Silicon

In this section a brief description is given of solar cells that were fabricated using ion implantation and laser annealing techniques (9-11) from single and polycrystalline silicon. Single-crystal, solar-cell parameters in Table II indicate high efficiencies, which are as high as those obtained using any other technique that is presently available. The solar cell efficiencies of the order of 12% from polycrystalline silicon are particularly encouraging because most of the conventional techniques based upon solid state diffusion yield efficiencies < 10%. The reason for high efficiency in polycrystalline solar cells may be found in Fig. 7, which shows a comparison between quantum efficiencies of single and polycrystalline cells. The

Table II. Solar Cell Parameters of Ion-implanted (As$^+$ 10 keV, 2 x 10^{15} cm^{-2}) and Pulsed Laser-annealed Single Crystal (C) and Polycrystalline (P) Silicon.

Cell No.	Laser Energy Density (J cm^{-2})	Pulse Duration (ns)	MCDL (µm)	Open Circuit Voltage (mV)	Short Circuit Current (mA/cm^2)	Fill Factor	Solar Cell Efficiency, AM$_1$ (%)
C$_1$	1.2	15-20	260	580	35.2	0.78	15.9
C$_2$	1.2	15-20	250	580	35.0	0.77	15.6
P$_4$	1.2	15-20	100	566	28.7	0.76	12.3
P$_5$	1.2	15-20	105	565	28.0	0.75	11.9

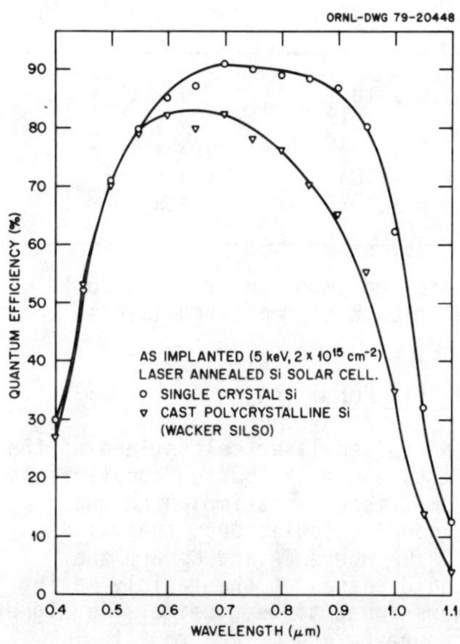

Fig. 7 Comparison of quantum efficiency spectra of ion-implanted, laser-annealed single and poly-crystalline Si solar cells.

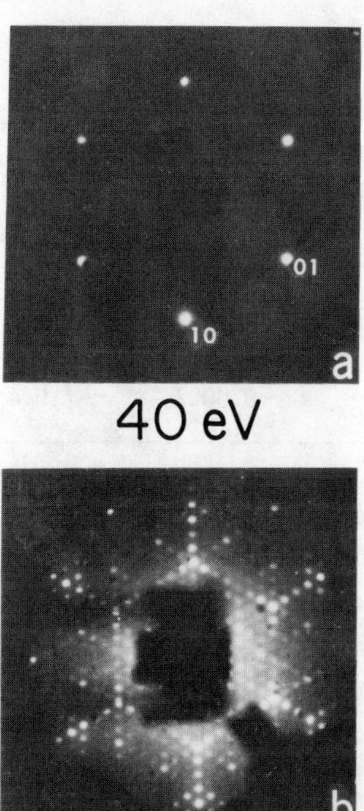

Fig. 8 Low-energy electron dif-fraction (LEED) patterns: (a) (111) pattern after laser annealing (E ≈ 2.0 Jcm^{-2}, τ =15 ns); (b) the pattern after sputter-cleaning and annealing (3).

almost identical quantum efficiencies in the short-wavelength region are indicative of similar electrical characteristics in the emitter regions of single and polycrystalline solar cells. Most of the light with longer wave-length is absorbed in the substrate, where the polycrystalline quantum effi-ciency is lower than that of single crystals due to the presence of electrically active defects.

<u>Surface Ordering and Cleaning</u>

Laser pulses can be used to atomically clean and restructure near sur-face layers (3). The techniques of Auger electron spectroscopy (AES) have been used to monitor the levels of impurities present in the surface regions of specimens, which were irradiated with laser pulses under a high vacuum environment (~2 x 10^{-10} Torr). These specimens, after pulsed laser melting, had O/Si and C/Si ratios of less than ~10^{-4}, indicating surface con-centrations of O and C impurities $\leq$ 0.1% of a monolayer. A LEED (low energy electron diffraction) pattern obtained from a Si(111) crystal subsequent to laser irradiation is shown in Fig. 8(a). This (1x1) pattern is indicative of a filled outermost (111) surface layer, i.e., surface has the same struc-ture as the bulk. This pattern is to be contrasted with that shown in Fig. 8(b) obtained from a (111) sample following a conventional sputter-annealing treatment. The observed (7x7) pattern indicates a reconstructed surface layer. The pattern in Fig. 8(a) transforms to that in Fig. 8(b) after heating the laser annealed specimen at ~700 K. Atomically clean and

Table III. Distribution Coefficients and Solubility Limits

Dopant	k_0	k'	Retrograde Maxima (cm^{-3})	Pulsed Laser Annealing (cm^{-3})
As	0.3	1	1.5×10^{21}	6×10^{21}
Sb	0.023	0.7	7×10^{19}	1.3×10^{21}
Ga	0.008	0.2	4.5×10^{19}	4.5×10^{20}
In	0.0004	0.15	8×10^{17}	1.5×10^{20}
Bi	0.0007	0.4	8×10^{17}	4×10^{20}

metastable, unreconstructed surfaces may play an important role in solid state devices and in solid phase epitaxial growth of deposited layers.

Metastable Alloying and Cell Formation

Since crystal growth velocities during pulsed-laser melting are of the order of meters per second, solubility limits and distribution coefficients are of considerable interest. From the comparison of as-implanted and laser-annealed dopant profiles, and using model calculations, the values of modified distribution coefficients ($k' = C_S'/C_L'$ where C_S' and C_L' are the dopant concentrations in the solid and liquid phases at the rapidly moving interface) have been determined. These were found to be considerable higher than the equilibrium values ($k_0 = C_S/C_L$, where C_S and C_L are equilibrium dopant concentrations in the solid and liquid phases) as determined from an equilibrium phase diagram (5).

For dopants such as B, P, and As in silicon, k_0 is ~0.5 (see Table III), and the k' values extracted from laser-annealed profiles approach unity. However, in the case of Sb, In, Bi, Ga, and Fe, k_0 is much less than unity, and even at such high velocities, the k' values are less than unity. As a result, a significant segregation near the surface is observed because the surface is the last region to solidify. If the implanted doses of these impurities are high enough, one can determine the maximum concentrations of these impurities which can be incorporated substitutionally at these solidification velocities (5). As shown in Fig. 9, C_S^{max} represents the maximum concentration of In in silicon at a solidification velocity of ~4 ms^{-1}. Above this concentration, the planar solid-liquid interface becomes unstable and breaks into cellular structures (7).

The cells are observed in the top region (as indicated in Fig. 9) and the amount of In incorporated into the cell walls corresponds to the difference between the total and substitutional concentrations. Figure 10(a) shows cells in a (100) silicon specimen. The average size of these cells is 500 Å with the cell wall width ~50 Å. A diffraction pattern of this specimen [Fig. 10(b)] contained, in addition to 220 and 400 silicon spots, weak 110 indium diffraction spots indicating that indium in the cell walls has grown epitaxially on the silicon substrate. A dark-field micrograph obtained using one of the In spots is shown in Fig. 10(c), which indicates that most of the indium is segregated at the cell walls. The interior of each cell was found to be a defect-free column of epitaxial silicon with an In concentration of ~1.5×10^{20} cm^{-3} (see Table III) trapped in substitutional lattice sites. The cells in (111) specimens, under similar conditions, exhibited a different morphology as shown in Fig. 10(d), and the average cell size was found to increase from 500 Å in (100) orientation to 700 Å in (111) orientation. This increase in the average cell size is interpreted as being due to a decrease in the solidification velocity in the (111) orientation compared to the (100) orientation. A TEM image of a

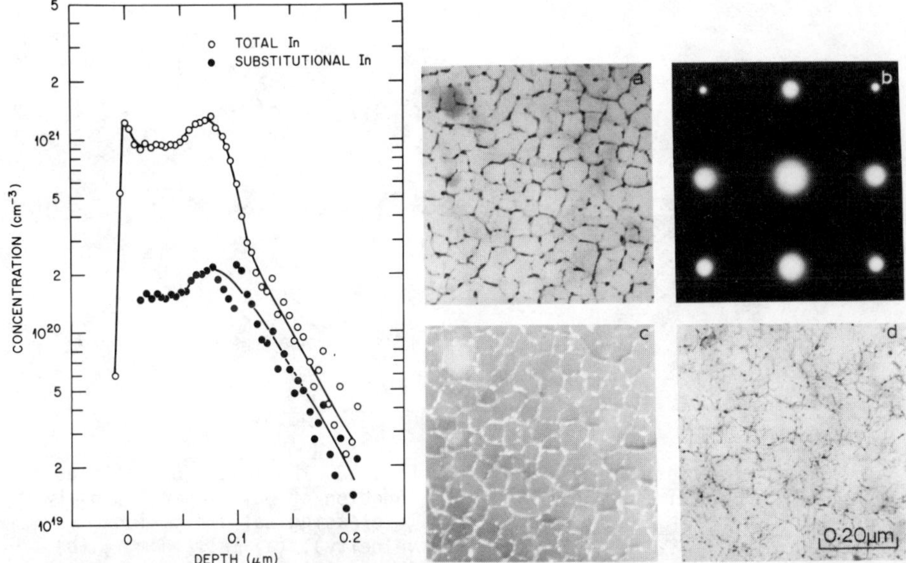

<table>
<tr><td>

Fig. 9 Comparison of total and substitutional concentrations for ^{115}In (125 keV, 1.25×10^{16} cm^{-2}) after laser annealing (after White et al., Ref. 2).

</td><td>

Fig. 10 Cell formation in laser-annealed silicon: (a) TEM images of cells in (100) silicon, (b) (100) diffraction pattern, (c) dark-field micrograph using one of the In spots, (d) cells in (111) silicon.

</td></tr>
</table>

x-section (Fig. 11) showed that the cells extend from the surface to a depth ~1000 Å, which is consistent with the In concentration profiles of Fig. 9. The average cell size was also found to increase from 350 Å to 750 Å when the pulse energy density was increased from 1.1 to 2.1 Jcm^{-2}, as shown in Fig. 12. The calculations shown in Fig. 1 indicated that as the pulse energy density increased, the velocity of solidification decreased. In the following, the results of a perturbation theory are presented, which predict the increase in cell size as the velocity of solidification decreases, consistent with the observations of Fig. 12. Results on cell formation similar to that observed in indium doped silicon have been obtained for Ga, Fe, and Bi doped silicon as well.

The size of cells, and the concentrations of impurities above which interface instability set in, can be calculated using the perturbation theory originally developed by Mullins and Sekerka (12). In their theory, we have incorporated the dependence of distribution coefficients on velocity, and good agreement has been obtained between the calculations and the experimental results.

The perturbation theory considers constitutional supercooling as the primary driving force of interfacial instability (similar to the classical theory), but it includes (unlike the classical theory) surface tension, latent heat and transport of heat from the interface. The latter terms produce a stabilizing influence on the interface; therefore, the classical theory considering only constitutional supercooling, predicts instability rather prematurely. According to the theory, the limiting concentration C_S^l is given by,

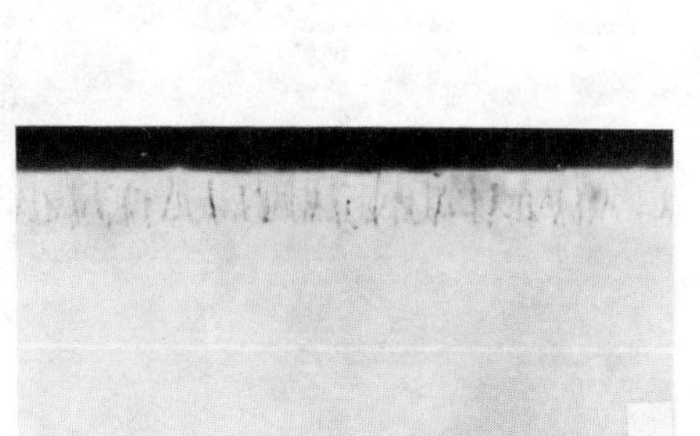

Fig. 11 TEM x-section image from 1.6 Jcm^{-2} laser annealed specimen.

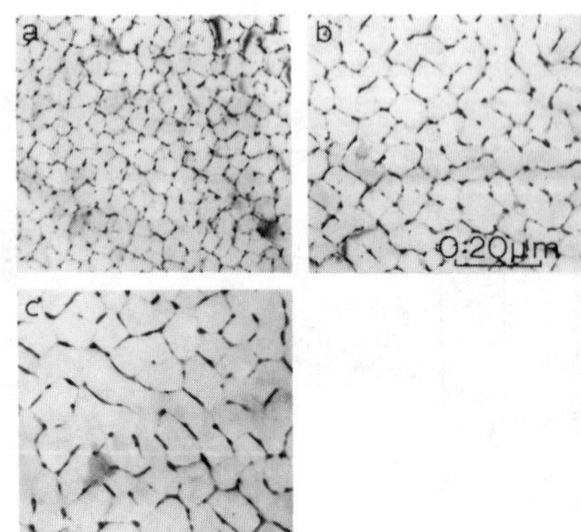

Fig. 12 TEM images of cells as a function of pulse energy density (decreasing solidification velocity): (a) E=1.1 Jcm^{-2}, (b) 1.6 Jcm^{-2}, (c) 2.1 Jcm^{-2}.

$$C_S' = \frac{D_L k'}{mV\bar{\alpha}(k'-1)} \left[\frac{k_L G_L}{k_S + k_L} \left(\alpha_L - \frac{V}{K_L}\right) + \frac{k_S G_S}{k_S + k_L} \left(\alpha_S + \frac{V}{K_S}\right) \right.$$

$$\left. + \bar{\alpha}\, T_m\, \Gamma\, \omega^2 \right] \left[1 + \frac{2 k'}{(1 + (2D_L\omega/V)^2)^{1/2} - 1} \right], \tag{1}$$

with

$$\alpha_L = (V/2K_L) + [(V/2K_L)^2 + \omega^2]^{1/2}$$

$$\alpha_S = -(V/2K_S) + [(V/2K_S)^2 + \omega^2]^{1/2}$$

$$\bar{\alpha} = (k_S\,\alpha_S + k_L\alpha_\ell) / (k_S + k_L)$$

where D_L is the impurity diffusion coefficient in liquid, V is the velocity of solidification, m is the liquidus slope in the phase diagram, k_L and k_S are the thermal conductivities, and K_L and K_S are the thermal diffusivities, of the liquid and solid, respectively, G_L and G_S are the thermal gradients in the liquid and solid respectively, T_m is the melting point

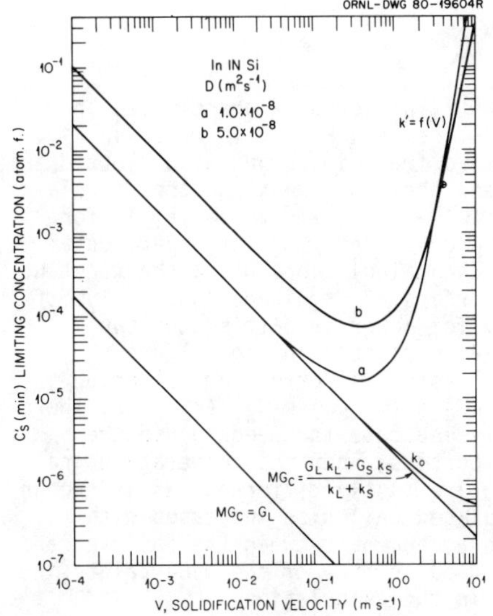

Fig. 13 Limiting concentration of instability versus velocity of solidifications: curve (a) D_L = 1.0 x 10⁻⁸ m²s⁻¹, curve (b) D_L = 5.0 x 10⁻⁸ m²s⁻¹ (k'=f(v) for both curves), k_0 curve for a constant value of k'=k_0. Experimental value is plotted as $\oplus$.	Fig. 14 Cell size as a function of solidification velocity: a, b, and k_0 curves correspond to those in Fig. 13, the curve with k'=0.15 (the value deduced at V=4.5ms⁻¹) is also included in the figure. $\oplus$ denotes the experimental value.

of the planar interface in the absence of solute, Γ is the ratio of the solid-liquid surface energy to the latent heat of fusion per unit volume, and ω is the spatial frequency associated with the instability. The following functional form (recently derived by Wood) for the dependence of k' upon solidification velocity was incorporated in equation (2):

$$k' = k_0 \exp\left(- RT \ln k_0\right)\left(1 - \exp - (V/V_0)\right)/ RT) \qquad (2)$$

where k_0 is the equilibrium distribution coefficient, and V_0 is a constant which can be determined experimentally. The value of V_0 was determined by plugging in the experimentally deduced value of k' = 0.15 at 4.0 ms⁻¹.

We have solved for $C_S^!$ (ω) as a function of ω for given values of V. The minimum value of $C_S^!$ (ω) (which is equal to C_S^{max}) gives the stability-instability demarcation. λ_{min} = $2\pi/\omega_{min}$ then represents the cell size where ω_{min} is the value corresponding to the minimum value of C_S. The following values of the materials constants were used in the calculations

$$k_L = 70 \text{ Jm}^{-1}\text{s}^{-1}\text{K}^{-1}, \quad k_S = 22 \text{ Jm}^{-1}\text{s}^{-1}\text{k}^{-1}$$

$$D_L = 1.0 \times 10^{-8} \text{m}^2\text{s}^{-1} \text{ (curve a)} \quad 5.0 \times 10^{-8} \text{m}^2\text{s}^{-1} \text{ (curve b)},$$

$$k_0 = 4.0 \times 10^{-4}, \quad V_0 = 2.83 \text{ ms}^{-1}$$

$$T_m = 1.3 \times 10^{-7} \text{ km}$$

$$K_L = 3.0 \times 10^{-5} \text{ m}^2\text{s}^{-1}, \quad K_S = 9.4 \times 10^{-6} \text{ m}^2\text{s}^{-1},$$

$$G_s = 6.57 \times 10^8 \text{ Km}^{-1}, \quad G_L = 1.33 \times 10^6 \text{ Km}^{-1}$$

$$m = -400 \text{ K (at. fract.)}^{-1} .$$

The results of the calculations of $C'_{S(min)}$ versus V are plotted in Fig. 13, and the corresponding plot of cell size versus V is given in Fig. 14. The values of $C'_{S(min)}$ or C_S^{max} obtained from Fig. 9 are also plotted in Fig. 13, indicating good agreement between the experiment and the calculations. The curve (a) is for $D_L = 1.0 \times 10^{-8} \text{ m}^2\text{s}^{-1}$, and curve (b) is for $D_L = 5.0 \times 10^{-8} \text{ m}^2\text{s}^{-1}$, these values of D_L cover the range of experimental D_L values available in the literature. The calculations using the constant value of k_0, however, give about a four orders of magnitude lower value of $C'_{S(min)}$ at $V \approx 4 \text{ ms}^{-1}$; but, as expected, for lower velocities the two calculations provide the same results. Also plotted in Fig. 11 are limiting concentrations obtained using classical supercooling criterion $mG_C = G_L$, and the modified CS criterion $mG_C = (k_L G_L + k_S G_S)/(k_L + k_S)$ where $G_C = VC'_S(k'-1)/D_L k'$. The results clearly indicate the need to use the modified CS criterion, even at lower velocities, when the temperature gradients in the liquid (G_L) and solid (G_S) are vastly different, as is encountered during laser annealing. The calculated cell size increases with decreasing solidification velocity. The agreement between the observed and the calculated cell sizes is considered good in view of the uncertainty in the materials and laser parameters used in the calculations.

Conclusions

In conclusion, pulsed laser melting can be used to crystallize amorphous and damaged layers in semiconductors, and the resulting laser annealed layers have been found to be remarkably defect-free. Melting of the surface region followed by liquid phase epitaxial growth can be used to remove defects such as precipitates and dislocation loops and to improve the electrical characteristics of p-n junctions. A new technique based upon laser-induced diffusion (surface alloying) has been developed for the formation of p-n junctions and solar cells. The electrical characteristics of p-n junction diodes and solar cells obtained by ion implantation and laser annealing are superior to that achieved by thermal annealing techniques, because the electrical properties of the substrates (such as minority carrier lifetime) are not degraded. The values of interfacial distribution coefficients at such high velocities of solidification are considerably higher than the equilbrium values. The maximum concentrations of dopants that can be incorporated into substitutional lattice sites, are found to exceed equilibrium solubility limits by factors that range from ~4 (for As) to ~500 (for Bi). Above these concentrations, a planar solid-liquid interface becomes unstable and breaks into cellular structures. The perturbation theory, which takes into account surface tension, latent heat of evolution, and temperature gradients in both the liquid and solid (whereas the CS criterion considers the temperature gradient only in the liquid), correctly predicts solute concentrations $C'_{S(min)}$ and wavelengths (cell sizes) associated with instabilities when the dependence of interfacial distribution coefficient upon the velocity of solidification is included in the theory. These results also demonstrate that laser melting can be used to study, in a controlled way, the difficult problem of metastable alloying and rapid solidification.

References

1. J. Narayan, R. T. Young, and C. W. White, "A Comparative Study of Laser
and Thermal Annealing of Boron-Implanted Silicon," J. Appl. Phys. 49 (7)
(1978) pp. 3912-3917; C. W. White, J. Narayan, and R. T. Young, "Laser
Annealing of Ion-Implanted Semiconductors," Science, 204 (1979) pp.
461-468; J. Narayan, "Laser and Electron Beam Processing of Silicon and
Gallium Arsenide," J. Electrochem. Soc. 80 (1) (1980) pp. 294-332; J.
Narayan, C. W. White, and R. T. Young, U.S. Patent, 4,181,538, Jan. 1, 1980.
2. J. C. Wang, R. F. Wood, and P. P. Pronko, "Theoretical Analysis of
Thermal and Mass Transport in Ion-Implanted, Laser-Annealed Silicon," Appl.
Phys. Lett. 33 (1978) pp. 455-458; R. F. Wood and G. E. Giles,
"Microscopic Theory of Pulsed Laser Annealing," Phys. Rev. B (in press).
3. D. M. Zehner, C. W. White and G. W. Ownby, "Preparation of Atomically
Clean Silicon Surfaces by Pulsed Laser Irradiation," Appl. Phys. Lett. 36,
(1980) pp. 56-58; "Silicon Surface Structures After Pulsed Laser
Annealing," Surf. Sci. 92, (1980) p. L67.
4. J. Narayan, R. T. Young, R. F. Wood, and W. H. Christie, "p-n Junction
Formation in Boron-Deposited Silicon by Laser-Induced Diffusion," Appl.
Phys. Lett. 33 (1978) p. 338; J. Narayan and R. T. Young, U.S. Patent
4,147,563, April 3, 1979.
5. C. W. White, S. R. Wilson, B. R. Appleton, and F. W. Young, Jr.,
"Supersaturated Substitutional Alloys Formed by Ion Implantations and
Pulsed Laser-Annealing of Group III-V Dopants in Silicon," J. Appl. Phys.
51 (1980) pp. 738-749.
6. C. W. White, J. Narayan, B. R. Appleton, and S. R. Wilson, "Impurity
Segregation by Pulsed Laser Irradiation," J. Appl. Phys. 50 (1979) pp.
2967-2969.
7. J. Narayan, "Interface Instability and Cell Formation in Ion Implanted
and Laser Annealed Silicon," J. Appl. Phys. (May 1981, in press).
8. R. T. Young and J. Narayan, "Laser Annealing of Diffusion-Induced
Imperfections in Silicon," Appl. Phys. Lett. 33 (1978) pp. 14-16.
9. R. T. Young, R. F. Wood, J. Narayan, and C. W. White, "Laser Processing
of Polycrystalline Solar Cells," pp. 651-659 in Laser and Electron Beam
Processing of Materials, ed. by C. W. White and P. S. Peercy, Academic
press, (1980, New York).
10. R. T. Young, R. F. Wood, J. Narayan, C. W. White and W. H. Christie,
"Pulsed Laser Techniques for Solar Cell Processing," IEEE Transactions
(Electron Devices) ED-27 (1980) pp. 807-815.
11. L. A. Ladd, K. V. Ravi, and J. Narayan, "Ion Implanted and Laser
Processed Solar Cells Made from EFG Ribbon," J. Electrochem. Soc. 80-5
(1980) pp. 122-130.
12. W. W. Mullins and R. F. Sekerka, "Stability of Planar Interface During
Solidification of a Dilute Binary Alloy," J. Appl. Phys. 35 (1964) pp.
444-451.

HEAT FLOW DURING CW LASER MATERIALS PROCESSING

T. Chande and J. Mazumder

Department of Mechanical and Industrial Engineering
1206 West Green Street
University of Illinois at Urbana-Champaign
Urbana, IL 61801

A three dimensional heat transfer model for laser materials processing
with a moving Gaussian heat source is developed using finite difference
numerical techniques.

This paper describes a moving heat source numerical model for three
dimensional heat conduction in the substrate during laser material
processing. It is capable of calculating the melt width and the depth of
penetration. The model allows for keyholing, laser beam mode structure and
surface heat losses from a semi-infinite slab of finite thickness and
width. It accounts for variable thermal properties and latent heat
effects. Hence, chemical reactions may be simulated as uniform heat sources
or sinks. In addition, a transformation monitor has been built in to track
the thermal response of the slab.

Introduction

The laser has added a new dimension to materials processing, being at once a very intense, well defined and precisely maneuvrable heat source. It has successfully made the transition from a laboratory to the shop floor and is now commercially used for welding, cutting, and surface treatment. A better understanding of how it works in these processes is necessary before its potential is fully tapped or its applications extended.

Laser material interactions being complex events, considerable time and experimental effort would be required to analyse them thoroughly. The principal variables are the substrate thermal and optical properties, the laser beam's total power, power distribution, diameter and traverse speed and the manner and rate of impingement of the cooling gas.

A mathematical model offers a powerful alternative to a purely experimental approach in analyzing this problem. A model may be used to check our physical concept of the process, estimate unmeasureable parameters and to study the effects of varying a parameter. Typically, the thermal cycle at each location in the fusion and heat affected zones, the temperature distribution and the cooling rates may be computed. These determine the extent of phase transformations and the possible occurrence of metastable phases.

A model for laser materials processing must account for the keyhole effect, whereby the intense laser beam vapourizes a fine hole into the substrate. This region has been modeled by Klemens [1] by a remarkably successful semi-quantitative approach that assumes only a one-dimensional flow along the quadrant position in the keyhole. A stationary heat source model, [2] moving line source model with non-uniform heat input, [3] and moving heat source with heat flow in one or two dimensions with variable properties [4] represent earlier attempts to tackle this problem. A numerical solution is needed if the non-linear terms of surface re-radiation and keyholing are to be included.

This paper describes a moving heat source numerical model for three dimensional heat conduction in the substrate during laser material processing. It is capable of calculating the melt width and the depth of penetration. The model allows for keyholing, laser beam mode structure and surface heat losses from a semi-infinite slab of finite thickness and width. It accounts for variable thermal properties and latent heat effects. Hence, chemical reactions may be simulated as uniform heat sources or sinks. In addition, a transformation monitor has been built in to track the thermal response of the slab.

1. Physical Definition of the Process to be Modeled

In order to define the mathematical model the process is physically defined as follows:

A laser beam having a defined power distribution strikes the surface of an opaque substrate having a finite width and thickness and moving in the positive x direction (along the length) with a uniform velocity. Some of the incident radiation is reflected, the rest absorbed. The reflectivity is considered to be zero if the temperature exceeds the boiling point, since it is assumed that a keyhole produced by vaporization acts as a blackbody. Some of the absorbed energy is lost by re-radiation and convection from the surface while the rest is conducted into the substrate. The convective heat

166

transfer is enhanced on the upper surface due to concentric gas jet used for shielding.

The three-dimensional system is assumed to be in a quasi-steady state condition after the keyhole initiation period, which for most practical purposes may be regarded as instantaneous, and will only occur at the beginning of a run.

This leads to the following assumptions:

1. The laser beam remains stationary relative to the earth and is incident at right angles at the center of the substrate width. The substrate moves in the positive x direction (along the length) with a uniform velocity U. The system is diagrammatically shown in Fig. 1.

2. The workpiece is infinite in length, but has finite width and thickness.

3. Quasi-steady state is assumed.

4. The power distribution in the beam is Gaussian (not a necessary assumption, but a precise power distribution is required and a gaussian distribution approximates that from a well-tuned laser with a stable cavity).

5. The thermal conductivity, specific heat, thermal diffusivity and the surface reflectivity are temperature dependent. The density is assumed to be independent of temperature.

6. The latent heats of the transformations and the resultant influence on the temperature profile are accounted for in the enthalpy calculations.

7. The surface relectvity is considered to be zero when the temperature exceeds the boiling point.

8. There is a radiation loss from the upper and lower surfaces of the workpieces.

9. There is a convective heat loss due to the shelding gas flow.

10. When any location exceeds the boiling point, it is considered to have evaporated.

11. Radiation penetrating the substrate is absorbed according to the Beer-Lambert's law

$$P(L) = P(s) \cdot e^{(-\beta L)}$$

where 'β' is the absorbing coefficient (m^{-1}) and is considered independent of position within the keyhole, and P(L) is the absorbed power density at depth L which was of value P(s) at the surface.

12. Matrix points considered to have "evaporated" remain in the conducting network at a fictitiously high temperature to simulate the high convection and radiation transfer effects from the fast moving plasma within the keyhole.

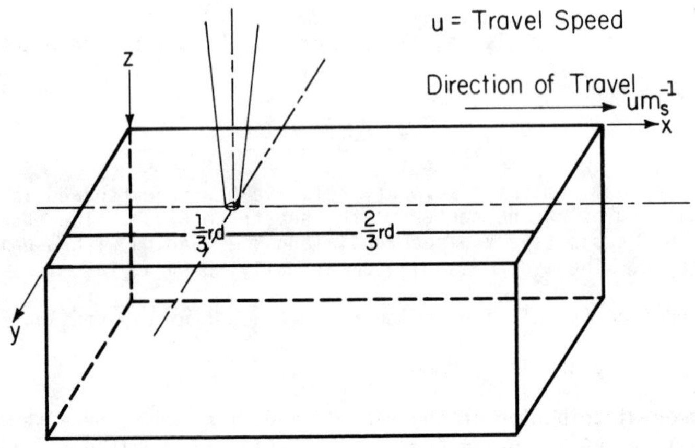

Fig. 1   Geometry of the total calculation volume with respect to the laser beam ('from Ref. 13)

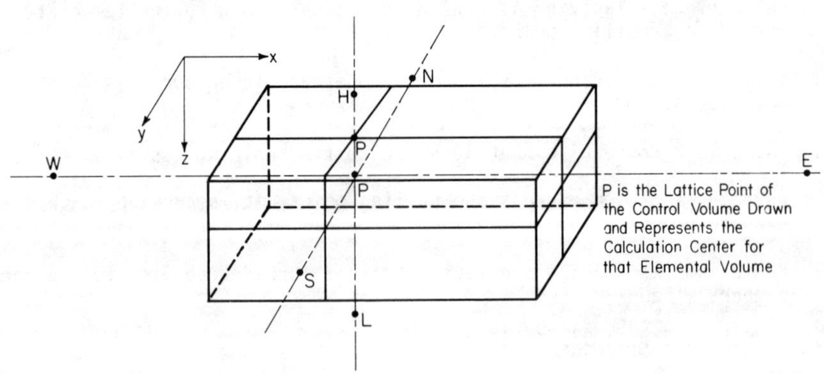

Fig. 2   Nomenclature used in and around an elemental volume (from Ref. 13)

2. Mathematical Definition of the Process to be Modeled

A large number of techniques have been reported [5-10] to solve partial differential equations. Since the present problem involves a steady state situation a 'relaxation' technique was chosen.

An exponentially expanding grid is used to provide optimal coverage of the slab and meet computer hardware constraints. Once the spatial limits beyond which the model predicts no temperature changes are established with such a grid, a linear grid is used to zero-in on this critical region. This maximizes the yield of useful information.

The heat balance on an asymmetric control volume within the body of the experimental grid, as shown in Fig. 2 can be stated in cartesian coordinates as

$$-K \, \delta y \, \delta z (\partial T/\partial x)_{PW} - K \, \delta x \, \delta z (\partial T/\partial y)_{PN} - K \, \delta x \, \delta y (\partial T/\partial z)_{PH}$$

$$+K \, \delta y \, \delta z (\partial T/\partial x)_{EP} + K \, \delta x \, \delta z (\partial T/\partial y)_{SP} + K \, \delta x \, \delta y (\partial T/\partial x)_{LP} +$$

$$\rho \, u \, Cp \, \delta y \, \delta z ((T_W + T_p)/2) - \rho \, u \, Cp \, \delta y \, \delta z ((T_p + T_E)/2) = 0 \qquad (1)$$

where K is the thermal conductivity W/m·k, and the subscripts P,E,W,S,N,H,L refer to the location shown in Fig. 2 for temperature T. This equation is then put in a finite difference form for the computer model.

The temperature gradient at the surface may be calculated by an application of Fourier's law of heat conduction, the net surface heat flux into the substrate being the incident flux corrected for the radiant and convective losses. The incident flux at a surface node is derived from the Gaussian power distribution for a TEM_{oo} mode. The convective heat transfer coefficient is calculated from Gordon and Cobonpue [11] for a vertically impinging gas jet. The radiant heat transfer coefficient is derived from its definition using Stefan-Boltzman equation.

To simulate keyholing, a point was deemed to be transparent if its temperature exceeded the boiling point. The laser power would then be incident on the node below, after suffering some absorption in the keyhole. This occurrence would then be recognized in computing the temperature gradient at the point. The process continues down through the substrate while the transparent grid points keep their high temperature, as they would if the keyhole were filled with hot plasma.

To be able to calculate the thermo-physical properties at any temperature, one must know the temperature dependence of these properties. This data is scarce and incomplete. In the temperture range that is available [12], such data may be numerically fitted to a curve of the type

$$P(T) = A + B \cdot T + \left(\frac{C}{T^2}\right) + \left(\frac{D}{T^3}\right) \qquad (2)$$

where P(T) is the desired property and A, B, C, and D are constants to be determined. If the data has considerable scatter then a spline fit may be tried. Alternately, temperatures may be averaged over incremental temperature ranges and the appropriate value chosen within the program. The above relation may be cautiously extrapolated to higher regions with a suitable, weighting factor if the data available is insufficient.

Incorporation of temperature dependence of properties can have a significant influence on the calculation of thermal gradients and hence has a direct bearing on the profile generated. The thermal diffusivity is prominent in the basic equation and therefore, a temperature dependent thermal diffusivity should be incorporated.

The temperature dependence of the specific heat obtained as above is used for enthalpy calculations. These are performed to account for the enthalpies of transformation in the heat balance and for their effect on the temperature profile. The enthalpy at any temperature may be computed by integrating term by term the temperature dependent specific heat, and adding the appropriate heats of transformation when necessary. To implement this, we need to set temperature traps about the transformation temperatures such that, when the temperature of any node falls within this range, the heat of transformation is added to the enthalpy calculations and the temperature of this node is reset to the known transformation temperature. The implicit assumptions are that the calculated temperatures of the node rise slowly and that the heat of transformation is supplied instantaneously by the laser beam. The latter is true for all practical purposes and the former was checked using a transformation monitor. Each time the temperature of a node is "trapped" and suitably reset, a tally mark is put against the identity of the node. The number of tally marks increases as the number of transformation ranges crossed increases. When the node is above the boiling point, only the tally marks are incremented.

The array of tally marks reveals at a glance the molten and heat affected zones and the number of iterations for which the maximum temperature was less than the boiling point. In effect it models the thermal response of the substrate and acts as a transformation monitor.

The listing has been recoded for optimality. The damping scheme is now execution dependent and this has enhanced its reliability. The damping coefficient is incremented after a predetermined number of iterations so that the fluctuations are not over damped. Further, this scheme gives accelerated convergence and is stable for values of the heat transfer coefficient of up to 32,000 $wm^{-2} k^{-1}$ and for laser beam radius up to 3 mm. There is no evidence to suggest that it would be unstable for values beyond those mentioned above, only, these represent the limits tried so far.

3. Operation of the Model

The temperature profile predicted by the model is plotted as generalized curve using the following dimensionless groups.

$$T^* = \frac{\Delta T\, k\, D_b}{P_{tot}(1 - r_f)}; \quad X^* = \frac{X}{D_b}; \quad Y^* = \frac{Y}{D_b}; \quad Z^* = \frac{Z}{D_b}; \quad U^* = \frac{U D_b}{\alpha} \qquad (3)$$

where D_b is the incident beam diameter. These plots would enable reasonable predictions for different operating conditions without recourse to a computer. The symmetry of Figs. 3 and 4 and the exponential type decay of Fig. 5 are as expected.

The laser beam being small in diameter and very intense, the temperatures rise rapidly in and around the beam as can be seen in Fig. 6.

Thermal property data [12] were fitted to an equation of the form $P(T) = A + BT + C/T^2 + D/T^3$ using a non-linear least squares fit through an IMSL

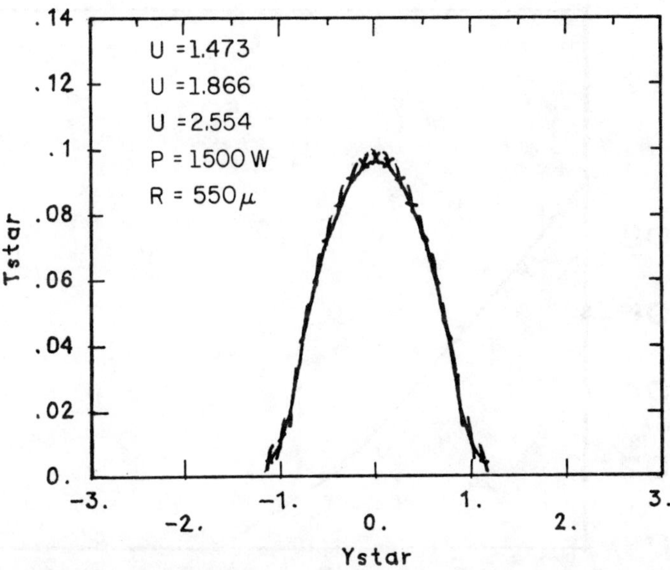

Fig. 3 Centerline dimensionless length (x*) versus temperature (T*)
through the point of interaction

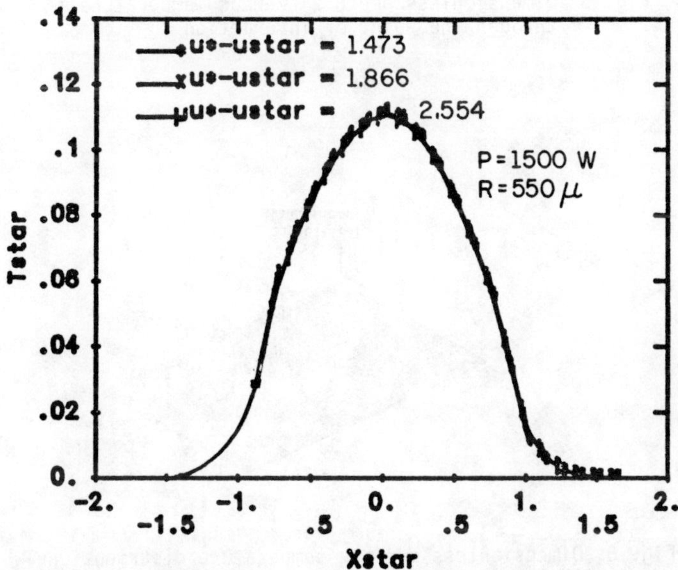

Fig. 4 Dimensionless width (Y*) versus temperature (T*) through
the point of interaction

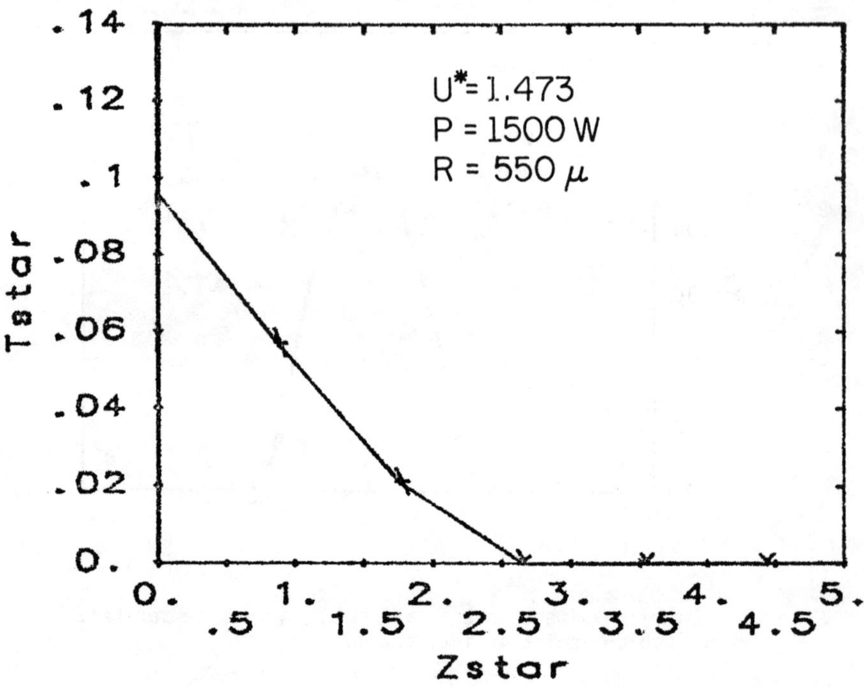

Fig. 5   Dimensionless depth z* versus temperature T*
through the point of interaction

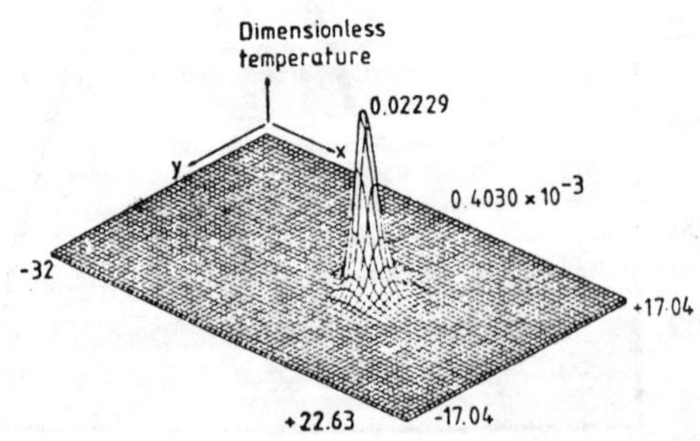

Fig. 6   Dimensionless surface temperature distribution T

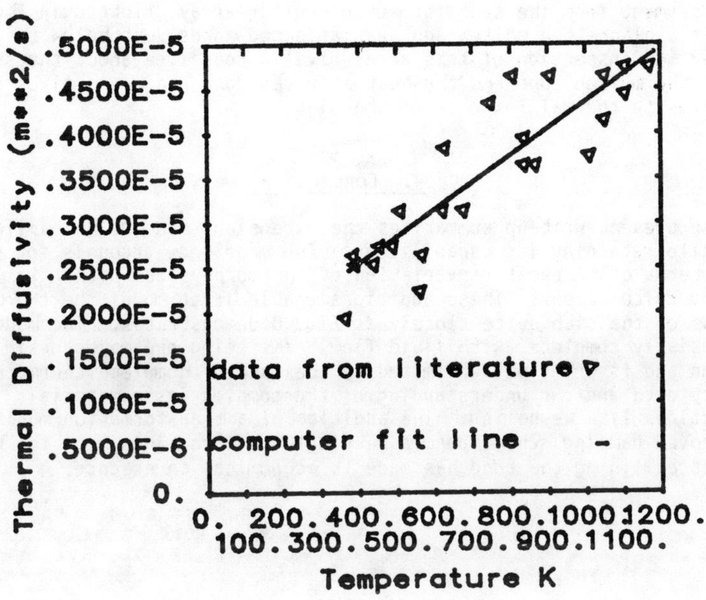

Fig. 7 Least squares computer fit of thermal property
data for Ti-6Al-4V

routine [15]. In particular, the thermal diffusivity data for Ti-6Al-4V
have more scatter than could be satisfactorily handled, and these were
fitted to a straight line, as shown in Fig. 7.

Significant information about the melt width, the melt depth and the
melt contour can be obtained from the temperature profile. The model is
sensitive to input parameters and with the appropriate values yield
excellent agreement with the experimentally plotted profile as can be seen
in Fig. 8.

A segment from the transformation monitor array, plotted in Fig. 9
shows at a glance the molten and heat affected zones just below the
surface. An inspection of this array gives a good idea about the shape or
size of the molten zone and the heat affected zone and hence gives some
insight as to thermal response of the slab.

4. Conclusions

The present writeup summarizes the extensions made to an earlier model
(13) while retaining its capabilities. The model now accounts for non-
linear terms of temperature variation of thermo-physical properties and
enthalpy calculations. These additions enable it to model the thermal
response of the slab quite closely as Fig. 8 demonstrates. The model is
substantially complete, with fluid flow terms being the major desirable
addition and is principally limited by the lack of complete thermophysical
property data and our understanding of the complex laser-material
interactions like keyholing. The addition of a transformation monitor and
an improved damping scheme permit meaningful interactions with the model
while streamlining the code has made it economical to execute.

(a)

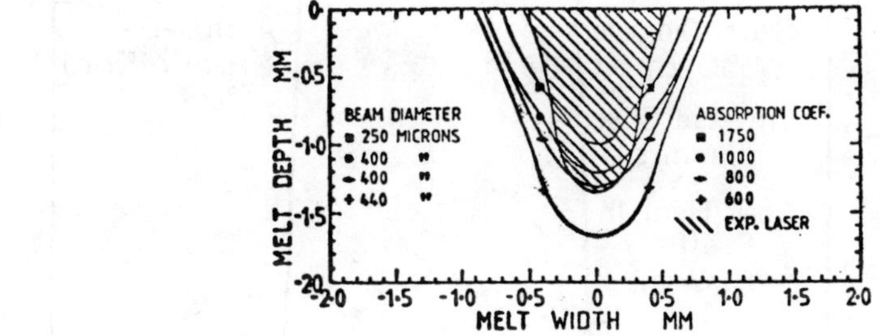

(b)

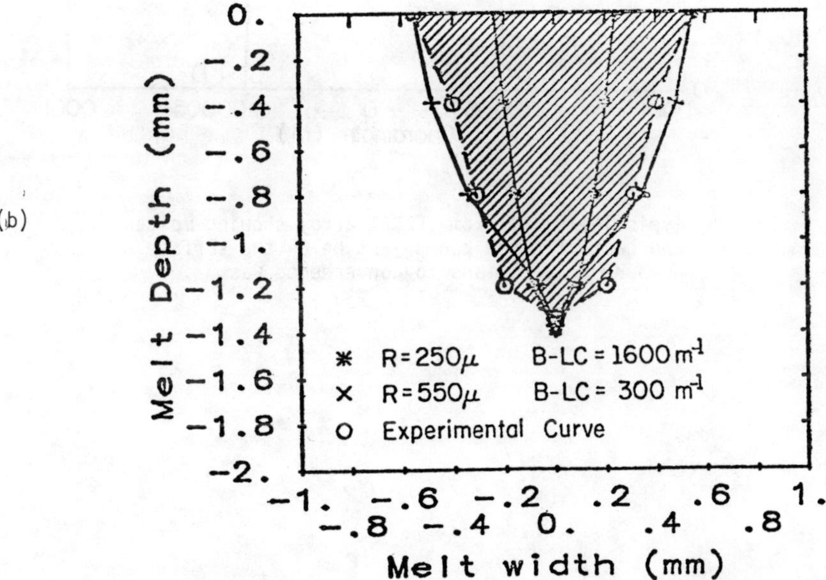

Fig. 8 Variation in laser melt widths with beam diameter
and absorption coefficient at a laser power of 1570W,
traverse speed of 35mm/s and reflectivity of 0.80,
(a) from ref. 13; (b) after current revisions.

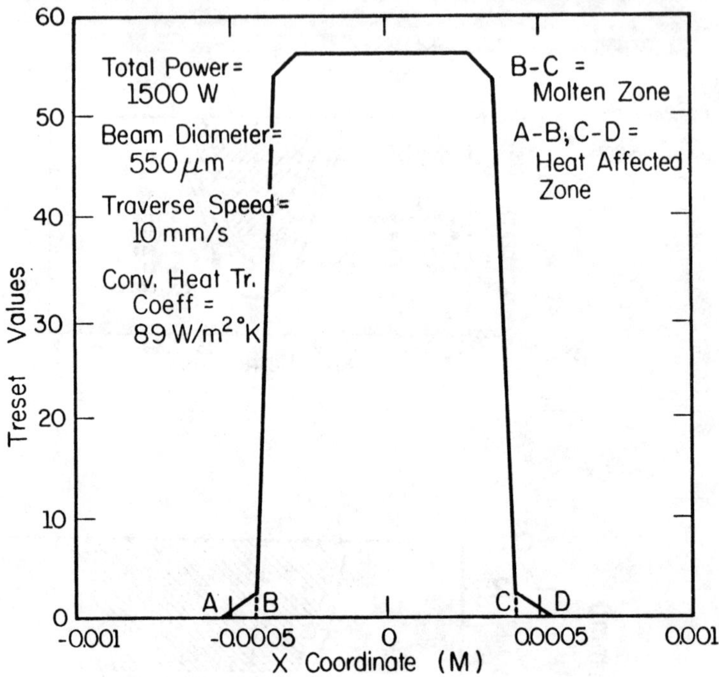

Fig. 9 Typical element from TRESET array showing molten
and heat affected zones just below the surface.
Number of iterations to convergence was 82.

References

1. Klemens, P. G., <u>J. Appl. Phys.</u> 47, 2165-2174, (1976).

2. Steen, W. M., <u>Lett. Heat Mass Trans.</u> 4, 167-178, (1977).

3. Swifthook, D. T. and A. E. F. Gick, <u>Welding J.</u> 52, 4925-4995, (1973).

4. O. Westby, The Technical University of Norway, (1968), (unpublished).

5. Dorn, V. S., and D. D. McCraken, Numerical Methods and Fortran Programming (Wiley, New York, 1964).

6. Dusinberre, G. M., Numerical Analysis of Heat Flow (McGraw-Hill, New York, 1949).

7. Scarborough, J. B., Numerical Mathematical Analysis, 6th ed. (Johns Hopkins Press, Baltimore, 1966).

8. Liebman, G., <u>Brit. J. Appl. Phys.</u>, 6, 129, (1955).

9. Crank, J. and P. Nicholson, <u>Proc. Cambridge Phil. Soc.</u>, 43, 50 (1947).

10. Richardson, L. F., Math Gazette, 12, 415, (1925).

11. Gordon, R. and J. Cobonpue, <u>Int. Heat Transfer Conf.</u> Pt. II, 454-460, (1961).

12. Touloukian, Y. S. (ed.), <u>The Thermophysical Properties of Matter</u>, Vol. 1,4,7 (1Fl/Plenum, New York, 1970).

13. Mazumder, J., and W. M. Steen, <u>J. Appl. Phys.</u>, 51 (2), 1980.

14. Edwards, J. B., E. E. Hucke, and J. J. Martin, <u>Metallurgical Reviews</u>, 120 (Pt. 1), 1968.

15. International Mathematical and Statistical Libraries, Inc., Edition 8, 1980, Subroutine ZXSSQ.

OPTICAL PROPERTIES OF METAL SURFACES

DURING LASER IRRADIATION

W. T. Walter, N. Solimene, K. Park. T. H. Kim and K. Mukherjee
Departments of Electrical Engineering/Computer Science and
Physical Metallurgy and Microwave Research Institute
Polytechnic Institute of New York
Route 110
Farmingdale, New York 11735

Direct real-time measurements of the target material's optical properties with subnanosecond resolution may provide crucial, process-revealing signatures in following the interaction of an intense laser beam with a metal as the surface progressively undergoes heating, plastic deformation, slip, vaporization, ejection of liquid metal, plasma formation, etc.

Three classes of physical processes have been proposed to account for a substantial decrease in reflectance observed during the interaction of an intense laser pulse with a metal surface: (1) deformation of the surface, (2) plasma formation, and (3) a nonlinear process causing enhanced absorption within the metal. Thus far we may conclude that specular reflectance is a sensitive indicator of surface deformation. Total reflectance measurements, on the other hand, indicate that until the surface temperature of a metal target reaches the vicinity of the boiling point, the total reflectance does not differ significantly from that given by a Drude-type free-electron model which is not large enough to account for a substantial reflectance decrease. In addition to reflectance measurements, spectral measurements have begun on the plasma which forms in front of a copper surface at incident laser power densities above 3×10^8 W/cm^2. Spectral emission lines have been observed from neutral copper atoms, ions and dimers (Cu$_2$) in the plasma when the surface temperature of copper exceeds the boiling point.

Real time monitoring of the material parameters during a laser-material interaction can improve our understanding of the physical processes involved and enable us to follow the transfer of energy from an incoming laser beam to the target material. Since bandwidths of optical detection systems exceed one gigahertz, optical measurements can be readily carried out on a subnanosecond time scale. The optical properties of a substance, therefore, merit special consideration because they are quantities that can be measured directly with subnanosecond resolution. Optical properties include the scalar quantities: absorptance, transmittance, and reflectance as well as the complex or vector quantities: dielectric constant, refractive index and conductivity. Although we would like to measure directly other properties such as surface and bulk temperatures on such a fast time scale, temperature is extremely difficult to measure directly even on a microsecond time scale. Therefore, in following the interaction of an intense laser beam with a target material, such as a metal, real-time measurements of the optical properties of the metal may differentiate and be process-revealing. Since for opaque targets the absorptance is one minus the reflectance, measurement of the time dependence of the metal surface reflectance will yield the laser power directly deposited into the metal. The absorbed power can then be used to calulate the metal's temperature history during and following laser irradiation. Our goal is to measure and then use the optical properties of metals to develop a better understanding of the laser-metal interaction from heating through to plasma formation. At present, polarimetric or ellipsometry measurements have not yet been carried out during laser interaction experiments. Hence in this paper, we will concentrate on the scalar optical properties of reflectance and absorptance and refractive index, plasma and electron-phonon collision frequencies derived from these measurements. Our long range goal is to measure these quantities directly as well as to carry out spectroscopic measurements of populations and temperatures in the plasma produced at the metal surface.

<u>Target Reflectance Behavior</u>

Experimentally, the reflectance of a metal surface has been observed to undergo a sharp and substantial decrease during an intense laser pulse (1-11). For incident laser power densities above threshold values which lie between 10^7 and $10^9 W/cm^2$, the measured reflectance of a metal surface drops significantly before partially or totally recovering as indicated by the curves in Figure 1a. This type of reflectance change was first reported by Bonch-Bruevich et al. (1) during individual spikes ~400-ns FWHM (Full Width at Half-Maximum intensity) within a millisecond 1.06-μ Nd glass laser pulse on targets of silver, copper, aluminum, dural and steel as indicated.

There are three distinct regions within a generalized reflectance curve as illustrated in Figure 1b: first, AB – an initial steep decrease to ~ one-half of the initial reflectance; second, BC or BC' – a plateau region during which the reflectance remains approximately constant; and third, C'D' – a complete reflectance recovery, or for higher intensity laser pulses CD, a further decrease to ~ one-tenth of the initial reflectance followed by a partial recovery.

Zavecz, Saifi and Notis (4) reported similar reflectance curves during Q-switched 1.06-μ Nd:YAG laser pulses ~60-ns FWHM on single crystal copper and tantalum targets as shown in Figure 1c. The remarkably close agreement between Zavecz's 0.64 plateau reflectance value and the 0.66 value of Bonch-Bruevich for copper, the only metal common to both investigations, is evident in Figure 1. Plateau features have also been reported

by Dymshits (8), using a 30 ~ns FWHM 1.06-μ Nd glass single laser pulse
on thin aluminum films, and Walters and Clauer (10), using gain-switched
50 to 70-ns 10.6-μ TEA CO_2 laser pulses on polished polycrystalline rods.

Bonch-Bruevich et al. (1) associated the initial steep decrease in the
reflectance curve, AB, with the heating of the metal surface to melting and
with the drop in electrical conductivity which accompanies melting. The
plateau BC or BC', they suggested, indicated that the temperature of the
molten layer remained constant while the absorbed radiant energy propagated a melting wave into the solid. The further reflectance drop during
region CD was attributed to a decrease in the amount of energy conducted
to the liquefaction wavefront and, therefore, indicated a second region of
increasing surface temperature.

Zavecz et al. (4) tested Bonch-Bruevich's conjecture that the metal
was melting during region BC. A single-crystal copper target was irradiated with a laser intensity ($3.2 \times 10^8 W/cm^2$) which was sufficient to attain
the plateau reflectance BC, as shown in Fig. 1c, but not high enough to
cause the second reflectance decrease. Examination of the target by
scanning electron beam microscopy revealed no evidence of melting. Melting occurred during the second decrease in reflectance, CD of Fig. 1b. A
major difference between the two experiments was the use of an integrating
sphere by Bonch-Bruevich and, therefore, a total reflectance measurement
while Zavecz et al. measured only the specular reflectance. In ensuing experiments to Zavecz's, Koo and Slusher (5) proposed that the reflected beam
could have been deflected into a wide cone of angles by a transient grating
structure produced on the metal target surface. Ready (6), using a 10.6-μ
CO_2 laser, and von Allmen (9) using a 1.06-μ Nd laser on metal targets,
found substantial and permanent changes from specular to diffuse reflectance
during the laser pulse. Ready used a probe laser to subtract plasma absorption and concluded that no recovery of reflectance occurred during or
after the pulse. An ellipsoidal light collector was employed by von Allmen

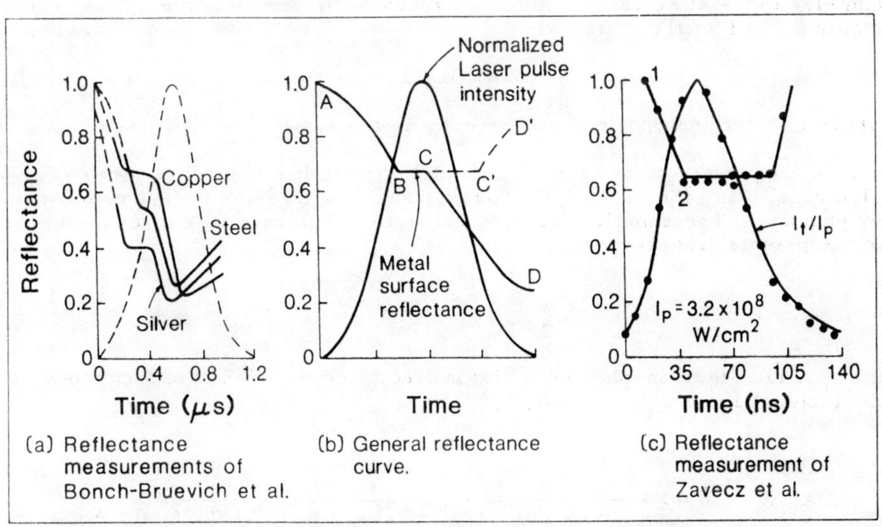

Fig. 1 - Metal Reflectance Behavior during an
intense laser pulse.

in front of a copper target in air. He found no significant change in the total reflectance until the metal surface exceeded the boiling point. Prokhorov et al. (3) suggested the formation of a liquid dielectric layer causing an enhanced absorption. More recently, Dymshits (8) observed Fig. 1-type reflectance changes during a specular reflectance measurement with the metal target film in vacuum. He suggested that the reflection takes place from a plasma formation in front of the metal surface. Walters and Clauer (10) report a sharp decrease to ~0.35 in specular reflectance at the melting point of Aℓ followed by nearly complete recovery. No luminosity was observed and no significant permanent loss of the high specular reflectivity was detected.

An understanding of the metal surface reflectance behavior during an intense laser pulse is lacking at present. In addition to Bonch Bruevich's suggestion that the initial steep decrease in reflectance is associated with the heating of the metal surface to melting, three classes of alternate explanations have been offered for the steep decrease in metal target reflectance during an intense laser pulse: 1) surface deformation, 2) plasma formation, or 3) nonlinear process. Before discussing these, we shall first examine the change in reflectance expected when a metal is heated to its melting point.

<u>Free Electron Model</u>

The Fresnel expression for the reflectivity, R, of radiant intensity at normal incidence on the surface of a homogeneous substance is (12, 13)

$$R = \frac{|\tilde{n} - 1|^2}{|\tilde{n} + 1|^2} \tag{1}$$

where $\tilde{n}$ is the complex refractive index. From the viewpoint of optics, the complex index of refraction may be expressed in terms of the optical constants n and k of the material

$$\tilde{n} = n - ik \tag{2}$$

where n is the index of refraction and k is the extinction coefficient

In the Drude or free electron model, a metal is viewed as a gas of free electrons which interacts with a background positive-ion lattice represented by phonons. The complex index of refraction then may be expressed in terms of the plasma frequency

$$\omega_p \equiv \sqrt{4\pi N e^2 / m^*} \quad , \tag{3}$$

the average electron-phonon collision frequency $\equiv \nu_c$ and the frequency of the incident light $\omega_{light} \equiv 2\pi c/\lambda$,

$$\tilde{n} = 1 - \frac{\omega_p^2}{\omega_{light}^2 + \nu_c^2}\left(1 + i\frac{\nu_c}{\omega_{light}}\right) = 1 - \frac{1}{\omega^2 + \nu^2}(1 + i\frac{\nu}{\omega}) \tag{4}$$

where $\omega \equiv \omega_{light}/\omega_p$, and $\nu \equiv \nu_c/\omega_p$. Here N is the free electron density, e is the electronic charge, m^* is the effective mass of the free electron in the metallic lattice and $i \equiv \sqrt{-1}$. For visible light incident on a good conductor

182

$$\omega_p \approx 10^{16} \text{ sec}^{-1} > \omega_{\text{laser}} \approx 10^{15} \text{ sec}^{-1} > \nu_c \approx 10^{14} \text{ sec}^{-1}. \quad (5)$$

Therefore, $\omega \sim 10^{-1}$ and $\nu \sim 10^{-2}$.

When the temperature is raised to the melting point, no significant change in the free electron density is expected for a metal whose bandgap is much greater than the thermal energy change. Therefore, no appreciable change is expected in the plasma frequency ω_p. The phonon density, however, may increase significantly and hence also may the electron-phonon collision frequency and ν. Figure 2 shows the dependence of the Fresnel reflectivity on the collision frequency for neodymium and ruby laser frequencies normally incident on a copper surface.

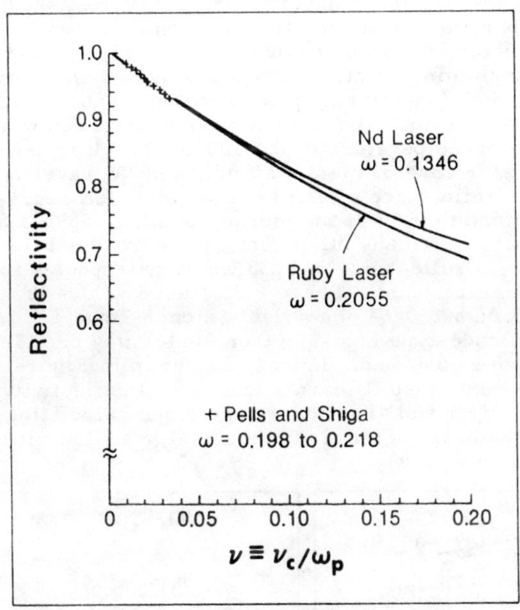

Fig. 2 - Reflectivity of Copper as a function of the electron-phonon collision frequency ν_c for two incident laser frequencies ω.

There has been considerable variation in the room-temperature reflectance measurements of many metals. Since the absorption depth, $\lambda/2\pi k$, is only a few hundred angstroms at most for metals in the visible-near infrared spectral region, the measured reflectances of metals are very sensitive to the method of preparation of the surface as well as to the history of the surface from preparation to measurement. When the metallic surface is produced by polishing, mechanical stress can deform the surface by producing an amorphous layer which may be several hundred angstroms thick (14). Even more detrimental is the embedding of polishing-compound particles which is more likely to take place in soft metals: copper, aluminum, etc. If an oxide layer is allowed to form, the reflectance measured will not be that of the pure metal. For metals which are good conductors and therefore also possess high reflectivities, processes which modify the metal-air or metal-vacuum interface generally reduce the reflectance. The highest reflectances of metals have been measured for evaporated coatings prepared in a high vacuum at a high rate of deposition of the metal and measured immediately after preparation while still in a high vacuum (15). Reflectances of Ag, Cu and Al measured under these conditions at 1.06μ are .994, .985, and .944, respectively.

Figure 2 indicates that for a room-temperature copper reflectivity of 0.985 at 1.06μ, the ratio of the electron-phonon collision frequency to the plasma frequency $\nu = 0.00749$. According to the free electron model then, for the reflectivity of copper to decrease to 75%, for example, ν must increase so that $\nu > 0.15$. This means an increase of more than 20 times in the electron-phonon collision frequency.

Ujihara (13) has examined the temperature dependence of a Drude-type free-electron model for high conductivity metals. He asked whether the increase with temperature of the electron-phonon collision frequency can account for a reflectance decrease to the plateau value by the time that the metal reaches its melting point. Ujihara assumed a Debye model for the phonon spectrum and N-scattering processes on a spherical Fermi surface. Both the Debye temperature and the electron distribution were considered to be independent of temperature in the 300°K to melting point region. Ujihara's results for several metals at ruby and Nd wavelengths are shown in Figure 3. The reflectance of solid copper at 1.06μ decreases from 0.95 at room temperature to 0.73 at the melting point, 1356°K. Since the conductivity of copper, like many other metals, decreases by a factor of 2 upon melting (16), a reflectance of 0.58 would be expected for liquid copper at 1.06μ at the melting point. These reflectance values bracket the plateau reflectances of 0.66 and 0.64 observed by Bonch-Bruevich and Zavecz, and suggest that the Drude-type free-electron model may be able to explain the reflectance behavior observed. Indeed, as shown in Figure 4, Chan et al. (7) claim to have confirmed Ujihara's theoretical reflectivity curves using a ruby laser on copper and aluminum targets and calculating the surface temperature on the basis of a one-dimensional heat flow model.

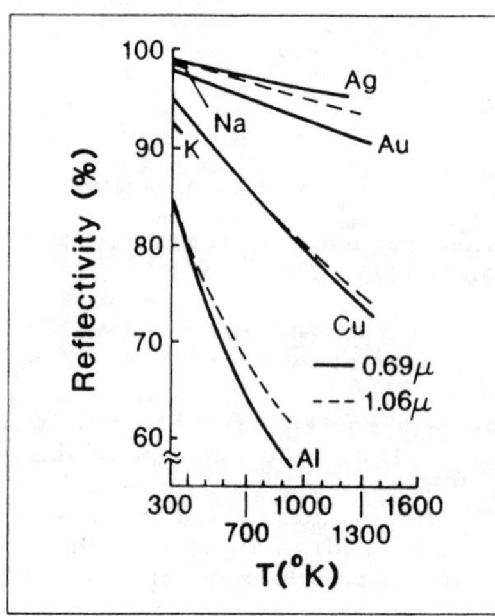

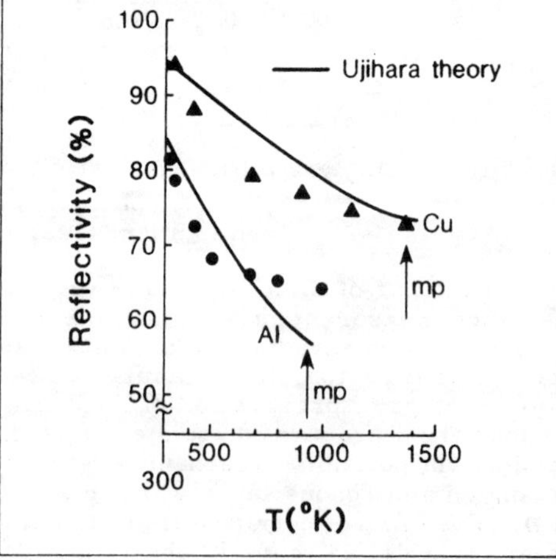

Fig. 3 - Ujihara's Calculated Reflectivities (13) for several metals at two laser frequencies.

Fig. 4 - Chan's Measured Reflectances (7) of copper (triangles) and aluminum (dots) at 0.69μ.

Ujihara's analysis has been examined by Walter (17), who pointed out that Ujihara, in his determination of the plasma frequency ω_p, used the 1913 effective-mass measurements (18). For copper, the old value is almost two times the more recent measurements. When the resulting low value for the plasma frequency was used with values of the optical constants, n and k, the room-temperature value Ujihara calculated for the electron-phonon collision frequency was 2.5 times too high and the room temperature reflectivity at 1.06μ was 95.2% instead of 98.5%. With these incorrect initial values, his analysis then predicted substantial changes in the collision frequency and in the reflectivity which are not confirmed by experiment, as indicated in Figure 5.

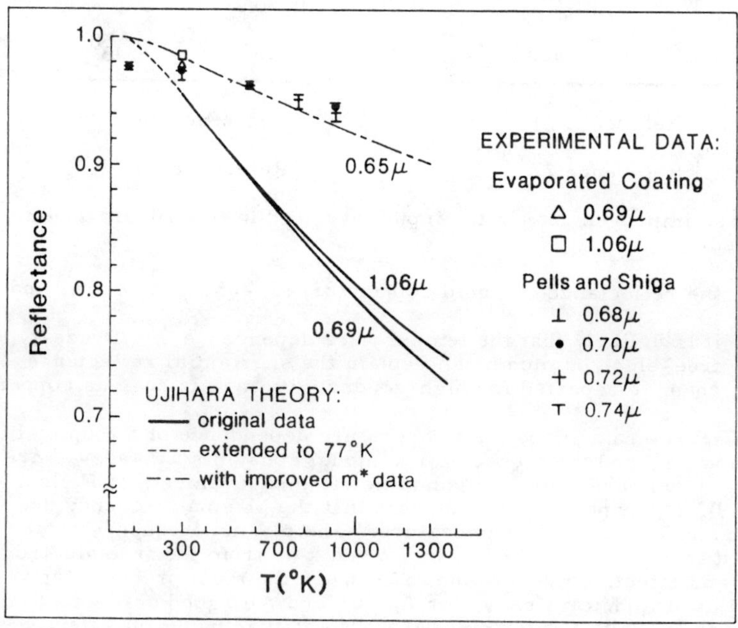

Fig. 5 - Temperature Dependence of the Reflectance of Copper.

The solid curve in Figure 5 shows the temperature dependence of the reflectivity of copper calculated by Ujihara. The dotted curve extends Ujihara's analysis down to 77°K. The optical properties of copper have been measured polarimetrically in ultra-high vacuum ($\sim 10^{-9}$ Torr) by Pells and Shiga (19) over the temperature range 77 to 920°K. Reflectances computed from Pells and Shiga's experimental values of n and k are also indicated in Figure 5. The discrepancy between theory and experiment is substantial. However, when the improved data listed in Table I is used, then the theory of Ujihara is in much better agreement with Pells and Shiga's experimental results. The dominant change responsible for most of the improvement is Schulz's careful determination (20) of the effective electron mass in copper including a correction for the anomalous skin effect.

Table I. Input Data for Ujihara Theory

Parameter		Original Ujihara Analysis	Improved Data of Reference 17
Index of refraction	n	0.150[a]	0.1074[b]
Extinction coefficient	k	4.049[a]	3.9104[b]
Wavelength	λ	0.70 μ[a]	0.65 μ[b]
Effective mass	m*/m	2.56[c]	1.45[d]

[a] Reference 21.

[b] Reference 22.

[c] Reference 18.

[d] Reference 20.

With these improvements in the input data used in the Ujihara theory, Walter concluded:

(1) the reflectance of liquid copper (23) is ~85%, not ~65%, and

(2) it is unlikely that the temperature dependence of a Drude-type free-electron model can explain the substantial reflectance changes reported for high-conductivity metals such as copper (17).

In Figures 6a and 6b, the temperature dependence of the optical constants n and k, and the free-electron model constants ω_p and ν_c, are displayed and compared with the polarimetric measurements of Pells and Shiga (19). As expected, it is evident that the plasma frequency does not change significantly with temperature. The plasma frequency can be determined from the optical constants n and k or from the free electron density and effective mass N and m^*. When the most precise n and k values (22) are used to determine ω_p for Ag, Cu and Au, good agreement is obtained with ω_p values determined from N and m^*, using the effective mass values of Schulz. Also, the effective mass for electrons in Cu determined by Schulz from optical properties at 2μ is nearly identical to the value from electronic specific heat determinations (20). It is also evident in Fig. 6b that the electron-phonon collision frequency undergoes a much smaller increase (~4) when copper is heated from room temperature to its 1356°K melting point than that required ($\geq$20) to produce the substantial reflectance drop to Fig. 1-type plateau values.

Finally, it is clear from the plots of n and ν_c in Figure 6 that even with the improved input data listed in Table I, the theory of Ujihara does not have the correct functional dependence with temperature for either the index of refraction or the electron-phonon collision frequency for T < 500°K. Refinements such as a departure from sphericity of the Fermi surface and more than one electron-phonon collision frequency could be considered to improve the Ujihara treatment.

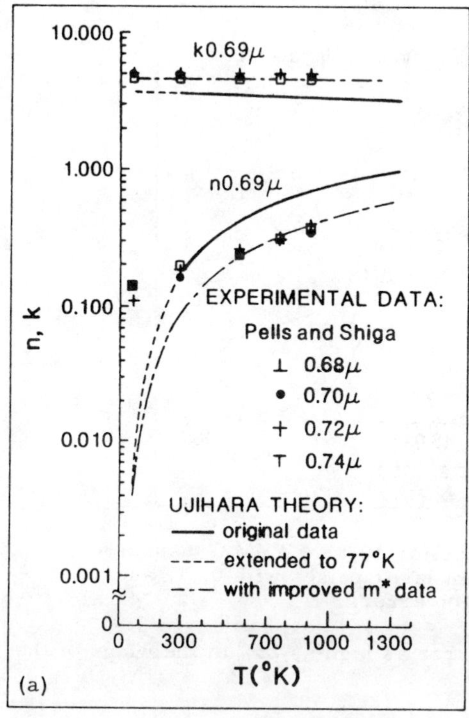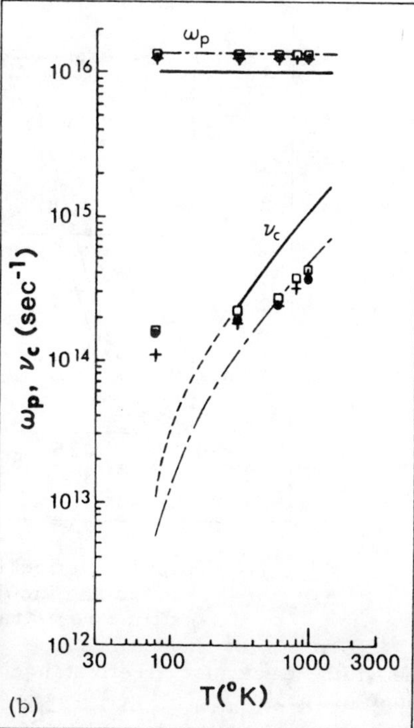

Fig. 6 - Temperature Dependence in Copper of (a) Optical Material Properties: index of refraction n and extinction coefficient k and (b) Free-Electron Material Properties: plasma frequency ω_p and electron-phonon collision frequency ν_c.

Reflectance Measurements

Other explanations for the reflectance behavior which is shown in Figure 1 fall into three categories: (1) a surface deformation which directs the reflected beam away from the detector, (2) plasma formation which absorbs or scatters the reflected light, or (3) a nonlinear process causing enhanced absorption within the metal. The first two categories involve processes which produce a reduction in light reaching the detector that could be misinterpreted as a decrease in the reflectance of the solid target surface. The importance of the first category, surface deformation, may be evaluated by comparing specular reflectance with total reflectance. The initial experiments of Bonch-Bruevich (Figure 1a) apparently measured total reflectance while those of Zavecz (Figure 1c) were specular reflectance measurements.

Experimentally, the reflectance of the metal surface is obtained by dividing the reflected pulse by the incident laser pulse. For a 30-ns laser pulse, even a time registration error as small as 1 ns between these two pulses can cause a significant error in the measured reflectance. This is indicated in Figure 7 where the solid pulse curve represents 30-ns FWHM Gaussian incident and reflected pulses, and the solid horizontal line is the corresponding reflectance. Time registration errors of 1 and 3 ns are shown by the dashed curves with the dashed reflected pulse now preceding the solid incident laser pulse. A shift of only 1 ns, corresponding to 3% of

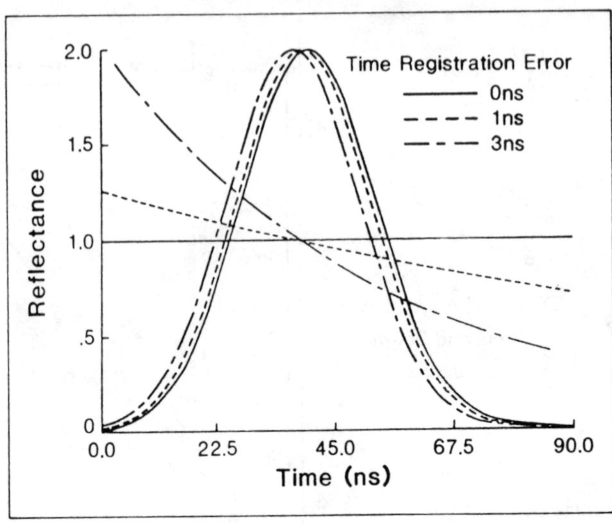

Fig. 7 - Computed Reflectance for 30-ns FWHM Gaussian
reflected and incident laser pulses with 0, 1, and
3-ns time registration errors.

the width, can cause a reflectance error as high as 25% in the wings of the
pulse shape.

To avoid a time registration error, Park and Walter (11) used a single
detector for both pulses and sent the sample of the incident laser pulse
through an air delay path. They carried out both specular and total reflectance measurements on copper targets. The experimental arrangement is
sketched in Figure 8. A TRG model 104 ruby laser operating in the Q-
switched mode produced 0. 6J in a 30-ns FWHM pulse. A one to four beam
expander reduced the measured beam divergence to ~ 1 mrad. A high-
intensity beamsplitter provided a sample of the incident pulse which was
delayed through an air path of ~ 20m for specular reflectance measurements and ~ 50m for total reflectance measurements. The targets were
located in a vacuum chamber, and a mechanical pump reduced the chamber
pressure below 0. 1 Torr to avoid air breakdown at the target surface.

The copper target surfaces were produced by vacuum-deposition on a
glass plate of good optical quality. Electroplating was used to back up the
~ 2000 Å thick copper film on the glass plate and produce a mechanically-
strong copper target a few mm thick. Just before the reflectance measurements were made, the vacuum-deposited target was removed from the glass
plate and placed in the vacuum chamber. In this way an optical-quality
copper surface could be prepared and protected until the measurements were
taken.

At the focused spot ~ 0. 5-mm diam on the copper target, a maximum
intensity of 10^9 W/cm^2 was produced. This was determined from a measured
energy of 75 mJ at the target location. Corning glass filters were used as
attenuators. For the total reflectance measurements indicated in Figure 8b,
an integrating sphere consisting of two plastic hemispheres coated with
Eastman White Reflectance coating (barium sulfate) was placed in the vacuum
chamber. Not only does an integrating sphere carry out a spatial integration
of all the light reflected from the target surface, but it delays the arrival of

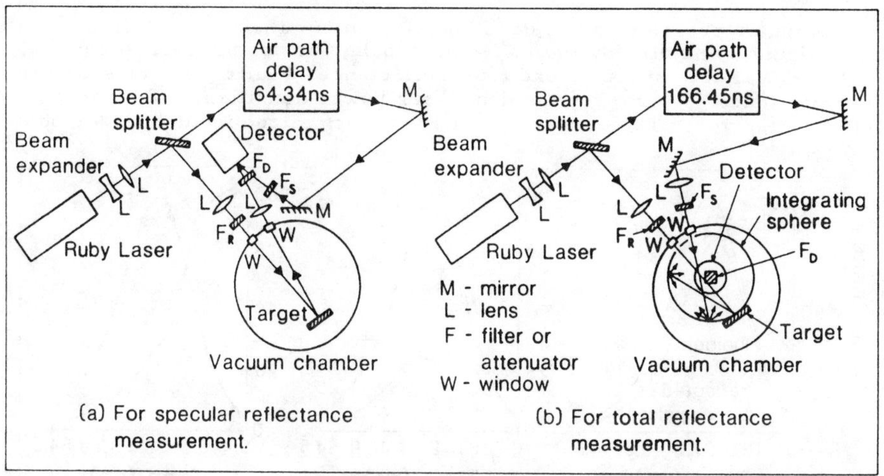

Fig. 8 - Experimental arrangement. M = mirror, L = lens, F = filter or attenuator and W = window.

the light at the detector by varying times depending on the path within the sphere. The 30-ns FWHM ruby laser pulse was lengthened by ~10 ns when detected by a fast vacuum photodiode located at the top of the sphere. Because of the pulse-lengthening property of the integrating sphere, a longer air delay path, ~50m, was required to avoid overlapping the reflected and sample laser pulses. As indicated in Figure 8b, the sample of the incident laser pulse was directed through a second entrance window into the integrating sphere, but was not focused on the surface of the sphere.

For each reflectance measurement, a Polaroid picture was taken of the Tektronix 519 oscilloscope trace and enlarged as shown in Figure 9 for digitization of the reflected and sample pulse shapes into a PDP 11 mini-computer.

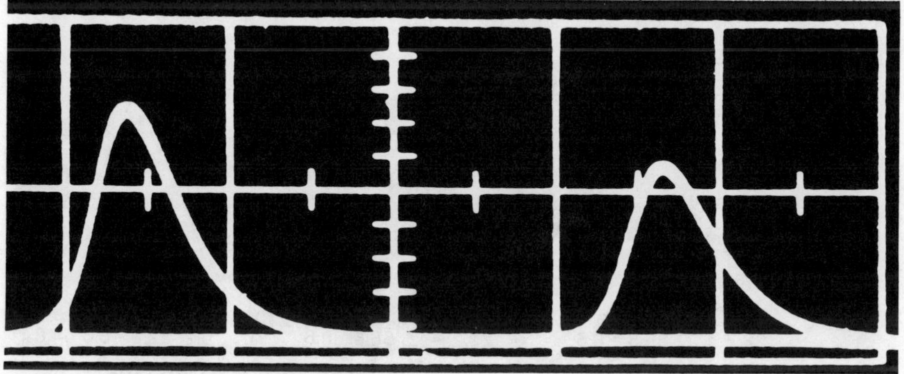

Fig. 9 - Enlargement of an oscilloscope trace showing the specularly-reflected pulse from an evaporated-copper target surface and a sample of the incident 3.15×10^8 W/cm^2 ruby laser pulse. The horizontal time scale is 50 ns/large division.

Computer programs were developed to remove the air-path time delay, calculate and display the reflectance behavior during the laser pulse.
Park and Walter's specular and total reflectance results (11) are shown in
Figures 10 and 11 for three incident laser power densities, ~ 2, 5 and
9×10^8 watts/cm^2. The curves shown are a normalized incident laser pulse,

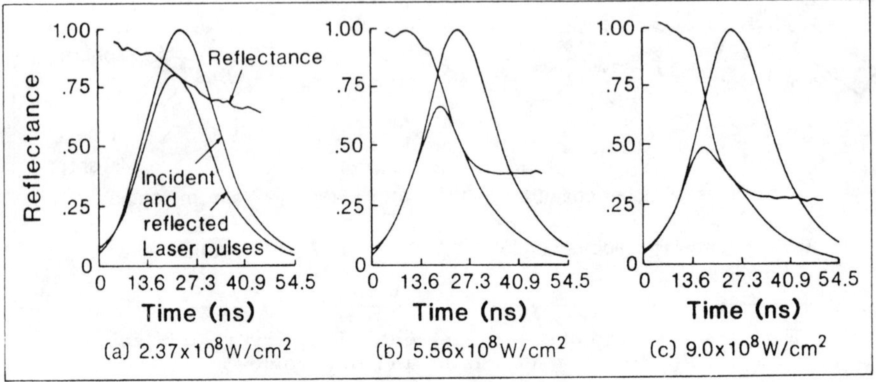

Fig. 10 - Specular Reflectance of Copper.

the reflected pulse and the metal target's reflectance behavior which is the
ratio of the two pulse curves. The laser energy absorbed was calculated
from these results and the temperature history of the metal surface determined using a one-dimensional heat-conduction approach.

The measured total reflectances of evaporated-copper target surfaces
show a slow decrease with increasing incident laser intensity or increasing
surface temperature up to the boiling point of the metal. At low incident
laser power densities (i.e., $< 3 \times 10^8$ W/cm^2) the specular and total reflectances of the metal target surface are similar (cf. Figure 10a with Figure
11a). The one-dimensional heat-conduction calculations indicate that the
copper surface did not reach the melting point with this incident laser

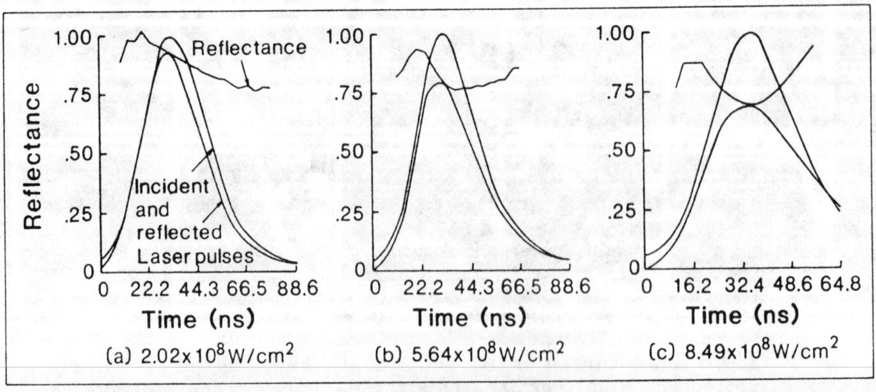

Fig. 11 - Total Reflectance of Copper.

power density; namely, $2 \times 10^8 W/cm^2$. No permanent damage was observed
in the copper. The copper target was examined under 200x magnification
and no trace of melting was observed. The lack of any permanent observable
damage is consistent with the calculated maximum surface temperature of
$1040^\circ K$ and with the equality of the specular and total reflectances.

At an incident laser power density of $5.6 \times 10^8 W/cm^2$, the specular
reflectance undergoes a sharp drop which is not observed in the total reflectance (cf. Figure 10b with Figure 11b). The one-dimensional heat-
conduction calculation indicates that the maximum temperature attained by
the copper surface was in excess of the boiling point. Permanent damage
was observed at the interaction site. Once again the permanent changes
observed in the target are consistent with the surface temperatures calculated and the reflectances measured which reveal a decrease in the specular
accompanied by an increase in the diffuse component of the reflectance.

An additional increase in incident laser power density to $\sim 9 \times 10^8 W/cm^2$
continues the decrease in specular reflectance (Figure 10c compared with
10b). Permanent target surface damage is more severe and the crater site
is more extensive. The recovery in the total reflectance behavior which is
evident in Figure 11c is caused by a "light flash" produced by emission of
light from a plasma that is formed at or in front of the metal surface. The
spectral content of this light includes sharp emission lines from neutral
copper atoms, copper ions and copper dimers (Cu_2 molecules) as indicated
in Figure 12. These sharp emission features are superimposed upon a continuous background. They were recorded on Polaroid film using a 0.5m
Jarrell-Ash Ebert spectrometer. Our current research program includes
improvement of the spectral resolution to identify the type and state of ex-
citation of the emitting species. Temporal resolution is also being used to
identify the order in which the different species are excited and contribute
to the "light flash."

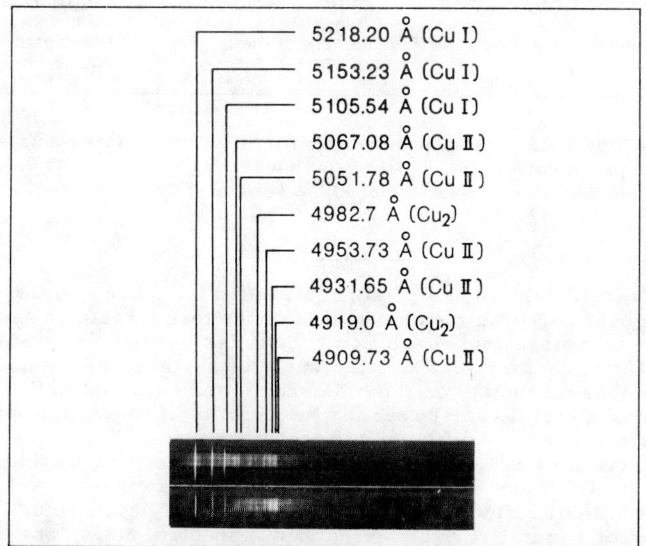

Fig. 12 - Spectral Characteristics of the "Light Flash"
in the blue-green region.

When an interference filter is inserted in front of the TRG_o105B vacuum photodiode thereby restricting the light detected to $\pm$ 50 Å around the 6943 Å ruby laser wavelength, then the recovery in total reflectance behavior so evident in Figure 11c and just noticeable in Figure 11b disappears. Our most recent reflectance behavior results are shown in Figure 13 for three different surface preparations of the copper target: vacuum-deposited, diamond-turned and mechanically-polished polycrystalline. These results were obtained with a ruby-laser wavelength interference filter included in the detection system. Notice that we are begining to see differences in the reflectance behavior for different surface preparations. This provides some preliminary confirmation for the monitoring of optical properties of target materials to identify the relevant processes taking place during the laser-material interaction.

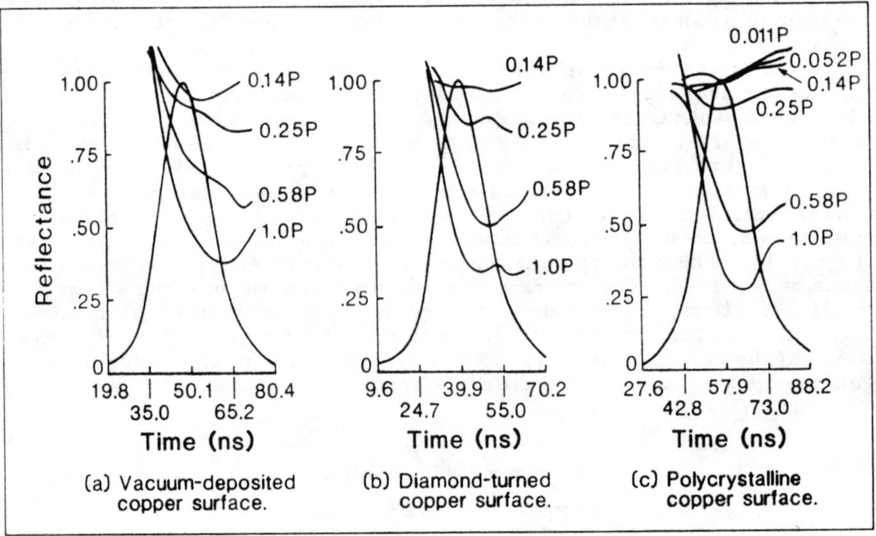

Fig. 13 - Total Reflectance Behavior for different copper surface preparations. P corresponds to an incident ruby laser power density of 1.5×10^9 W/cm^2.

Conclusions

Surface deformation and plasma formation have been observed by means of direct real-time measurements of the target material's optical properties with nanosecond resolution. Thus far we may conclude that specular reflectance is a sensitive indicator of surface deformation. Total reflectance measurements, on the other hand, indicate that until the surface temperature of a metal target reaches the vicinity of the boiling point, the total reflectance does not differ significantly from that given by a Drude-type free-electron model. These results are in agreement with those of von Allmen et al. (9) but contradict the conclusion of others that a decrease of ~ 50% in the total reflectance occurs at or possibly even before the melting point of a metal surface is reached. Spectral emission lines have been observed from neutral atoms, ions and dimers in the plasma formed as the surface temperature of copper exceeds the boiling point.

Acknowledgments

This research has been supported in part by the Office of Naval Research under Contract No. N00014-78-C-0678 and by the Air Force Office of Scientific Research under the Joint Services Electronics Program, Contract No. F44620-74-C-0056, and Grant No. AFOSR-78-3557. The helpful collaboration of N. Marcuvitz and M. Newstein, measurement assistance of W. Woturski and R. DiFazio, target preparation by M. Eschwei and S. Bielaczy, and computer programming of F. Richter and P. Levin are gratefully acknowledged.

References

1. A. M. Bonch-Bruevich, Ya. A. Imas, G. S. Romanov, M. N. Liebenson and L. N. Mal'tsev, "Effect of a Laser Pulse on the Reflecting Power of a Metal," Sov. Phys. -Tech. Phys., 13 (1968) pp. 640-643.

2. N. G. Basov, V. A. Boiko, O. N. Krokhin, O. G. Semenov and G. V. Sklizkov, "Reduction of Reflection Coefficient for Intense Laser Radiation on Solid Surfaces," Sov. Phys. -Tech. Phys., 13 (1969) pp. 1581-1582.

3. A. M. Prokhorov, V. A. Batanov, F. V. Bunkin and V. B. Federov, "Metal Evaporation under Powerful Optical Radiation," IEEE J. Quantum Electronics, QE-9 (1973) pp. 503-510.

4. T. E. Zavecz, M. A. Saifi and M. Notis, "Metal Reflectivity under High-Intensity Optical Radiation," Appl. Phys. Lett., 26 (1975) pp. 165-168.

5. J. C. Koo and R. E. Slusher, "Diffraction from Laser-Induced Deformation on Relfective Surfaces," Appl. Phys. Lett., 28 (1976) pp. 614-616.

6. J. F. Ready, "Change of Reflectivity of Metallic Surfaces During Irradiation by CO_2-TEA Laser Pulses," IEEE J. Quantum Electronics, QE-12 (1976) pp. 137-142.

7. P. W. Chan, Y. W. Chan and H. S. Ng, "Reflectivity of Metals at High Temperatures Heated by Pulsed Laser," Phys. Lett., 61A (1977) pp. 151-153.

8. Yu. I. Dymshits, "Reflection of Intense Radiation from a Thin Metal Film," Sov. Phys. -Tech. Phys., 22 (1977) pp. 901-902.

9. M. von Allmen, P. Blaser, K. Affolter and E. Stürmer, "Absorption Phenomena in Metal Drilling with Nd-Lasers," IEEE J. Quantum Electronics, QE-14 (1978) pp. 85-88.

10. C. T. Walters and A. H. Clauer, "Transient Reflectivity Behavior of Pure Aluminum at 10.6μm," Appl. Phys. Lett., 33 (1978) pp. 713-715.

11. K. Park and W. T. Walter, "Reflectance Change of a Copper Surface During Intense Laser Irradiation," pp. 274-282 in Proceedings of the International Conference on LASERS '78, V. J. Corcoran, ed.; Orlando, Florida, December 1978, and "Metal Reflectance Changes During Intense Laser Irradiation," pp. 21-31 in Applications of Lasers in Materials Processing, E. A. Metzbower, ed.; American Society for Metals, Metals Park, Ohio, 1979.

12. N. F. Mott and H. Jones, The Theory of the Properties of Metals and Alloys, Clarendon Press, Oxford, England (1936).

13. K. Ujihara, "Reflectivity of Metals at High Temperature, " J. Appl. Phys., 43 (1972) pp. 2376-2383.

14. N. F. Mott and H. Jones, op cit p. 116.

15. G. Hass et al., 1965, as quoted in American Institute of Physics Handbook, D. E. Gray, ed.; McGraw Hill, New York (1972) pp. 6-157.

16. N. F. Mott and H. Jones, op cit p. 278.

17. W. T. Walter, "Reflectance Changes of Metals During Laser Irradiation, " pp. 109-117 in Laser Applications in Materials Processing, J. F. Ready and C. B. Shaw, Jr., ed.; Proc. of SPIE, 198 (1980).

18. K. Forsterling and V. Freedericksz, Ann. Physik, (1913) 40 p. 200, as quoted in M. P. Givens, "Optical Properties of Metals, " p. 324 in Solid State Physics 6 Academic Press, New York (1958).

19. G. P. Pells and M. Shiga, "The Optical Properties of Copper and Gold as a Function of Temperature, " J. Phys. C. (Solid St. Phys.), 2 (1969) pp. 1835-1846; and G. P. Pells, n and k data furnished by private correspondence as indicated in this reference.

20. L. G. Schulz, "An Experimental Confirmation of the Drude Free Electron Theory of the Optical Properties of Metals for Silver, Gold and Copper in the Near Infrared, " J. Opt. Soc. Am., 44 (1954) pp. 540-545.

21. K. Weiss, Z. Naturforsch, 3a (1948) p. 143, as quoted in American Institute of Physics Handbook, D. E. Gray, ed.; McGraw Hill, New York (1972) pp. 6-133.

22. S. Roberts, Phys. Rev., 118 (1960) p. 1509, as quoted in American Institute of Physics Handbook, D. E. Gray, ed.; McGraw Hill, New York (1972) pp. 6-133.

23. F. P. Gagliano and V. J. Zaleckas, "Laser Processing Fundamentals, " p. 146 in Lasers in Industry, S. S. Charschan, ed.; Van Nostrand Reinhold, New York (1972).

THERMAL COUPLING OF CO_2 LASER RADIATION TO METALS

C. T. Walters, T. R. Tucker, S. L. Ream,
A. H. Clauer, and D. J. Gallant
Battelle,Columbus Laboratories
505 King Avenue
Columbus, Ohio 43201

Thermal coupling coefficients are reported for the first time for TEA
CO_2 laser pulses (10.6 µm) on pure iron and iron -4% carbon alloy surfaces.
Large increases in the pure iron thermal coupling were observed with
increasing fluence that were not attributable to plasma enhancement.
Thermal modeling results are also reported which indicate that the details
of the increase with increasing fluence cannot be correlated with a simple
free-electron model of optical absorption. Iron alloy thermal coupling
results are also presented which indicate less dependence on fluence but
significant surface preparation effects.

The pulsed CO_2 laser has great potential for application to surface modification of metals. While pulsed device development lags that of continuous CO_2 lasers, it has been demonstrated that rapidly solidified surface layers can be produced in the pulsed mode which have an appearance similar to those produced by 5000 W continuous scanned beams. An example is shown in Figure 1 which presents a micrograph of a section through a rapidly solidified surface layer (white region 6 µm thick) produced on as-cast Fe-4 wt% C by single TEA-CO_2 laser pulses incident at 71 J/cm^2.

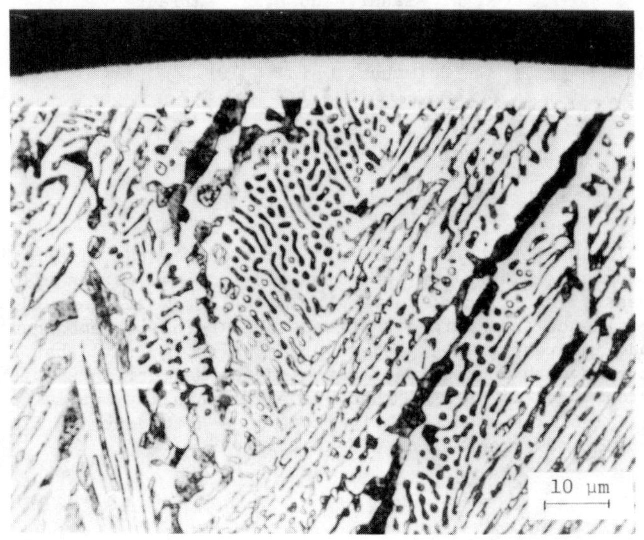

Fig. 1 – Cross section of Fe-4C sample which has been
surface melted with TEA laser pulses.

An important key to development of pulsed CO_2 laser processing of metals is an understanding of thermal coupling mechanisms operative on the surface of metals under various fluence and power density conditions. Thermal coupling is characterized by the thermal coupling coefficient which is simply the total energy retained in a target sample during a laser pulse divided by the incident energy. Research has been undertaken at Battelle (1-4) and elsewhere (5-12) to elucidate the details of thermal coupling, plasma effects, reflectivity transients, and microstructure changes in several materials. This paper presents the first experimental data on thermal coupling of CO_2 laser pulses to pure iron and an iron-carbon alloy. In prior work (3) it was discovered that pure iron exhibited an anomalous specular reflectivity transient similar to that discovered for aluminum (1) when subjected to intense CO_2 laser pulses in vacuum. In this transient, the specular reflectivity dropped to 35% of the initial value during a 70 ns pulse and recovered to full reflectivity prior to the end of the pulse. Open shutter photography revealed no surface luminosity indicative of plasma. To assess the extent to which the anomaly can be related to increased target absorption, thermal coupling measurements were initiated under conditions similar to those reported earlier (3).

Laser irradiations were conducted with 1.4 cm x 1.4 cm x 2-3 mm thick target samples mounted in a vacuum chamber as illustrated schematically in Figure 2. The beam from a Lumonics Model 301 TEA CO_2 laser was focused

SPOT-LOADED IRRADIANCE GEOMETRY

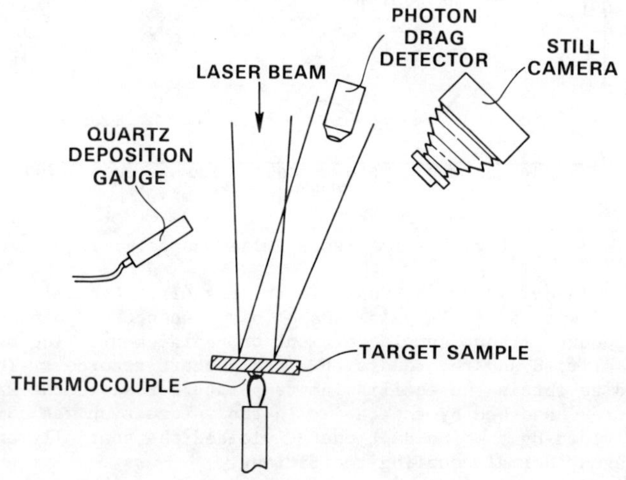

Fig. 2 – Experimental arrangement for thermal coupling measurements with spot-loaded irradiance geometry.

with a 1 m focal length germanium lens and was directed into the chamber through a sodium chloride optical flat which provided reflections for beam calorimetry external to the chamber. The target samples were instrumented with rear surface 30 gauge copper/constantan thermocouples which also served to support the sample with a minimum of thermal conduction loss.

The target samples were located slightly behind the focal plane where the beam spot was completely on the target (spot-loading conditions). The effective 1/e diameter of the spot was 0.7 cm (0.35 cm^2 effective area). The beam energy density profile was not uniform but could be characterized as a distorted Gaussian profile. The principal advantages of this arrangement were that the measurement of thermal coupling could be performed without sensitivity to beam alignment errors and without instrumentation difficulties associated with a small target which would be required for flood-loading geometry. Additionally, all luminosity in this arrangement could unambiguously be attributed to target effects. The disadvantage of this arrangement was the spatial integration of the measured effects as a result of the nonuniform energy density profile.

Sequences of CO_2 laser irradiations were conducted in vacuum for several iron and iron alloy samples with two different temporal pulse shapes as illustrated in Figure 3. The long pulse is the normal TEA laser pulse consisting of a gain-switched spike (70-100 ns FWHM) followed by a long low intensity tail lasting about 2 μs. The short pulse was obtained

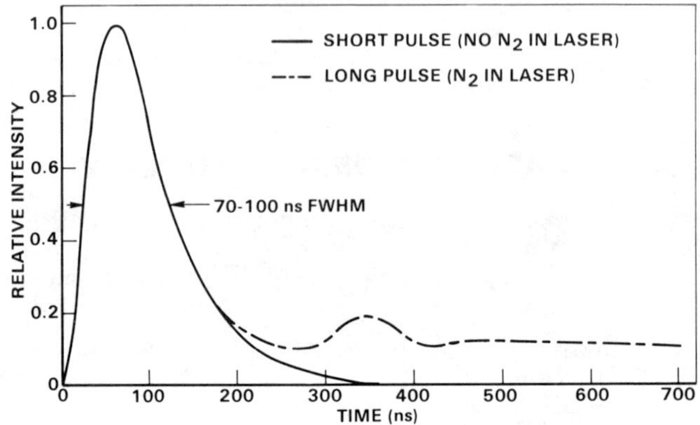

Fig. 3 - TEA laser pulse shapes.

by running the laser without nitrogen in the gas mix. Generally, samples
were irradiated with a single pulse shape in a sequence of pulses of
increasing fluence without sample movement or replacement. Thermocouple
output was amplified and recorded with a strip chart recorder. The data
were analyzed to obtain the equilibrium temperature rise in the sample from
which the energy absorbed by the target in the interaction was computed.
This value divided by the incident energy yielded the spatially and tem-
porally averaged thermal coupling coefficient.

In addition to thermocouple data, simultaneous observations of plasma
luminosity, reflectivity transients, and mass removal were recorded. Plasma
luminosity was recorded with an open shutter camera (f/4.7) and ASA 400 speed
Polaroid film. A photon drag detector was aligned in the specularly reflected beam to view the interaction area but definitive results were not
anticipated because of spatial averaging of the reflected beam. An oscil-
lating-quartz-disc thin-film deposition gauge was employed to detect vaporization of target material.

<u>Experimental Results</u>

Thermal coupling coefficients measured for mechanically polished pure
iron samples are presented in Figures 4 and 5 for the long and short TEA CO_2
laser pulses, respectively. It is important to note that these coefficients
are spatial averages for the irradiance pattern, but are plotted as a function of the spatial peak fluence. In Figure 4, all data were taken at the
same surface location of the same sample with the long pulse. The open
circles indicate the coefficient measured for the first pulse at a given
fluence level in the sequence of increasing fluence levels while the solid
circles indicate repeat pulses after the sample had been exposed to fluences
higher than that indicated. Good agreement is noted between initial and
repeat data at a given fluence, indicating a minimal effect of damage accu-
mulation on thermal coupling. The dashed line is included to guide the eye
along what is believed to be the trend in the data. Random scatter in the
data is fairly small, but systematic error bounds due to calorimetry and
profiling inaccuracy are estimated at ±30% for fluence and ±20% for
coupling coefficient. The low fluence limit for thermal coupling appears
to be near 2.2% which is somewhat lower than that indicated by the limited
room temperature 10.6 μm iron absorptance data available from the literature

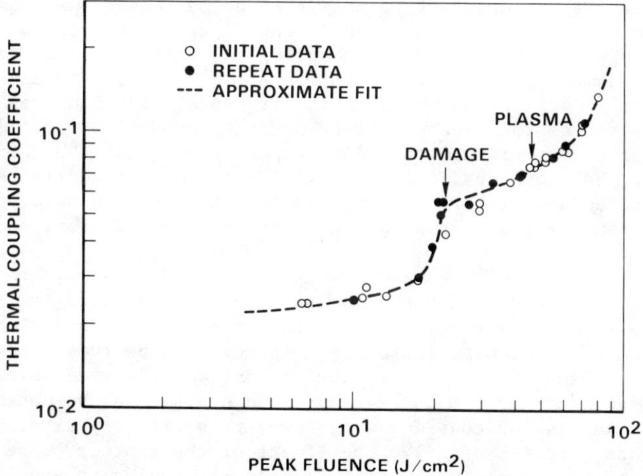

Fig. 4 – Thermal coupling coefficient for polished
pure iron exposed to long TEA CO_2 laser
pulse (10.6 μm) in vacuum.

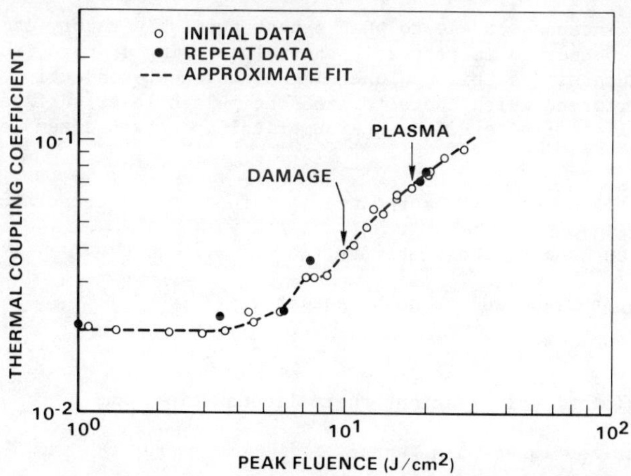

Fig. 5 – Thermal coupling coefficient for polished
pure iron exposed to short TEA CO_2 laser
pulse (10.6 μm) in vacuum.

(3.0-4.5%) (13). The high degree of polish, sample cleaning, and vacuum
environment may account for some of the difference. As peak fluence was
increased the thermal coupling coefficient increased monotonically as might
be expected simply from thermal effects as manifested in increased electron
scattering. Near 19 J/cm^2, however, a dramatic rise in thermal coupling with
increasing fluence was observed. An arrow in the figure indicates where
damage was first observed (21 J/cm^2). Damage in this case was a very subtle
change in the surface as indicated by visual observation of the specularity

of the surface after each pulse as compared to that of unirradiated portions
of the sample. It is important to note that no luminosity was observed to
be associated with the large increase in coupling. As fluence was increased
above the damage threshold, thermal coupling increased at a reduced slope.
At 45 J/cm^2 plasma luminosity was first observed. The plasma apparently
enhanced thermal coupling as noted in the figure at high fluences where the
coefficient rises sharply again. At 80 J/cm^2 the coupling reached 14% which
is a factor of 6 greater than the room temperature value. No definitive
indication of the anomalous reflectivity effect was observed in the spatially integrated specular reflectivity signal at fluences below plasma
threshold. Material removal was detected by the vapor deposition gauge at
60 J/cm^2.

Thermal coupling data for the short pulse are presented in Figure 5.
Again the low fluence coefficient was found to be near 2% and a monotonic
increase in coefficient with increasing fluence was observed. A small
deviation from a smooth increase in coefficient was observed below damage
threshold (9.8 J/cm^2) but this may have been scatter in the data. The
magnitude of increase of coupling was, however, still very significant below
the plasma threshold (18 J/cm^2). The effect of the shorter pulse appears to
be reduction of the damage and plasma fluence thresholds by about a factor
of two. This was expected because of the higher peak power density (at
constant fluence) and associated more rapid heating and shallower melt
depths.

Discussion of Results

It is of interest to see to what extent available theory can be used
to explain the observed increases in thermal coupling of pure iron with
increasing fluence. To this end, one-dimensional heat-conduction calculations were performed which characterized the pulsed laser/surface interaction. An explicit finite difference numerical model was assembled which
employed:

- prescribed optical absorptance as a function of temperature,

- constant thermal conductivity,

- temperature-dependent heat capacity,

- heat of fusion,

- one-dimensional transient thermal conduction, and

- linearized laser pulse shape.

The critical input model, of course, was the function describing the
optical absorptance variation with temperature. At low beam intensities it
is reasonable to assume that the simple Drude (free-electron) theory of
optical absorption in metals may be used as a starting point to describe
the interaction. As pointed out by Wieting and Schriempf (14), however,
even this assumption may be subject to question for metals such as nickel
and iron where bound electrons may play a role. Lacking any better theory,
free electrons were assumed for the calculation and the optical absorptivity,
α, was computed as (14)

$$\alpha = \frac{4n}{(n+1)^2 + k^2} \tag{1}$$

200

where n is the index of refraction and k is the extinction index for the metal. These parameters are in turn related to the real and imaginary parts of the electrical conductivity, σ_1 and σ_2, by

$$2n^2 = \left(1 - \frac{\sigma_2}{\omega\varepsilon_o}\right) + \left[\left(1 - \frac{\sigma_2}{\omega\varepsilon_o}\right)^2 + \left(\frac{\sigma_1}{\omega\varepsilon_o}\right)^2\right]^{1/2} \quad , \qquad (2)$$

$$2k^2 = -\left(1 - \frac{\sigma_2}{\omega\varepsilon_o}\right) + \left[\left(1 - \frac{\sigma_2}{\omega\varepsilon_o}\right)^2 + \left(\frac{\sigma_1}{\omega\varepsilon_o}\right)^2\right]^{1/2} \quad , \qquad (3)$$

where

$$\sigma_1 = \frac{\sigma_o}{(1+\omega^2\tau^2)} \quad , \qquad (4)$$

$$\sigma_2 = \frac{\omega\tau\sigma_o}{(1+\omega^2\tau^2)} \quad , \qquad (5)$$

and ω is the angular frequency of the incident radiation ($\omega = 1.78 \times 10^{14}\ s^{-1}$ for 10.6 μm laser beam), ε_o is the permittivity of free space ($\varepsilon_o = 8.85 \times 10^{-12}\ s/\Omega m$), τ is the relaxation time of the electrons due to collisions, and σ_o is the dc electrical conductivity. The latter two parameters are related by

$$\sigma_o = \frac{Ne^2\tau}{\beta m_o} \quad , \qquad (6)$$

where e is the electron charge, m_o is the electron mass, N is the electron number density and β is the ratio of effective electron mass in the lattice to actual electron mass. Following Reference (14), β/N was assumed to be independent of temperature and to have the value $5.89 \times 10^{-24}\ cm^3$ based on 2 free electrons per atom. The temperature dependence of α, then, is completely embodied in the relaxation parameter, τ. The functional dependence $\tau(T)$ was determined from literature values for the electrical resistivity of iron for temperatures varying up to and past the melt point (15,16).

The temperature dependence of α for pure iron as calculated by the Drude theory using published resistivity data is given in Figure 6 by the solid line. The function is characterized by a reduction in slope at the Curie point (770 C) and a small jump at the melt point (1536 C). This behavior is close to the square root of that of the resistivity function. The calculated room temperature value of α turned out to be 0.03 which was 50% greater than that measured in the pulsed laser experiments. To force agreement at lower temperatures in the heat conduction analysis, the Drude theory result was scaled by a constant factor (0.66) as shown by the dashed line in Figure 6. Independent data taken with a continuous CO_2 laser on a low alloy steel indicate reasonable agreement with the scaled theory as noted by the open circles in Figure 6. An additional curve in the figure indicates a function having a jump discontinuity at the Curie point, which, although included inadvertently in some of the modeling, gave improved agreement with the data as discussed below. There is no physical basis for such a jump at low beam intensities.

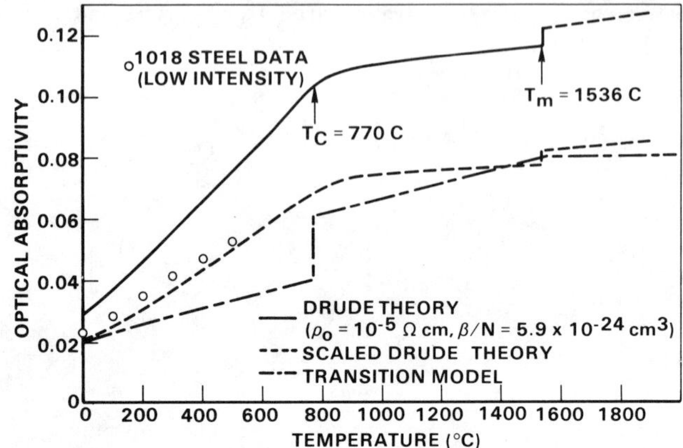

Fig. 6 – Optical absorptivity of pure iron at 10.6 μm.

Numerical calculations of the thermal response of pure iron samples to 10.6 μm laser pulses were performed using the prescribed optical absorptivity functions discussed above. Curves of effective coupling coefficient versus fluence were generated by repeating the calculation for various fluence levels and integrating the absorbed energy over the pulse length at each level. Results of the calculations are compared to the experimental data in Figures 7 and 8. The upper solid curve in Figure 7 gives the

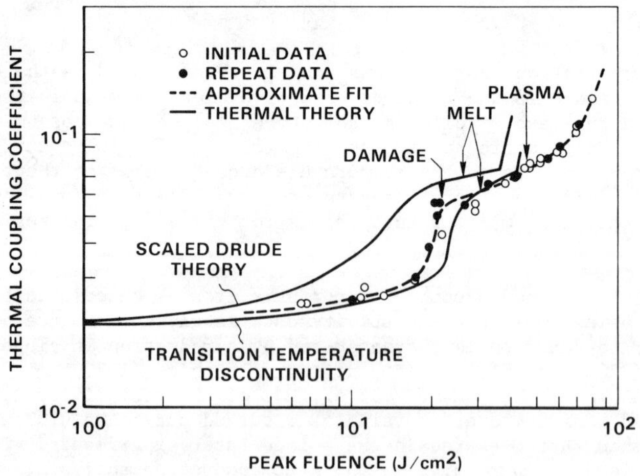

Fig. 7 – Thermal theory results for long pulse
on pure iron at 10.6 μm.

computed thermal coupling for the scaled Drude theory absorption. The arrow on the curve indicates the computed melt threshold which is somewhat higher than the observed damage threshold although probably within the combined error band of the theory and experiment. The shape of the curve does not agree well with the experimental result because the theoretical coupling does not rise sharply with increasing fluence in the "shelf" region. It can

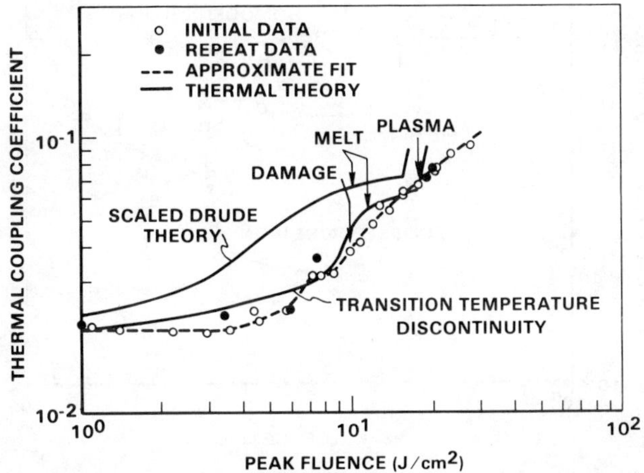

Fig. 8 - Thermal theory results for short pulse on
pure iron at 10.6 μm.

be argued that some of the disagreement might arise from the fact that the
theoretical result was not integrated over the spatial beam profile, however such an integration would be expected to smooth the curve rather than
sharpen transition features. The lower solid curve indicates the theoretical
result for the absorption model with a jump at the Curie point and surprising
agreement is noted. While the jump in absorption is unphysical at low beam
intensities, it might be possible for such a change to occur under intense
beam conditions typical of those occurring in the experiments. Results for
the short laser pulse are presented in Figure 8. Again, the theory with the
jump in absorption yielded better agreement with the experiment than did the
scaled Drude theory.

It is also of interest to compare the measured pure iron thermal coupling results with other experimental data to assess the extent to which the
observed sharp rise in thermal coupling occurs in alloys of iron. Figure 9
presents long pulse thermal coupling data for two iron alloy samples (Fe-4.2
wt% C) taken under conditions similar to those of the pure iron irradiations.
The open squares indicate data taken for a polished alloy sample. The low
fluence coupling in this case is near 9% which indicates the strong effect
of the alloying constituent on absorptivity. The collision frequency of the
free electrons is apparently increased greatly by the presence of the carbon
atoms. As fluence is increased, however, the coupling does not increase
greatly, indicating that the carbon atom scattering probably dominates the
pure thermal effect. As noted in the figure, the damage and plasma thresholds are lower for the alloy than for the pure material as expected. Also
shown in the figure are data for a sample of the same alloy which was
continuous laser homogenized from the as-cast and surface ground condition.
The surface of this sample was oxidized and not free of defects as in the
polished surface case. For this alloy sample (open triangles) the low
fluence coupling was twice as great as that for the polished alloy. Furthermore, above the damage threshold coupling decreased with increasing fluence
indicating that cleanup of the surface probably dominated any thermal
effects. Also it was noted that plasma initiated below the visual damage
threshold, which is typical of defect initiated plasma.

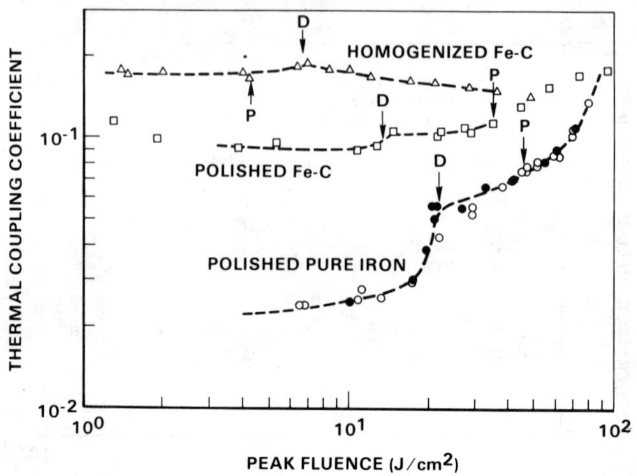

Fig. 9 – Comparison of pure iron and iron alloy
thermal coupling at 10.6 μm (long pulse).

As a further check on the consistency of the alloy thermal coupling
measurements, the low fluence results are compared to continuous laser
measurements made on the same alloy under various conditions in Table I.

Table I. Thermal Coupling to Iron Alloy (4% C) at 10.6 μm

Condition	CW Beam (<200 C)	Pulsed Beam (Low Intensity)
Polished	0.083	0.093
Homogenized	----	0.18
Grit Blasted	0.25	----
Oxidized	0.38	----

Good agreement is noted for the polished condition. It is also clear that
higher coupling is possible by surface alteration as indicated for a grit-
blasted and a heavily oxidized sample. Details of the continuous laser
measurements will be presented in a forthcoming paper.

Conclusions

Based on the research reported herein, it is clear that the thermal
coupling coefficient for TEA CO_2 laser pulses (10.6 μm) on pure iron rises
significantly with increasing fluence. This increase was more than a factor
of three greater than the low fluence value for fluences less than the
plasma threshold and cannot, therefore, be attributed to plasma enhancement.
The details of the variation of coupling with fluence do not correlate well
with a thermal model based on a simple (free electron) Drude theory but
seem to require a nonlinear effect to be invoked. The sharp rise in coupling to iron with increased fluence appears to be masked by impurity

scattering when alloying constituents are added. For the alloy case, coupling is higher and much less fluence dependent.

Acknowledgments

The authors wish to thank Bernerd E. Campbell and Gary L. Long for assistance in preparation of the samples and conduct of the experiments. This research was sponsored by the U.S. Army Research Office.

References

1. C. T. Walters and A. H. Clauer, "Transient Reflectivity Behavior of Pure Aluminum at 10.6 µm," _Applied Physics Letters_, _33_ (8) (October, 1978) pp. 713-715.

2. C. T. Walters and A. H. Clauer, "Time-Resolved Specular Reflectivity of Metals Subjected to 10.6-µm Laser Pulses," pp. 67-72 in _Laser-Solid Interactions and Laser Processing - 1978_, S. D. Ferris, H. J. Leamy, and J. M. Poate, eds.; AIP Conference Proceedings (50), Boston, Massachusetts, 1978.

3. C. T. Walters, "Anomalous Behavior of the Reflectivity of Metals Irradiated with 10.6 µm Laser Pulses," pp. 98-108 in _Laser Applications in Materials Processing_, J. F. Ready, ed.; SPIE _198_, San Diego, California, August, 1979.

4. C. T. Walters, A. H. Clauer, and B. P. Fairand, "Pulsed Laser Surface Melting of Fe-Base Alloys," pp. 241-245 in _Rapid Solidification Processing, Principles and Technologies II_, R. Mehrabian, B. H. Kear, and M. Cohen, eds.; Second International Conference on Rapid Solidification Processing, Reston, Virginia, March, 1980.

5. J. F. Ready, "Change of Reflectivity of Metallic Surfaces During Irradiation by CO_2-TEA Laser Pulses," _IEEE Journal of Quantum Electronics_, _QE-12_ (2) (February, 1976) pp. 137-142.

6. K. Ujihara, "Reflectivity of Metals at High Temperatures," _Journal of Applied Physics_, _43_ (5) (May, 1972) pp. 2376-2383.

7. J. A. McMordie and P. D. Roberts, "The Interaction of Pulsed CO_2 Laser Radiation with Aluminum," _Journal of Physics D: Applied Physics_, _8_ (1975) pp. 768-781.

8. T. J. Wieting and J. T. Schriempf, "Infrared Absorptances of Partially Ordered Alloys at Elevated Temperatures," _Journal of Applied Physics_, _47_ (9) (September, 1976) pp. 4009-4011.

9. T. J. Wieting and J. L. DeRosa, "Effects of Surface Condition on the Infrared Absorptivity of 304 Stainless Steel," _Journal of Applied Physics_, _50_ (2) (February, 1979) pp. 1071-1078.

10. K. Park and W. T. Walter, "Reflectance Change of a Copper Surface During Intense Laser Irradiation," pp. 274-282 in _Proceedings of the International Conference on Lasers '78_, Vincent J. Corcoran, ed., Orlando, Florida, December, 1978.

11. V. P. Ageev, A. I. Barchukov, F. V. Bunkin, V. I. Konov, S. B. Puzhaev, A. S. Silenok, and N. I. Chapliev, "Heating of Metals by CO_2 Laser Radiation Pulses," _Soviet Journal of Quantum Electronics_, __9__ (1) (January, 1979) pp. 43–47.

12. J. A. McKay and J. T. Schriempf, "Anomalous Infrared Absorptance of Aluminum Under Pulsed 10.6 μm Laser Irradiation in Vacuum," _Applied Physics Letters_, __35__ (6) (September, 1979) pp. 433–434.

13. Y. S. Touloukian and D. P. Dewitt, _Thermophysical Properties of Matter_, Vol 7, _Thermal Radiative Properties: Metallic Elements and Alloys_, p 1644; Plenum Publishing Corporation, New York, N.Y., 1970.

14. T. J. Wieting and J. T. Schriempf, "Free-Electron Theory and Laser Interactions with Metals," Report of NRL Progress, pp. 1–13, Naval Research Laboratory, Washington, D.C., June, 1972.

15. R. W. Powell, "LXXX. The Electrical Resistivity of Liquid Iron," _Philosophical Magazine_, Ser. 7, __44__ (354) (July, 1953) pp. 772–775.

16. H. J. Güntherodt, E. Hauser, H. U. Künzi, and R. Müller, "The Electrical Resistivity of Liquid Fe, Co, Ni and Pd," _Physics Letters_, __54A__ (4) (September, 1975) pp. 291–292.

17. D. J. Gallant, "Temperature Dependence of the Optical Absorption in Metals," thesis presented to the Faculty of the Graduate School of the University of Southern California, June, 1980 (unpublished).

SURFACE MELTING OF AN ALLOY UNDER

STEADY STATE CONDITIONS

J. A. Sekhar*, R. Mehrabian** and H. L. Fraser*
*Department of Metallurgy and Mining Engineering
University of Illinois
Urbana, IL
&
**Metallurgy Division
National Bureau of Standards
Washington, DC

A combined theoretical and experimental study is described for the
surface melting of an Al-4.5 wt% Cu alloy substrate subjected to a high
intensity stationary heat flux. Both the calculations and the experiments were done under steady state conditions. That is, the heat flux
absorbed, through the circular region on the bounding surface of the
substrate, is exactly balanced by conduction of heat into the substrate -
thermal profiles remain the same after an initial transient. The heat flow
model is based on a new formulation and solution methodology of the heat flow
equation for the two free moving boundary problem at hand. The experiments
were carried out on an electron beam welding apparatus especially modified
for rapid solidification studies. Agreement between theory and experiment is
shown to be reasonably good considering the limitations of the former due to
a number of assumptions.

Introduction

The increasing acceptance of the various lasers and electron beam
systems for the study of rapid solidification (surface modification by the
self quenching techniques) has prompted the need for heat flow models to
adequately characterize the melt depths, heat affected zones, thermal gradients and the velocities of the melting and solidifying interfaces. Various
reports have subsequently appeared detailing the heat flow in a substrate
heated by a stationary or moving beam directed high energy [1-7]. A major
criticism of these models are the use of temperature independent thermal
properties [1-7], neglect of the heat of fusion [2,6], use of temperature
independent absorptivity and the inability to model an alloy solidification
problem with a moving "mushy" zone and heat of fusion liberated over a range
of temperatures. This paper presents a review of our recent work on
solidification of a binary alloy and compares the calculations to experiments on an Al-4.5wt%Cu alloy substrate. These comparisons are only for the
range of power inputs that result in steady state conditions. That is, the
temperature on the surface of the stationary melt remains below the vaporization temperature of the alloy for very long pulse times (of the order of
seconds) - the absorbed power is exactly offset by conduction of heat in the
substrate resulting in a steady state thermal field.

Development of Heat Flow Model and Computer Solutions for Melting and Solidification of a Binary Alloy Substrate

All mathematical models to date have assumed that the substrate has a
discrete melting point. We have addressed the problem of a substrate that
melts and solidifies over a range of temperatures [8]. In this model, it
was assumed that the substrate was a binary alloy and that the phase diagram
contained a eutectic. Furthermore, the substrate had previously been subjected to conventional solidification and was segregated with the weight
fraction of the eutectic being determined by the modified Scheil equation [9].

$$f_E = (\frac{C_E}{C_0})^{-\frac{1}{1-k'}} \tag{1}$$

Where f_E is the weight fraction of eutectic, k' if the equilibrium partition
coefficient; C_0 and C_E are the initial and eutectic solute contents, respectively.

Now consider the rapid melting and subsequent solidification of the
surface layer of the conventionally solidified semi-infinite alloy of average

composition C_o. The substrate is subjected to a stationary high intensity
heat flux over a circular region on its bounding surface. The surface outside the heat region is considered adiabatic. The generalized expressions
previously derived [10] for the determination of the coefficients in the
finite difference equations governing the heat transfer within the discretized spatial domains were used [8]. The heat conduction equation for the
axisymmatric system was put in oblate spheroidal coordinates which is more
compatible with the axisymmetric heat flow problem under consideration.

The finite difference equations were rewritten in terms of dimensionless
nodal enthalpy and temperature in a manner similar to that previously described [4] except that appropriate modifications are made to account for
melting and solidification over a range of temperatures [8]. In this way,
both dimensionless temperature and enthalpy were used to formulate a single
energy conservation equation for each discretized spatial domain regardless
of whether it is in the solid state, the "mushy" state, the liquid state or
contains the liquidus or eutectic isotherms.

The finite difference representaion of the heat conduction equation
was put in terms of dimensionless nodal enthalpy, ψ, and dimensionless nodal
temperature, θ [8]. These two dependent variables are defined as:

$$\psi = \frac{1}{\rho V} \int_V \rho \frac{(H - H_{sE}^*)}{\Delta H_{s\ell}} \, dV = \frac{H - H_{sE}^*}{\Delta H_{s\ell}}$$

$$\theta = \frac{k}{k_s} \frac{C_{ps} (T - T_E)}{\Delta H_{s\ell}} \tag{2}$$

$$\text{where: } H_{sE}^* = \frac{C_{ps} (T_E - T_0)}{\Delta H_{s\ell}}$$

C_p refers to the heat capacity, k refers to the thermal conductivity and the
subscripts s, ℓ, E and sℓ refer to solid, liquid, eutectic and solid plus
liquid ("mushy") phases, respectively. T_0 is the ambient temperature. In
general H in equation (2) refers to the specific enthalpy of the discretized
space volume and assumes different forms when the node is in the liquid, the
solid, the "mushy" region or contains the eutectic isotherm. H_{sE}^* is the
specific enthalpy of the solid at the eutectic temperature. The conductivity,
k, assumes values of the liquid, the "mushy" zone or the solid. In this
formulation the following assumptions were made [8]. The density of the
solid and the liquid are equal, the conductivities and heat capacities of the
solid and liquid are independent of temperature, but different from each

other. For the "mushy" region, the heat capacity of the liquid and an average thermal conductivity are used. Finally, in derivation of equation (2) it is assumed that the heat of fusion, $\Delta H_{s\ell}$, is independent of composition.

<u>In the solid region</u>, the dimensionless nodal enthalpy, ψ, is negative and is equal to the dimensionless nodal temperature, Θ:

$$\psi = 0 = \frac{C_{ps}(T - T_E)}{\Delta H_{s\ell}} < 0, \ T < T_E \tag{3}$$

<u>In the region containing primary solid phase and melted eutectic</u> the dimensionless enthalpy is:

$$\psi = \frac{C_{ps}(T_E - T_0) + f_{\ell E} \, H_{s\ell} - H_{sE}^{*}}{\Delta H_{s\ell}} = f_{\ell E} \tag{4}$$

$$0 \leq f_{\ell E} \leq f_E, \ T = T_E$$

where $f_{\ell E}$ is the fraction of the eutectic that has melted and f_E is given by equation (1).

The dimensionless nodal temperature is:

$$\theta = \frac{k}{k_s} \ \frac{C_{ps}(T_E - T_E)}{\Delta H_{s\ell}} = 0 \tag{5}$$

The conditions for this region then becomes:

$$0 \leq \psi \leq f_E \tag{6}$$

$$\theta = 0$$

<u>In the solid plus liquid "mushy" region</u> the temperature is between the eutectic temperature T_E, and the liquidus temperature, T_L. The assumption of the heat capacity equal to that of the liquid* for this region leads to the following expression for the dimensionless enthalpy.

$$\psi = \frac{C_{p\ell}(T - T_E)}{\Delta H_{s\ell}} + f_\ell \tag{7}$$

$$\text{where } f_E < f_\ell \leq 1.0, \ T_E < T \leq T_L$$

---

*   In metals such as Al, Fe, and Ni etc. the heat capacity of the solid and liquid are close enough that this assumption should not introduce any discernable error in the computation.

Assuming an average thermal conductivity $\bar{k}$ for this region given by:

$$\bar{k} = k_s f_s + k_\ell f_\ell \tag{8}$$

The equations for the nodal enthalpy and temperature become:

$$f_\ell \leq \psi \leq 1 + \frac{C_{pL}(T_L - T_E)}{\Delta H_{s\ell}}$$

$$\theta = (\psi - f_\ell) \left[\frac{C_{ps}}{C_{p\ell}} \cdot \frac{\bar{k}}{k_s} \right] > 0 \tag{9}$$

$$1.0 \geq f_\ell = f_E + \left(\frac{T - T_E}{T_L - T_E}\right)(1 - f_E) > f_E$$

Equation (9) thus assumes a linear distribution of weight fraction liquid in the "mushy" zone, between the eutectic (the weight fraction of which is calculated using equation (1)) and the liquidus temperatures.

In the superheated region, the dimensionless enthalpy and temperature become:

$$\psi = 1 + \frac{C_{p\ell}(T - T_E)}{\Delta H_{s\ell}} > 1 + \frac{C_{p\ell}(T_L - T_E)}{\Delta H_{s\ell}} \tag{10}$$

$$\Theta = (\psi - 1) \frac{C_{ps}}{C_{p\ell}} \cdot \frac{k_\ell}{k_s}, \quad T > T_L$$

The appropriate form of equation (2) for any volume element is substituted into the finite difference representation of the heat conduction equation given in reference [8]. After some manipulation and multiplication by different factors a common implicit heat balance equation similar to that in reference [4] is obtained for all the cases.

Results of Computer Simulations

Calculations for the Al-4.5 wt% Cu alloy verified earlier findings [4] that for the same value of $\underline{qa}$* (W/m) isotherms, including the liquidus (923K) and eutectic (821K) isotherms delineating the "mushy" zone, are located at the same dimensionless distance for identical values of temperature at T(0,0) or identical Fourier numbers.** The shape, location and velocity of the

---

* $\underline{a}$ is the radius of the circular region through which a uniform heat flux $\underline{q}$ is absorbed. $\underline{qa}$ physically indicated power per unit length.
** Fourier number, $Fo = \alpha_s t/a^2$ where α_s is the thermal diffusivity of the solid and t is time.

liquidus and eutectic isotherm can be conntinuously monitored for different
values of $\underline{qa}$ as described in Reference [8]. Calculations also showed that
minimum values of $\underline{qa} \simeq 9.44 \times 10^4$, and $\simeq 3.4 \times 10^5$ W/m are needed if the
surface temperature $T(0,0)$ is to reach T_E (the eutectic temperature), and T_v
(the vaporization temperature), respectively [8]. Clearly then, for the
establishment of steady state melt depths $qa \leq 3.4 \times 10^5$ W/m. Furthermore,
steady state is achieved at long times [8], or for values of $a/2\sqrt{\alpha_s t} \lesssim 0.25$.
Figure 1 compares the steady state shapes and locations of the "mushy" zone
for two values of the product $\underline{qa}$. Both values are within the range noted
above. While the size of the "mushy" zone does not significantly change
with increasing values of $\underline{qa}$, Figure 1 shows that the melt depth more than
doubles as this product is increased from 2×10^5 to 3×10^5 W/m. The
corresponding maximum calculated temperatures $T(0,0)$ are 1400K and 2356K,
respectively.

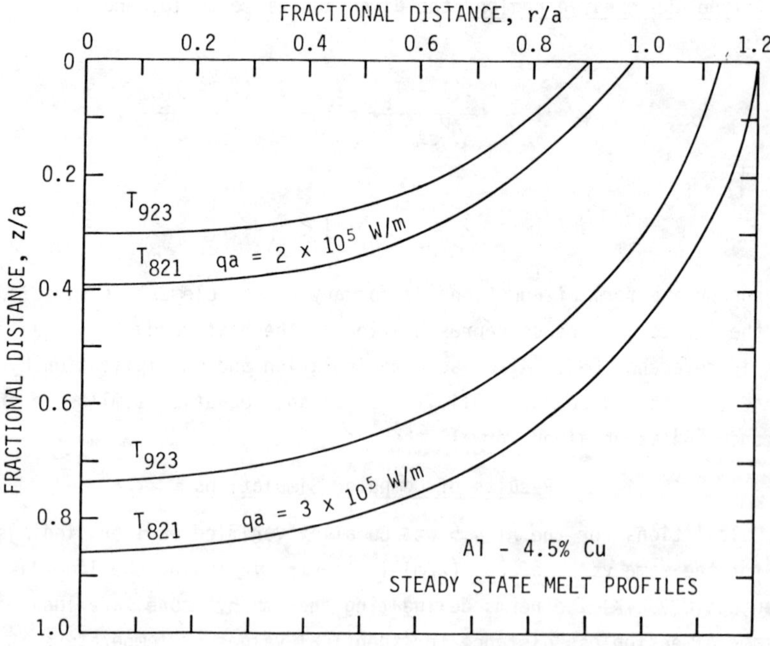

Figure 1. Steady state shapes and locations of the "mushy" zone in an
Al-4.5 wt% Cu substrate subjected to two different values of the
product $\underline{qa}$.

Figure 2 shows the size and location of the "mushy" zone as a function of dimensionless time for the product $\underline{qa}$ = 2.0 x 10^5 W/m. These curves were generated by turning off the beam power once a steady state thermal profile was established. Vertical distance between the curves denotes the size of the "mushy" zone. The data show that loss of superheat and shallower temperature gradients during solidification result in significantly large "mushy" zone sizes. For example, the "mushy" zone is almost as large as the maximum depth of the liquid metal pool when the substrate surface reaches its liquidus temperature during solidification. In other words, during initial stages of solidification the velocity of the eutectic isotherm along the z-axis is significantly slower than the liquidus isotherm. Average cooling rates during

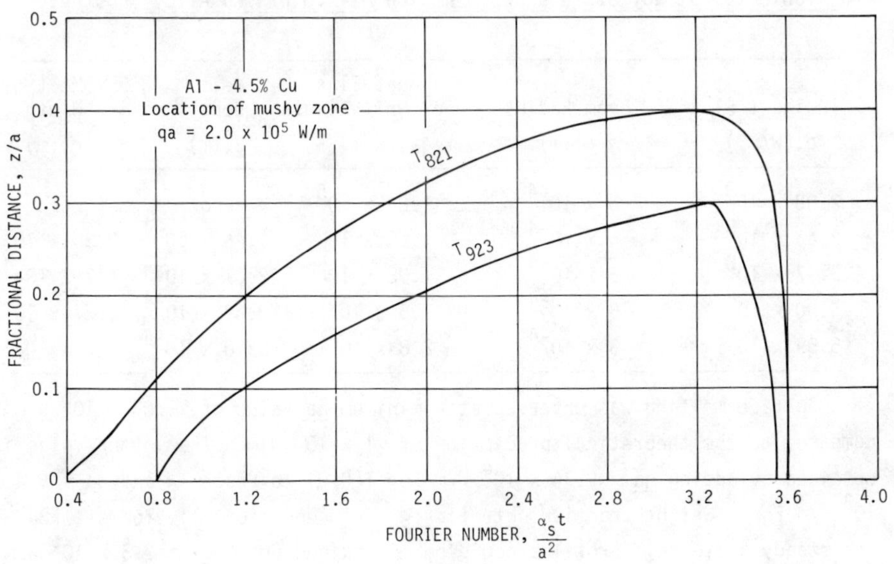

Figure 2. The dimensionless size and location of the "mushy" zone along the z-axis versus time for the product $\underline{qa}$ = 2.0 x 10^5 W/m. T_{923} and T_{821} denote the locations of the liquidus and solidus isotherms, respectively.

solidification are ready deduced from those type of computer calculations. For example, for this value of $\underline{qa}$ when $\underline{a}$ = 500 µm ($\underline{q}$ = 4 x 10^8 W/m^2) the calculated average cooling rates, deduced from Figure 2, at the bottom (z = 150 µm) and the top (z = 0) of the metal pool are ≈1.2 x 10^5 K/s and ≈6.2 x 10^5 K/s, respectively. These cooling rates are calculated by dividing the solidification temperature range of the alloy $\Delta T_{s\ell}$ = 102K by the difference in time from Figure 2 for the passage of both isotherms across a given location. Note that cooling rate increases as solidification proceeds toward the surface, therefore, finer structures are expected at this location.

Comparison of Experiments to Theory

The experiments were conducted on a 10kW electron beam machine. For these experiments, the incident powers were chosen such that steady state melt profiles could be obtained for long times. A limiting condition to steady state was obtained when the melt surface reached and far exceeded the vaporization temperature of the Al-4.5 wt% Cu alloy substrates under the 10^{-5} torr vacuum. Table I lists the various experiments conducted with the heat input maintained in the calculated steady state regime.

Table I. Steady State Experiments on Al-4.5 wt% Cu Alloy Substrate

Incident Flux $\underline{q}$,(W/m^2)	Beam Radius $\underline{a}$,(m)	Heat Flux Per Unit Length $\underline{qa}$,(W/m)	Melt Depth $\underline{z}$,(m)	Melt Width $\underline{r}$,(m)
2.08 x 10^8	5 x 10^{-4}	1.04 x 10^5	Zero	Zero
2.43 x 10^8	5 x 10^{-4}	1.22 x 10^5	1.46 x 10^{-4}	2.38 x 10^{-4}
3.97 x 10^8	5 x 10^{-4}	1.98 x 10^5	2.01 x 10^{-4}	4.7 x 10^{-4}
5.00 x 10^8	5 x 10^{-4}	2.5 x 10^5	2.69 x 10^{-4}	5.77 x 10^{-4}
5.59 x 10^8	5 x 10^{-4}	2.8 x 10^5	23.0 x 10^{-4}	7.7 x 10^{-4}

Surface melting was observed at a minimum $\underline{qa}$ value of ∿1.05 x 10^5 W/m compared to the theoretical prediction of ∿1 x 10^5 W/m. The latter value is obtained by adding $\underline{qa}$ = 9.44 x 10^4 W/m for T(0,0) to reach T_E and $\underline{qa}$ = 5.4 x 10^3 W/m for a weight fraction eutectic f_E, from equation (1), to melt [8]. The steady state melt profiles could be maintained up to $\underline{qa}$ ∿ 2.5 x 10^5 W/m before excessive vaporization led to a deep penetration condition. Figure 3 shows a typical steady state melt profile obtained for $\underline{qa}$ = 1.22 x 10^5 W/m.

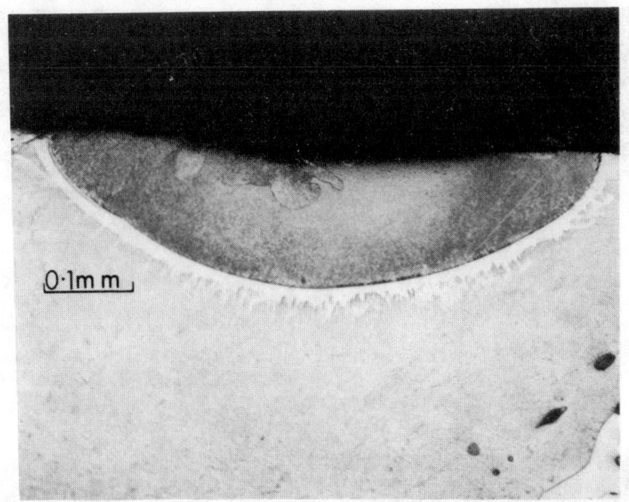

Figure 3. Macrograph showing the steauy state melt profile obtained on the
surface of an Al-4.5 wt% Cu alloy substrate. The E.B. conditions
were, q = 2.43 x 10^8 W/m^2, a = 500 μm, qa = 1.22 x 10^5 W/m.

Microstructural features include an uneven, saw-tooth, melt zone indicating
melting of the substrate over a range of temperatures. Of course, the existing
"mushy" zone at steady state is partially obscured by the solidification
microstructure. The light gray area outlining the melt profile appears
featureless at this magnification. Some macroscopic perturbation of the melt
surface is noted. This is probably due to a combination of surface evaporation, surface tension gradients and the induced pressure from the beam of
electrons.

Figure 4 shows photographs of the melt profiles obtained under the
various qa conditions listed in Table I. First, increasing the qa values
results in larger melt pools while the melt geometry remains essentially the
same, Figures 4(a) to 4(c). The relative size of these steady state melt
geometries, are plotted in Figure 5. Figure 4(d) shows the deep penetration
mode achieved at the higher power input.

Experimental steady state melt profiles are compared with theoretical
predictions in Figure 6 for two different values of the product qa. Agreement
between the two is reasonably good considering the various assumptions of the
calculations including constant but different thermophysical properties for
the solid, the "mushy" zone and the liquid. The calculations are based on
complete absorption of the incident energy by the substrate. This is a reasonable assumption for electron beam surface melting.

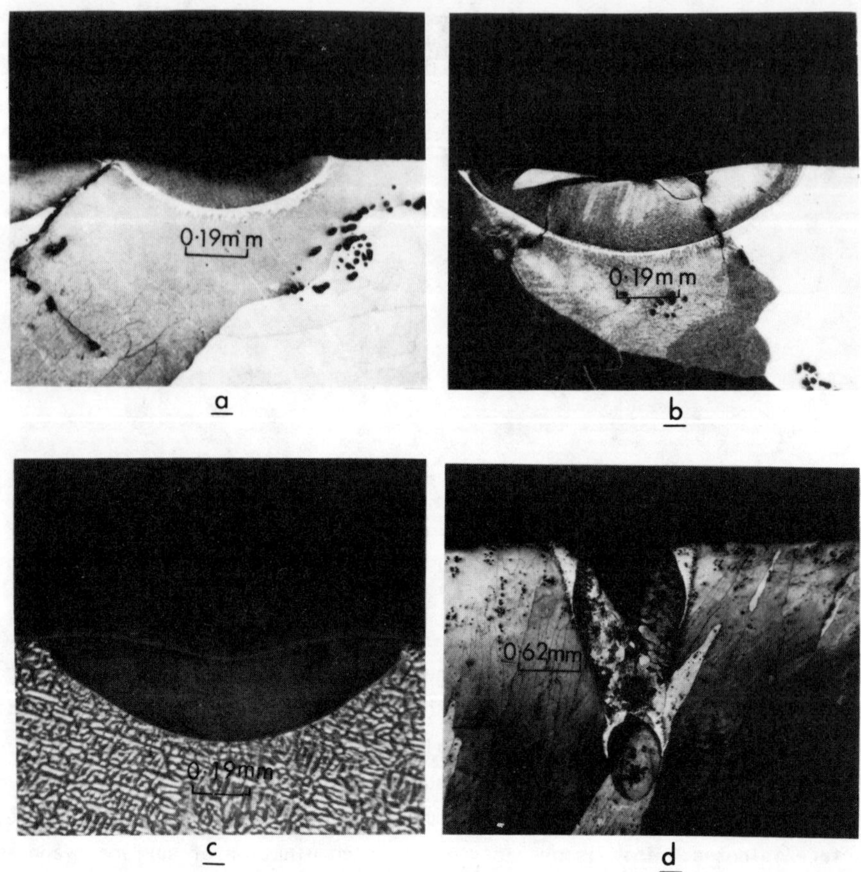

Figure 4. Macrographs showing steady state and deep penetration melt geometries in the Al-4.5 wt% Cu substrate. Conditions for the E.B. were those listed in Table I; (a) $\underline{qa}$ = 1.22 x 10^5, (b) $\underline{qa}$ = 1.98 x 10^5, (c) $\underline{qa}$ = 2.5 x 10^5, and (d) $\underline{qa}$ = 2.8 x 10^5 W/m.

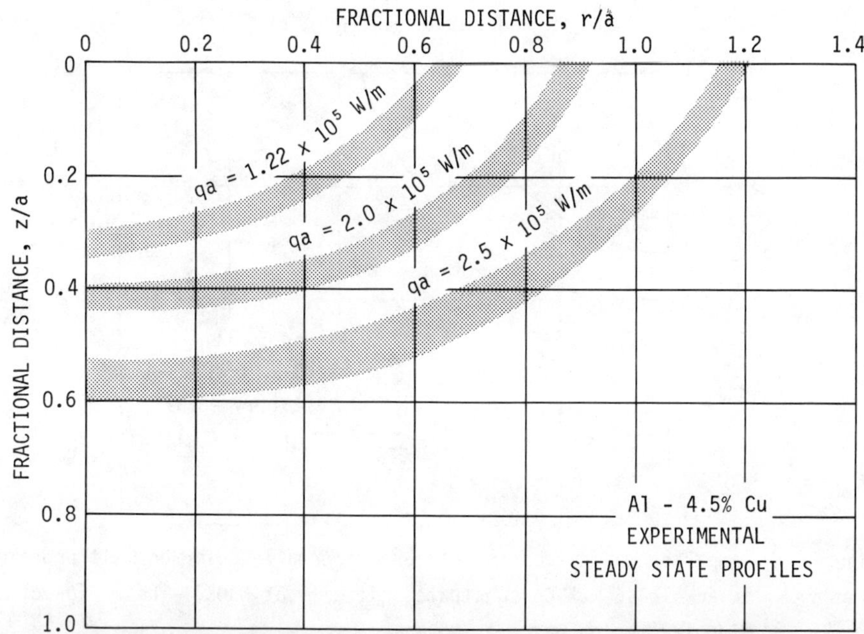

Figure 5. Comparison of the various melt profiles and partially observable "mushy" zones obtained on the Al-4.5 wt% Cu alloy.

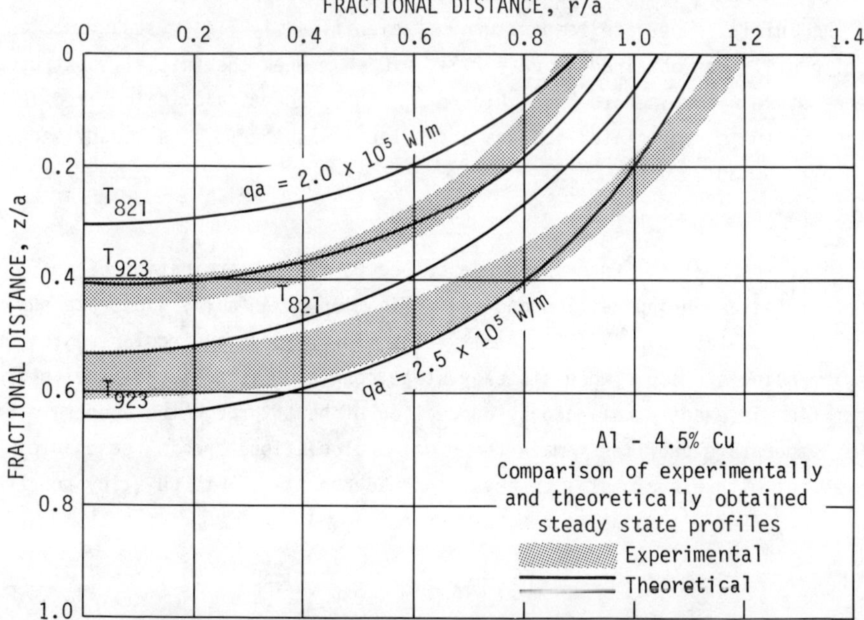

Figure 6. Comparison of calculated and experimentally observed steady state melt profiles in Al-4.5 wt% Cu alloy substrate. Substrates were initially at room temperature and $\underline{a} \simeq 500\mu$m.

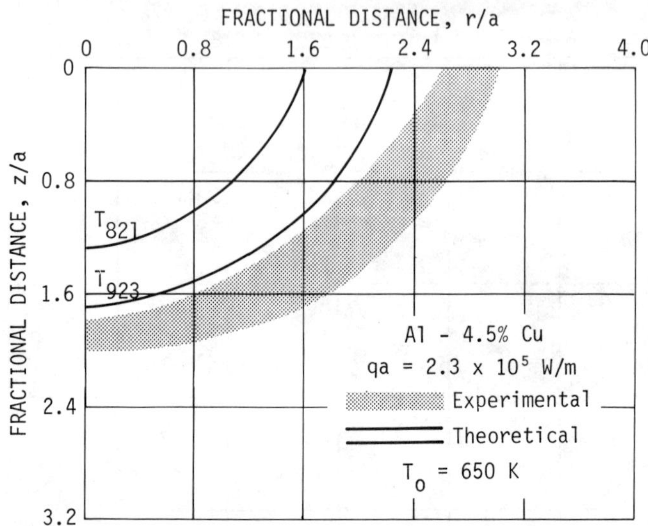

Figure 7.   Comparison of calculated and experimentally observed melt profiles
in an Al-4.5 wt% Cu substrate initially at 650K. The radius of the
circular region was $\underline{a}$ = 500µm.

An increase in the steady state melt depth is expected with an increase
in the initial substrate temperature. A significantly larger steady state
pool profile was obtained at $\underline{qa}$ = 2.31 x 10^5 W/m when the initial substrate
temperature was raised to 650K, Figure 7. The theoretically calculated locations of the steady state liquidus and eutectic isotherms, the "mushy" zone,
are off by ∿30%.

Summary

The mathematical formulation described permits determination of the
thermal fields during melting and solidification of an alloy substrate subjected to a stationary high intensity heat flux through a circular region on
its bounding surface. When a steady state condition is achieved, the absorbed
heat flux is exactly balanced by conduction of heat through the substrate and
the temperature profiles remain the same. Calculations are in reasonable
agreement with experimental observations made on Al-4.5 wt% Cu alloy substrates
surface melted with an electron beam source.

Acknowledgements

This work was jointly supported by the National Bureau of Standards and the Defense Advanced Research Projects Agency under DARPA Order No. 4275. Technical monitor of the latter contract is Lt. Col. L. A. Jacobson. The authors thank Dr. R. J. Schaefer and Mr. F. S. Biancaniello of the National Bureau of Standards for their valuable guidance and help in preparation of the electron beam melted specimens.

References

1. M. I. Cohen, *J. Franklin Inst.*, **283** (4) (1967) p. 271.
2. H. E. Cline and T. R. Anthony, *Journal of Applied Physics*, 48 (9) (1977).
3. S. C. Hsu, S. Chakravorty, and R. Mehrabian, *Metallurgical Transactions B*, **11B** (1978) p. 221.
4. S. C. Hsu, S. Kou and R. Mehrabian, *Metallurgical Transactions B*, **11B** (1980) p. 29.
5. D. S. Gnanamutha, C. B. Shaw, Jr., W. E. Lawrence, and M. R. Mitchell, *Laser-Solid Interactions and Laser Processing - 1978*, p. 173; AIP, 1979.
6. J. Mazumdar and W. M. Steen, *Journal of Applied Physics*, **51** (2), (1980).
7. S. Kou, S. C. Hsu, and R. Mehrabian, *Metallurgical Transactions B*, **12B** (1981) p. 33.
8. J. A. Sekhar, S. Kou, and R. Mehrabian, submitted for publiication in Metallurgical Transactions B.
9. M. C. Flemings, *Solidification Processing*, Chapter 2; McGraw-Hill, Inc., New York, NY, 1974.
10. G. E. Schneider, A. B. Strong, and M. M. Yovanovich, *Proc. of International Symposium on Computer Methods of Partial Differential Equations*, R. Vichnevetsky, ed.; pp. 312-317, AICA, New Brunswick, NJ, 1975.

LASER WELDING: STATE OF THE ART REVIEW

J. Mazumder

Department of Mechanical and Industrial Engineering
1206 W. Green St.
University of Illinois at Urbana-Champaign
Urbana, Illinois 61801

Abstract

From the initial development, the laser has been hailed as a
potentially useful welding tool for a variety of applications. The scope
for technically and commercially feasible laser welding applications has
increased greatly since the development of multikilowatt CW CO_2 lasers
around 1970. The laser's capability for generating a power density greater
than 10^6 watts/cm^2 is a primary factor in establishing its potential for
welding. Numerous experiments have shown that the laser permits precision
weld joints of a high quality only rivalled by an electron beam. The up to
date work on laser welding will be reviewed in order to determine the state
of the art.

Introduction

The scope for technically and commerically feasible laser welding applications has increased greatly since the development of multikilowatt CW CO_2 lasers around 1970.[1-3] Presently, commercially available CW CO_2 lasers range from a output power of few watts to 15 KW. Recently high power (up to 400 W CW) Neodymium-Yttrium aluminum garnet (Nd-YAG) laser are also commerically available. The availablity of the high power lasers and need for precision joining techniques for modern technology lead to the development of various deep penetration laser welding applications.

The potential advantages[4,5] of laser welding are the following.

(1) Light is inertialess; hence high processing speeds with very rapid stopping and starting becomes possible.
(2) Focused laser light can have high energy density.
(3) Laser welding can be used in room atmosphere.
(4) Difficult materials (e.g. titanium, quartz, etc.) can be handled.
(5) The workpiece need not be rigidly held.
(6) No electrode or filler materials are required.
(7) Narrow welds can be made.
(8) Very accurate welds are possible.
(9) Welds with little or no contamination can be produced.
(10) The heat affected zone (HAZ) adjacent to the cut or weld is very narrow.
(11) Intricate shapes can be cut or welded at high speed, using automatically controlled light deflection techniques.
(12) Time sharing of the laser beam can be achieved.

The localized heating obtained with laser sources was soon realized to be an important advantage. Anderson and Jackson[6] reported an interesting comparison between heating effects produced with a conventional arc source and those occurring with a pulsed laser. They show that not only is the heat affected zone small but also the laser source is more efficient since it requires only 37.8 J/cm^2 to produce melting to the required depth, while the arc source must deliver 246 J/cm^2 to the workpiece. Localized heating makes the laser ideal for welding electronic components such as printed circuit boards where high average temperatures even in small volumes surrounding the weld region cannot be tolerated.[7]

The penetration depth during laser welding is less than that observed during Electron Beam Welding (EBW). The maximum thickness welded by laser is reported to be by 2 in. thick steel plate using a 77 KW CO_2 laser[8], whereas EBW can produce welds up to several inches deep. However, the deep penetration of EBW extends only a short distance outside a vacuum. So far, optimum efficiency electron beam welding needs to be done in a vacuum. By contrast, laser beams can be transmitted for appreciable distance through the atmosphere without serious attenuation or optical degradation. The laser thus offers an easily maneuvered, chemically clean, high intensity, atmospheric welding process producing deep penetration welds (aspect ratio greater than 1:1) with narrow HAZ and subsequent low distortion.

Principles of Laser Welding

The mechanism of deep penetration welding by a laser beam is very similar to that encountered with an electron beam, i.e., energy transfer via "keyhole" formation.[7,9,10] This keyhole may be produced when a beam of sufficiently high power density causes vaporization of the substrate and the pressure produced by the vapor in the crater causes displacement of the

molten metal upwards along the walls of the hole. This hole acts as
a blackbody and aids the absorption of the laser beam as well as
distributing the heat deep in the material. On the other hand, in most
conventional welding processes the energy is deposited at the surface of the
workpiece and is brought into the interior by conduction.

The conditions of energy and material flow during beam welding were
investigated theoretically by Klemens.[11] According to Klemens[11] the
"keyhole" or cavity is formed only if the beam has sufficient power
density. The keyhole is filled with gas or vapor created by continuous
vaporization of wall material by the beam.[11] This cavity is surrounded by
liquid which in turn is surrounded by solid.

The flow of the liquid and surface tension tend to obliterate the
cavity while the vapor which is continuously generated tends to maintain the
cavity. There is a continuous flow of material out of the cavity at the
point where the beam enters. For a moving beam, this "keyhole" achieves a
steady state, i.e., the cavity and the beam associated molten zone move
forward at the speed set by the advance of the beam, while the material lost
by vaporization shows up as a depression in the solidified melt as porosity
or as an inward deformation of the workpiece, or possibly as a combination
of these effects. The requirement that sufficient vapor be produced to
maintain a steady state leads to a minimum advance speed for a steady
state. While the cavity moves through the solid and liquid material at a
speed determined by the motion of the beam, material must be moved
continuously from the region ahead of the cavity to the region behind it.
This is confirmed by experimental works of Sickman, et al.[12-13] and Arata,
et al.[14]

Fluid flow during penetration welding was also studied by Arata,
et al.[14] using laser beam irradiation of a glass with low viscosity at
high temperature. High speed photography of the phenomenon, taken at a
speed of 8,000 frames per second, clearly showed a melt flow and the motion
of the cavity. Molten fluid, formed at the front wall of the cavity, was
accelerated at an angle along the wall by the violent driving force of the
laser driven vaporization. In the process, a large vortex was formed behind
the cavity near the surface (Fig. 1). This is considered to be the cause of
the generation of so called "wine-glass" beads.

The transport of material is mainly by flow in the liquid. However,
part of the transport is in the vapor phase, and the vapor transport
provides the excess pressure which drives the liquid flow.

Laser Welding Variables

The major independent process variables for laser welding include the
following:

(1) incident laser beam power
(2) incident laser beam diameter
(3) absorptivity
(4) traverse speed of the laser beam across the substrate surface.

Parameters such as weld design, shielding gas, gap size for butt welds and
depth of focus with respect to the substrate also pay important roles.

The dependent variables are considered to be (a) depth of penetration,
(b) microstructure and metallurgical properties of the laser welded
joints. Effect of the some of the important variables are discussed in
brief.

(1) <u>Laser Beam Power</u>
        The depth of penetration during laser welding is directly related
to the power density of the laser beam and is a function of incident beam
power and beam diameter. Generally, for a constant beam diameter,
penetration increases with the increasing beam power. Lock, et al.[15] and
Baardsen, et al.[16] report that penetration increases almost linearly with
incident laser power. Generally it is observed that for a laser welding of
a particular thickness, a minimum threshold power is required.

        The effect of power (up to 77 KW) on butt weld penetration for 304
stainless steel is studied by Banas[8]. The qualitative observations made
by Banas[8] conforms the expression maximum penetration α Power[0.7].

(2) <u>Laser Beam Diameter</u>
        This parameter is one of the most important variables because it
determines the power density. However, it is very difficult to measure for
high power laser beams. This is partly due to the nature of the beam
diameter and partly due to the definition of what is to be measured. A
Gaussian beam diameter may be defined as the diameter where the power has
dropped to $1/e^2$ or $1/e$ of the central value. The beam diameter defined on
the basis of $1/e^2$ of the central value contains more than 80 percent of the
total power whereas the power contained for $1/e$ beam definition is slightly
over 60 percent [17]. Therefore, the $1/e^2$ beam diameter is recommended.

        Many techniques have been employed to measure the beam diameter, but
most are unsatisfactory in some respect or another. Single isotherm
contouring techniques such as charring paper and drilling acrylic or metal
plates, suffer from the fact that the particular isotherm they plot is both
power and exposure time dependent. They are also highly unlikely to
coincide with either the $1/e$ or $1/e^2$ diameters. Multiple isotherm
contouring techniques overcome these difficulties but are tedious to
interpret.

        One of the better methods for the measurement of beam diameter is the
photon drag detector[18-21], this has limitations too. To convert the
signal from this instrument, it is necessary to assume the beam is
axisymmetric and this might not always be strictly correct. However, this
assumption becomes critical if one makes an attempt to interpret the spacial
mode distribution of the beam. For simple beam diameter measurement, this
is not a problem. Although to date this device has only been used for TEM_{00}
or lower order mode beam it may also be used for beams with other types of
spacial distribution. Previous work by the author[18] revealed that the
response time for the photon drag dectector is fast enough to determine an
accurate beam diameter

(3) <u>Absorptivity</u>
        The efficiency of laser welding depends on the absorption of light
energy by the workpiece. Any heat transfer calculation for laser processing
is based on the energy absorbed by the workpiece.

        The infrared absorpiton of metals largely depends on conductive
absorption by free electrons. Therefore, absorptivity is a function of the
electrical resistivity of the substrate material. Arata, et al. [5,22]
measured the absorptivity of polished surfaces of various materials and
concluded that absorptivity is proportional to the square root of the
electrical resistivity. This agrees closely with the following formula.

$$A = 112.2 \sqrt{\rho_r} \tag{1}$$

where, A = absorptivity ρ_r = electrical resistivity.

A temperature dependent relationship between the electrical resistivity and emmissivity of the metal was derived by Bramson.[23] Bramson's formula can be used for theoretical calculation of the absorptivity from the electrical resistivity since absorptivity equals emmissivity. However, such a calculation will be valid only for metals heated in vacuum without a surface oxide layer. The presence of an oxide layer will increase the absorptivity. The relationship between the emmissivity and the electrical resistivity of a substrate, for perpendicular incidence of radiation of long wavelength, derived by Bramson is,

$$\epsilon_\lambda(T) = 0.365 \left[\frac{\rho_r(T)}{\lambda} \right]^{1/2} - 0.667 \left[\frac{\rho_r(T)}{\lambda} \right] + 0.006 \left[\frac{\rho_r(T)}{\lambda} \right]^{3/2} \tag{2}$$

where $\rho_r(T)$ is the electrical resistivity at absolute temperature T expressed in Ohm-cm.
$\epsilon_\lambda(T)$ is the emmissivity of the substrate at T°C temperature for radiation having a wavelength of λ.

The estimated absorptivity for Ti-6Al-4V at 300°C using Bramson's formula is around 15 percent. Experimental data published by Arata, et al.[5,22], indicate that the absorptivities of Al, Ag and Cu are 2 and 3 percent and those of stainless steel (304), Fe and Zr, are below 15 percent even in the molten state. This means that reflection losses are particularly great. Therefore, when sheet metal is welded various measures must be taken when required to avoid reflection. Applying absorbent powder or forming an anodized film, for example, are considered to be very effective.[5]

Absorptivity can also be increased by the use of reactive gas. Jørgensen[24] reported that the addition of 10 percent of oxygen to argon shielding gas gave an increase of up to 100 percent in welding depth. Jørgensen[24] also found that gas flow had no significant effect on weld depth but an increase in depth was associated with a decrease in reflectivity which was achieved by the addition of a small amount of oxygen.

Although metals are poor absorbers of infared energy at room temperature, above a certain threshold (approximately 10^6 - 10^7 W/cm^2) energy transfer via the "keyhole" leads to much higher effective absorptivity. Once a keyhole has formed absorptivity increases rapidly. Multiple internal reflections inside a "keyhole" are responsible for the deep penetration welding by the laser in spite of the large convergence angles for laser beams compared to electron beams.[2] However, the threshold energy required for keyhole formation in laser welding is higher than that required for an electron beam (1.5 x 10^5 W/cm^2) because of the poor absorptivity. Nevertheless, energy transfer by this "keyhole" mechanism permits laser to perform efficient welding of even highly reflective materials such as aluminum.

(4) <u>Traverse Speed</u>
The correlation of penetration depth with laser welding speed is discussed by Duley[7] (p. 241) and Locke, et al.[15] in comparison with that of electron beam welding. It is evident from Fig. 2 that the penetration in the laser weld is consistently less than that possible with an electron beam, but the difference between the two penetration depths diminishes as the welding speed increases. But Duley[7] (p. 241) found that this was

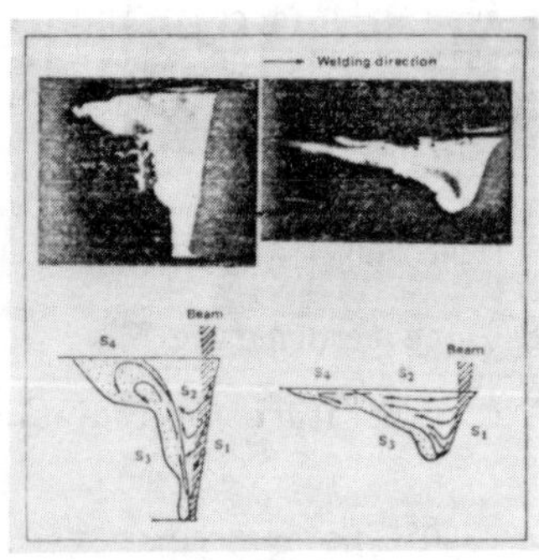

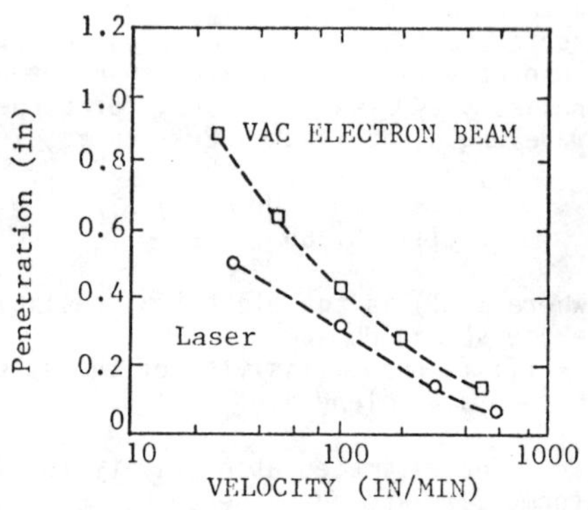

Fig. 1   The weld cavity and movement of the melt

Fig. 2   Weld penetration vs. weld speed for type 304 stainless steel welded with a $10kW-CO_2$ laser.(15)

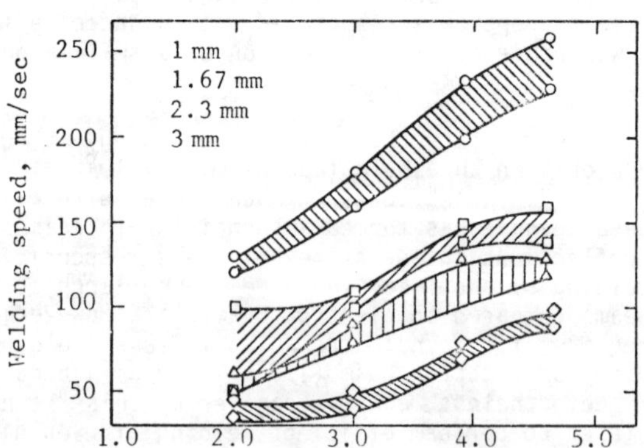

Fig. 3   Welding speed against power for Culham/ Ferranti 5 kW CO_2 (28)

somewhat surprising, since as pointed out by Baarsden, et al.[16] that the time to form a void or "keyhole" may become comparable to the illumination time for a particular area on the surface of the workpiece as the welding speed increases. When this occurs, the average power dissipated in the sheet is expected to drop because the keyhole is no longer a completely effective trap for the incident laser radiation. For an electron beam, the absorptivity of the material is independent of the shape and extent of the keyhole and hence the total power dissipated in the workpiece will be less strongly dependent on the welding speed. However, Crafer[25] reports that the keyhole penetration threshold for laser or electron beams of radius 0.1 mm incident on a steel surface of thermal diffusivity $\simeq$ 10 mm^2/sec. is achieved in 1 millisec. This could be regarded as instantaneous. Again, for very low welding speeds, the penetration depth of laser welds becomes significantly less than that attainable with the electron beam. According to Lock, et al.[26-27] this can be attributed for the formation of a plasma cloud, which attenuates the incident beam.

Figure 3 shows that for a particular power, welding can be performed over a range of thickness.[28] Beyond that range full penetration welding is not possible. Higher speeds may lead to improper penetration whereas lower speeds lead to excessive melting, loss of material and weld perforation. The range of welding speed (i.e., range between maximum and minimum welding speed for successful welding decreases with increase of thickness[28], as previously noted by Adams.[29]

Applications

A large variety of metals and alloys have been welded with pulsed or CW lasers. Until the first reported demonstration of deep penetration LBW by Brown and Banas,[33] most laser welds were performed with pulsed lasers. Several reviews of pulsed laser welding are available.[7,34,35,36]. However, penetration welding of metallurgical interest are reviewed here.

Penetration Welding

The scope of technically and commercially feasible laser welding applications has increased greatly since the demonstration of penetration welding using a multikilowatt CW CO_2 laser by Brown and Banas.[33] Numerous experiments have shown that the laser permit manufacture of precision weld joints of high quality for ferrous alloys, nickel alloys and Titanium alloys rivalled only by those made by EBW. Data describing the plate thickness of different laser welded materials are given in Table 1.

Aluminum and Its Alloys

Although the use of high power, CO_2 lasers for welding titanium, stainless steel, alloy steels is well researched, their application to aluminum welding did not attract the attention it deserved. To date, very limited data are available in open literature.[45-47] One of the reasons is that aluminum alloys have proven to be very difficult to weld is the high initial surface reflectivity for 10.6 μm radiation from CO_2 lasers.

The laser welding study of aluminum alloys 5456 and 5086 by Snow, et al.[47], revealed that these two aluminum alloys appear to differ in their welding response to a significant degree, both in ability to penetrate under a given set of conditions and also in bead appearance. Penetration in 5456 was substantially greater, the difference being primarily attributed to the 1.5% greater magnesium content of this alloy. Porosity was present in unacceptable amounts in all weld specimens Excessive "drop-through" of the

Material	Thickness		Laser power, kW
	in	mm	
Ship steel, grades A, B, C	1.125 butt	28.6	12.8
	1.0 butt	25.4	12.0
	0.75 butt	19.0	12.0
	0.625 butt	15.9	12.0
	0.50 butt	12.7	12.0
	0.375 butt	9.5	10.8
	0.375 to 0.5 tee	9.5 to 12.7	11.9
Low-alloy carbon steel	0.375 to 0.5 tee	9.5 to 12.7	7.5
AISI 4130	0.60 butt	15.2	14
Low-alloy high-strength steel 300M	0.75 butt	19.0	14
Arctic pipeline steel X-80	0.52 butt	13.2	12
D6AC steel	0.25 butt	6.4	15
	0.50 butt	12.7	15
HY-80	0.49	12.5	10.6
HY-130 steel	0.25 butt	6.4	5.5
HY-180 steel (HP9-4-20)	0.062 butt	1.6	5.5
	0.062	1.6	10.5
	0.64	16.3	5.5
Nickel-base alloy--Inco 718	0.57 butt	14.5	14
Stainless steel (AISI 304)	0.24	6	17
	0.49	12.5	17
	0.67	17	17
	0.65	16.5	17
	0.15	3.8	11.5
	0.22	5.6	11.5
	0.35	8.9	11.5
	0.48	12.3	9.5
	0.35	8.9	8
	0.8	20.3	20
	0.5	12.7	20
Stainless steel (AISI 321)	0.57 butt	14.5	14
Aluminum alloy:			
2219	0.50 butt	12.7	13
2219	0.25 butt	6.4	5
2219	0.25 butt	6.4	16
5456	0.125 butt	3.2	5.5
5456	0.187 butt	4.7	5.5
5456	0.49 butt	12.5	8
Titanium alloy 6A14V	0.60 butt	15.2	13.5
	0.125, 0.25	3.2, 6.4	5.5
	0.12	3	4.7
	0.49	12.5	11

bead was encountered in all full penetration welds.[47] This problem is related to liquid metal viscosity and surface tension. The interaction of shielding gas and/or plasma with the beam and the workpiece is very much a part of this phenomenon. Snow, et al.[47], concluded that aluminum alloys are very sensitive to the intensity of the input energy and the welding variables.

Breinan, et al., achieved some success in laser welding of aluminum alloys 2219 and 5456. Both of these materials have been welded in thicknesses of up to 9.5 mm with acceptable microstructure and bead profile but porosity has been a problem. Tensile tests on these welds resulted in failure by diagonal shear through the welds at an average strength of 342 MPa (49.7 ksi) where as the ultimate tensile strength value for the parent metal was 345 MPa (50 ksi).[48] Bend tests results of laser butt-welds were acceptable.[49] Bend test results for laser fillet weld joints of aluminum alloy 5456 (4.7 mm thick) exhibited sharper bend radii without fracture of the weld[47] when compared with those of similar gas metal welds (GMAW).

Metzbower and Moon[46] studied the mechanical properties of laser welded 12 mm thick aluminum alloy (5456). Tensile tests were perforemd on a standard ASTM round specimen 6.4 mm in diameter. The weld was transverse to the loading direction in the tests. All of the aluminum laser welds failed in the weld zone. The ductility of these specimens was low and the amount of porosity visible on the fracture surfaces was high. They used a non-standard 12 mm thick dynamic tear (DT) test specimens of laser welded 12 mm thick aluminum alloy (5456). Tensile tests were perforemd on a standard ASTM round specimen 6.4 mm in diameter. The weld was transverse to the loading direction in the tests. All of the aluminum laser welds failed in the weld zone. The ductility of these specimens was low and the amount of porosity visible on the fracture surfaces was high. The mechanical properties data obtained by Metzbower and Moon[46] are given in Table 2. They used a non-standard 12 mm thick dynamic tear (DT) test specimen with the notch at the weld for their fracture toughnes tests (the standard size is 16 mm). Porosity was also observed in DT fracture surfaces.

Mazumder[50] studied the feasibility of laser welding of three aluminum alloys (2036, 5182, 6009). A 1350 watt CW CO_2 laser was used for the bead on plate and lap welding of 1 mm thick aluminum plates. It was found that the composition of the alloy determined whether the irradiation parameters were critical. Mazumder[50] reportd that it is possible to produce crack-free laser welds with optimum irradiation parameters for aluminum alloy 2036. Aluminum alloy 5182 was found to be the easiest to weld compared to Al 2036 and Al 6009. Laser irradiation parameters were not very critical for successful welds but a pronounced loss of magnesium by evaporation was observed. Aluminum alloy 6009 demonstrated very poor weldability. Laser irradiation parameters required for successful weld seemed to be extremely critical. An excessive precipitation of Al-Mg-Si was observed and may be responsible for the cracking of these welds. However, additional cooling of the irradiated material by a chill plate facilitated successful welding.

A systematic study of Laser Welding of 1/4 in. thick Al-Mg alloy (Al-5083) is being carried out by Blake and Mazumder.[62] They used a 10 KW CO_2 laser and a gas shielding system where plasma formed during laser materials interaction is pushed into the "keyhole". This experimental approach successfully produced apparently porosity free welds (Fig. 4).

TABLE 2
Mechanical Properties of a Laser Welded Aluminum Alloy[46]

0.2% Yield Strength MPa	Ultimate Tensile Strength MPa	% Elongation	% Reduction in Area	Average Dynamic Tear (DT) Energy (Nm)	Range of DT Energy (Nm)
193	289	9.2	16.9	249	176-311

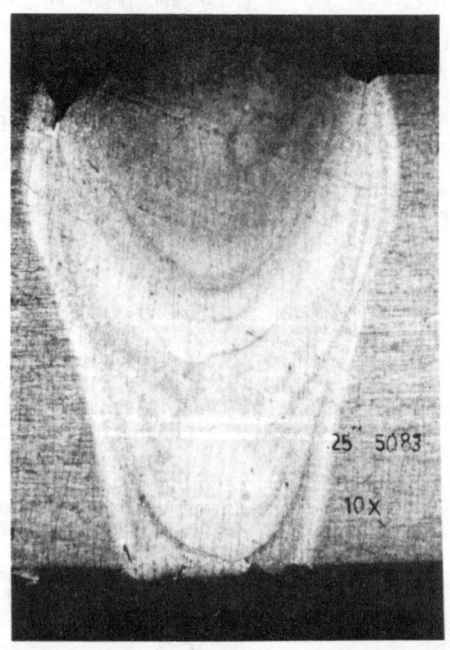

Fig. 4   Macrostructure of a laser welded 1/4 in.
thick aluminum alloy (Al-5083). Laser power
6 kW. Traverse speed 30 ipm. Gas nozzle
diameter 1/8 in. Nozzle angle 51°.

Steels

A great deal of laser welding effort has been expended on ferrous alloys. Most of the initial parametric laser welding studies were carried out on stainless steel.[15,26,33] This was investigated by various researchers[51,52] due to its importance in the power plant and chemical industries. Laser welding of rimmed steel sheet was evaluated for automotive applications by Baardsen, Schmatz, and Bisaro.[16] High speed laser welding of tin plate and tin free steel for container industries was reported by Mazumder and Steen.[18,53]

Results in HY-alloy steels were evaluated and reported by Banas[59] and Metzbower.[46] Laser welding of modified 4340 alloy was reported by Seaman and Hella.[59] Several other ferrous alloys studied for the applicability of laser welding include X-80 Arctic pipeline steel[55], tanker construction steels[56], D6AC[58], and high strength, low alloy (HSLA) grade steels.[58]

High-power laser welding experiments with 300 series stainless steel were conducted by Banas.[53] He reported that welds were formed in stainless steel with aspect ratios (depth-to-width ratios) as large as 12:1. Data were obtained in a series of bead-on-plate penetration tests conducted on samples exposed to the ambient atmosphere using laser powers up to 5.5 kW. It was concluded that the depth-of-penetration at cosntant laser power is a relatively weak function of welding speed.

The effect of power (up to 77 kW) on butt weld penetration for 304 stainless steel is studied by Banas.[8] The qualitative observation by Banas[8] is that the maximum penetration is porportional to the laser power to the 0.7 power.

Radiographic inspection of selected laser welds in stainless steel has shown that high-density, non-porous welds can be achieved.[59] Tensile tests of stainless steel welds performed by Banas[59] have shown that the joint strength can, with appropriate selection of weld parameters, equal that of the parent material. Similar observations were made by Wilgoss, et al.[51], for stainless steel 316 but data obtained for stainless steel 310 were less encouraging.

Wilgoss, et al.[51], performed tensile tests on laser butt welded stainless steel 316, 310, and ferritic steel DUCOL W30. Flat tensile specimens were used with the plane of the weld running transversely across the gauge length at its center. The room temperature uniaxial tensile tests were performed at a Crosshead velocity of 1.6 mm/sec corresponding to a strain rate of 2.8×10^{-2}/sec. Results (from Ref. 51) for the tensile tests are given in Table 3. Wilgoss, et al[51], evaluated laser welding of 6 mm thick stainless steel 316 and DUCOL W30 against TIG, plasma arc, and electron beam welding methods. Criteria for the comparison is given in Table 4. It is evident from the comparison that for high productivity and better weld quality electron beam and laser welding techqiues are to be preferred. However, one should keep it in mind that electron beams require a vacuum chamber which lasers do not.

Successful, autogenous, square-butt welds of X-80 Arctic pipeline steel using a high power CW CO_2 were reported by Nagler[60] and Breinan and Banas.[55] Both single pass and dual pass techniques are used to weld 13.2 mm (0.52 in.) material. Dual-pass welds exhibited smaller grain structure than single-pass welds; the upper shelf for dual-pass welds was also greater than 358J (264 ft-lb) and the transition temperature was below -51°C

TABLE 3 Tensile Test Results[51]

Material	Welding Conditions	Welding speed (mm s⁻¹) Power=5kW	Stress at ultimate tensile strength (MPa) Failure in parent (P) or weld (W)	Mean	Strain (%) 	Mean
316 stainless steel	Optimum	16 16	P 599 P 598	598	50.0 50.0	50.0
	Non-optimum	25 25 16 16	W 634 W 554 W 715 W 557	594 636	40.4 51.0 57.6 56.0	45.7 56.8
	Parent plate	-- --	P 592 P 590	591	54.0 54.0	54.0
310 stainless steel	Non-optimum (but fully penetrated)	25 25 16 16	W 410 W 446 W 506 W 510	428 508	16.8 22.6 31.0 26.4	19.7 28.7
Ducol W30 steel	optimum	16 16	P 766 P 769	767	16.0 15.0	15.5
	Non-optimum	25 25 16 16	W 666 W 754 P 774 P 740	710 757	13.2 8.0 22.8 21.0	10.6 21.9
	Parent plate	-- --	P 769 P 780	775	18.0 17.0	17.5

BS 1501 Standard Data

Material	Welding Conditions	Welding speed (mm s⁻¹) Power=5kW	Stress at ultimate tensile strength (MPa) Failure in parent (P) or weld (W)	Mean	Strain (%) 	Mean
316 (hardness)	Parent plate (143-212)	--	--	at least 520	--	at least 40
310 (hardness)	Parent plate (149-217)	--	--	at least 540	--	at least 40
Ducol W30 (hardness)	Parent plate (not quoted)	--	--	at least 590-700	--	at least 16

TABLE 4
Comparison of Processes (Plate Thickness 6 mm)[51]

PROCESS		LASER	TIG	PLASMA	ELECTRON BEAM
Power absorbed by workpiece		4 kW	2 kW	4 kW	5 kW
Total power used		50 kW	3 kW	6 kW	6 kW
Typical traverse speed		16 mm s^{-1}	2 mm s^{-1}	6.7 mm s^{-1}	40 mm s^{-1}
Alignment accuracy required		±0.5 mm	±1.0 mm	±1.0 mm	±0.3 mm
Energy input per unit length (absorbed by workpiece)		250 J mm^{-1}	1000 J mm^{-1}	600 J mm^{-1}	125 J mm^{-1}
Capital cost and availability		$250K for 5 kW output Available on a year's delivery in U.K. and U.S.A. $750K for 15 kW output available on a year's delivery in U.S.A.	~$16K per welding set Immediate delivery	~$20K per welding set Immediate delivery	~$100K for 6 kW machine On short delivery
Possibility of all positional welding		YES but requires optics to manipulate the beam. Optimum when moving workpiece	Serious penetration characteristic changes with attitude	Serious penetration characteristic changes with attitude	YES but requires mechanics to move gun. Optimum when moving workpiece
Distortion	Axial shrinkage	Minimal (small HAZ)	Significant ~1 mm on 5 mm plate	Significant ~1 mm on 5 mm plate	Minimal (small HAZ)
	Angular	Minimal parallel sidedness	Significant V shaped weld	Significant V shaped weld	Minimal parallel sidedness
Surface profile defects		Very fine flow lines	Underside protrusion held in by surface tension	Underside protrusion held in by surface tension	Produces ruffled swarf on backface
Special requirements for process operation		Safety interlock to guard against misplaced beam reflections	Normal light screening	Normal light screening	Vacuum chamber Local vacuum X-RAY SCREENING
Process END effects	Start	Slight surface protrusion	Smooth	Slight surface protrusion	Slight surface protrusion
	Finish	Smooth		Slight surface protrusion	Slight surface protrusion

(-60°F). The mechanical properties of the laser welds appear to be better than those of the base metal. Therefore, laser welding promises to be suitable for high-quality large diameter pipe welding application.

Successful laser welding of D6AC steel in thicknesses of 6.5 mm and 12.7 mm (0.25 in. and 0.5 in.) using a CW 15 kW laser is reported.[57] However, preheating up to 232°C (450F) was found to be necessary to eliminate porosity. The resultant welds exhibited consistent and reproduceable tensile test data with fracture appearing in the base metal demonstrating the quality of the weld.

The laser welds in HSLA steels exhibited properties equivalent to those of the base metal.[58] This is very difficult to obtain using conventional arc welding techniques.

Laser welded rimmed steel sheet was found to be acceptable for automobile sheet joining applications.[16] However, the welds possessed some gas porosity.

One of the principal commercial areas for the welding of thin material is that of can manufacture. Mazumder and Steen[18,53] evaluated the possibility of high speed welding of steels used in can making (tin plate and tin free steels) with a 2 kW CW CO_2 laser. A welding speed in excess of 19 meters/min was achieved for bead on plate welding of 0.2 mm tin plate using 1950 watts of laser output energy. The laser welding process was compared with other can making proceses and found to be the only method capable of welding tin free steel (with a 0.01 μm layer of chromium and a 0.04 μm layer of chromic oxide as a corrosion inhibitor) without auxiliary preparation. Although a 2 kW CW CO_2 laser by itself cannot reach the required welding speeds, arc augmentation of the 2 kW laser appears to be capable to do so.[53,61]

Mechanical properties of laser welds seem to be at least as good as those of the base material. All fractures during tensile tests were observed in the base material.[32] Laser welds were also found to be radiographically sound.[53] Simple bending fatigue tests revealed an endurance ratio of 0.45 to 0.5.[53] Data for corrosion rates of the laser welds using the Tafel extrapolation method has shown that the weld zone is at least as corrosion resistant as the base metal, if not better.[53] In conclusion[53], sound welds of good appearance and mechanical properties can be made using a laser in tin plate and tin free steel. The laser welds have a very narrow HAZ, and could be made through painted areas; they are autogeneous and so present no recycling problem as the solder does in lock seam soldered cans.

Titanium and Its Alloys

Until recently, the EBW technique has been the most popular method for the welding of Ti-6Al-4V, this alloy being widely used in the aerospace industries for its remarkable strength to weight ratio. However, the deep penetration of EBW can be obtained only up to a short distance under non-vacuum conditions, and for optimum efficiency electron beam welding is carried out in an evacuated chamber.[9] In contrast, CO_2 laser beams can be transmitted for appreciable distances through the atmosphere without serious attenuation or optical degradation. The laser thus offers an easily maneuvered, chemically clean, high intensity, atmospheric welding process, producing deep penetration welds (aspect ratio greater than 1:1) with narrow heat affected zone (HAZ) and subsequent low distortion.

The application of the laser technique to a metal such as a titanium alloy, which is difficult to weld, is not only of direct interest to the aerospace and chemical industries but also more generally a study into the welding of a chemcially sensitive metal with a complex temperature dependent structure. The importance and the need for better joining methods for titanium and its alloys resulted in several investigations of laser welding technqiues over various power ranges; Banas[59,69,65] (up to 5.5 kW), Seaman and Hella (up to 16 kW)[59], Adams[67] (up to 2 kW), and Mazumder and Steen (up to 5 kW).[18,63]

Mazumder and Steen[21,33,63] reported the relationship between laser welding parameters and the metallurgical and mechanical properties of laser welded Ti-6Al-4V and commerically pure titanium. As shown in Fig. 3, welding speeds in excess of 15 m/min were obtained by Mazumder, et al.[33], for 1 mm thick Ti-6Al-4V for 4.7 kW of laser power.

X-ray radiographs of successful laser butt-welds of Ti-6Al-4V and commercially pure (CP) titanium showed no cracks, porosity or inclusions.[63] Low porosity in laser welded titanium alloy was also observed by Seaman[66] and radiographically sound welds were also produced by Banas.[64] Undercutting was not prominent.

Tensile tests performed by Mazumder and Steen[18,63] on laser butt welds of titanium alloys revealed that laser welds are at least as strong as the base metal. Typical tensile test data[63] are given in Table 5.

Under simple bending fatigue, the endurance ratio for welded specimens (with a transverese central weld) was found to be 0.40 to 0.47 whereas that for unwelded specimens was 0.50.[63] Adams[67] reported that, under proper welding conditons, laser welds can be made in Ti-6Al-4V which exhibits the same fatigue characteristics as the base metal.

The variation of microstructure for a typical Ti-6Al-4V weld as seen in an optical photomacrograph in comparison with its macrostructure is represented in a composite (Fig. 5 after Ref. 63). The parent metal (Fig. 5A) shows dark β in a bright α marix which represents a typical annealed structure of α-β titanium. The heat affected zone consists of a mixture of martensitic α and primary α which corresponds to a structure quenched from the range 980 to 720°C.[68] The HAZ is shown in Fig. 5B. The right-hand side of the micrograph, Fig. 5B, is near the fusion zone. In that area, traces of α' (martensitic α) can be observed, whereas the left-hand side of the micrograph (i.e. further away from the fusion zone) shows a relative increase of primary α. This observation is confirmed by the scanning electron microscopy study.

The fusion zone consists mainly of α' (martensitic α). Both Figs. 5D and 5E represent the fusion zone of the same specimen. The abundance of α' (martensitic α) with the β-grain boundary is evident in these micrographs. This is confirmed by the structure revealed under both scanning and transmission electron microscopes. Such microstructure corresponds to a structure quenched from the β phase above 985°C[68] and is confirmed by comparison with the hardness values of others.[68]

Figure 6 shows the transmission electron micrograph (Fig. 6A) of the weld zone. It confirms the observations using optical microscopy that the fusion zone consists mainly of α' (martensitic α) of acicular morphology. this (acicular martensitic α) is characterized by large "primary" plates and smaller "secondary" plates. The primary plates extend for large distances across the parent β grain and effectively partition the untransformed β. The

TABLE 5
Typical mechanical properties of titanium alloys

MATERIALS	WELDED MATERIAL			ORIGINAL MATERIAL		
	UTS MN/m^2	0.2% proof stress MN/m^2	Elongation %	UTS MN/m^2	9.2% proof stress MN/m^2	Elongation %
AMS 4911-TA-10 (6Al-4V-Ti)	860-923	800-860	11-14	895-1004	834-895	10-15
Commercially pure titanium	≈530-573	≈460-503	≈27	>494	>416	27-28

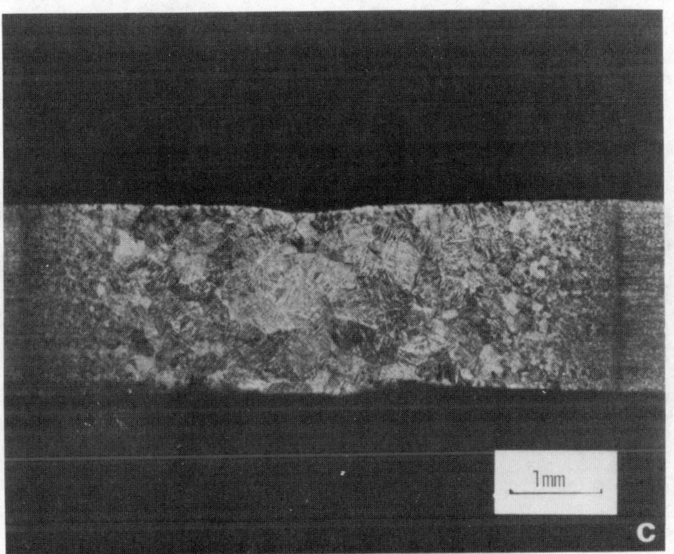

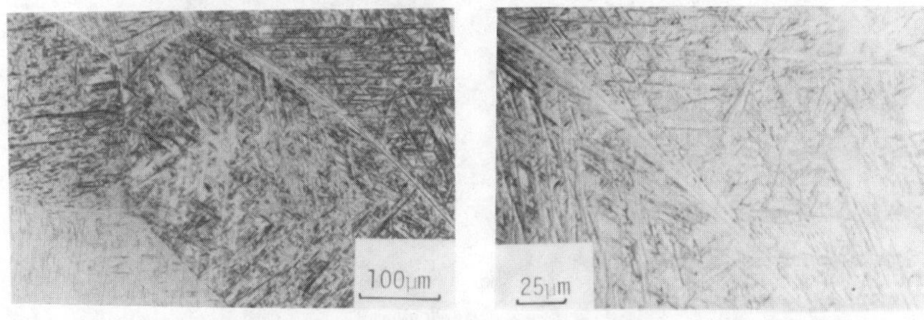

Fig. 5 Composite showing structural variation for 2 mm thick
6Al-4V-Ti alloy welded with CW CO_2 laser. Welding speed
7.5 mm/sec., laser power 1500W, (a) parent matrix; (b) HAZ;
(c) macrostructure of the weld; (d) and (e) fusion zones.

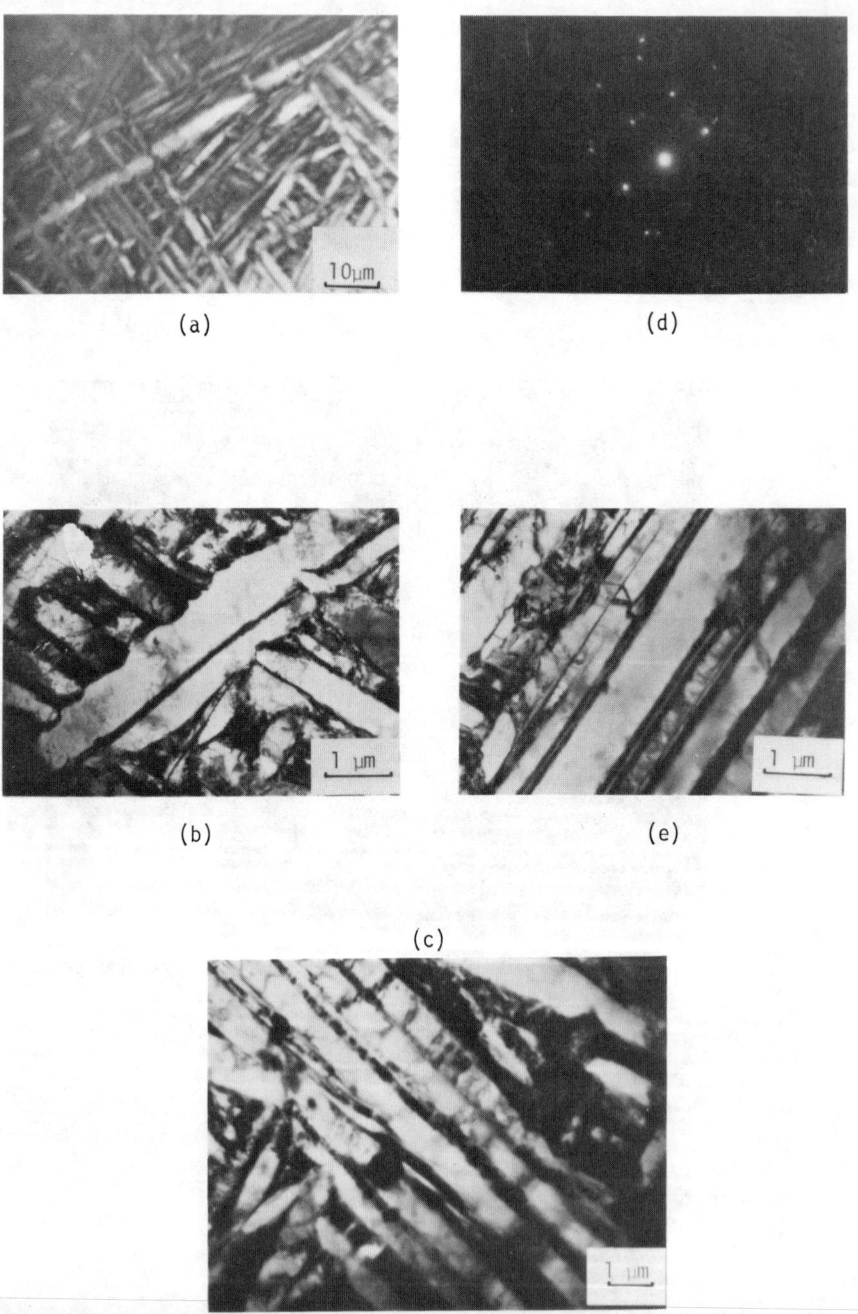

Fig. 6   Composite showing the transmission electron
micrographs of the 6Al-4V-Ti weld.

partitioned β subsequently transforms to a series of short, acicular, secondary plates of martensite (Fig. 6B and 6C). A similar microstructure was observed by Zaidi[69] for water quenched 6Al-4V-Ti from above 1000°C which also correlates with Hochied, et al.[68] The selected area diffraction pattern (Fig. 6D) of a martensite plate shown in Fig. 6B indicates that the zone axis is (10T1) hcp. But, in another sample, the zone axis of the martensite plate was found to be (0001).

This martensitic structure at the weld zone is responsible for the good tensile properties of the laser welded titanium alloy. By comparing the microstructure with the prediction of a three-dimensional transfer model[70] the cooling rate of the weld zone was estimated to be 10^4°C/sec.[63] This high cooling rate is responsible for the martensitic structure.

The total oxygen analysis of a weld sample indicates that there is no significant oxygen contamination taking place during laser welding when shielding gas is used.[63]

A comparative study of electron beam, laser, and plasma arc welds in Ti-6Al-4V alloy has been conducted by Banas.[64] Rodiographically sound welds were produced by all three technqiues.[64] The electron beam welds were quite narrow and exhibited a somewhat non-uniform radiographic appearance due to lower surface weld spatter whereas the arc welds were considerably broader but also quite uniform in density. Laser welds were narrower than arc welds and were comparable to but more uniform than EBW weld beads. Following stress relieving for two hours at 538°C (100°F), the welds produced by all three techniques had tensile strengths equivalent to or exceeding those of the base metal.

Iridium Alloys

High power CW CO_2 lasers are found to be an attractive tool for welding difficult alloys such as Thorium doped iridium alloys (DOP 14 and DOP 16). These alloys crack severely during GTA welding or welding with a highly defocussed electron beam[71]. Successful laser welds free from hot cracking have been reported in DOP-14 by David and Liu[71]. This is due to the characteristics of the highly concentrated heat source available with the laser and from the refinement in the fusion zone structure. The fusion zone structure is a strong function of the welding speed.[71] Typical microstructures of the DOP-14 laser welds made at 12.5 mm/sec. as shown in Fig. 7.

Laser welds with a refined fusion zone structure have also been reported in DOP-26 alloy[71]. The fusion zone structure of the laser welds compares well with the best structures obtained by arc welding using arc oscillation.

Conclusion

Numerous experiments since the development of multikilowatt CW CO_2 laser around 1970, have shown that the laser permits precision weld joints of a high quality only rivalled by an electron beam. Various commerical CO_2 laser systems lasting form power of few watts to 15 KW are available for multitude of laser processing applications. Now a days, high power (up to 400W average CW power) pulsed Nd-YAG lasers are also available for overlapping spot welding applications. Due to the shorter wave length (1.06 μm) Nd-YAG laser produces higher power density compared to CO_2 laser. Otherwise welding principles for Nd-YAG lasers are same as CO_2 lasers. However, the availability of high power lasers, limitations of

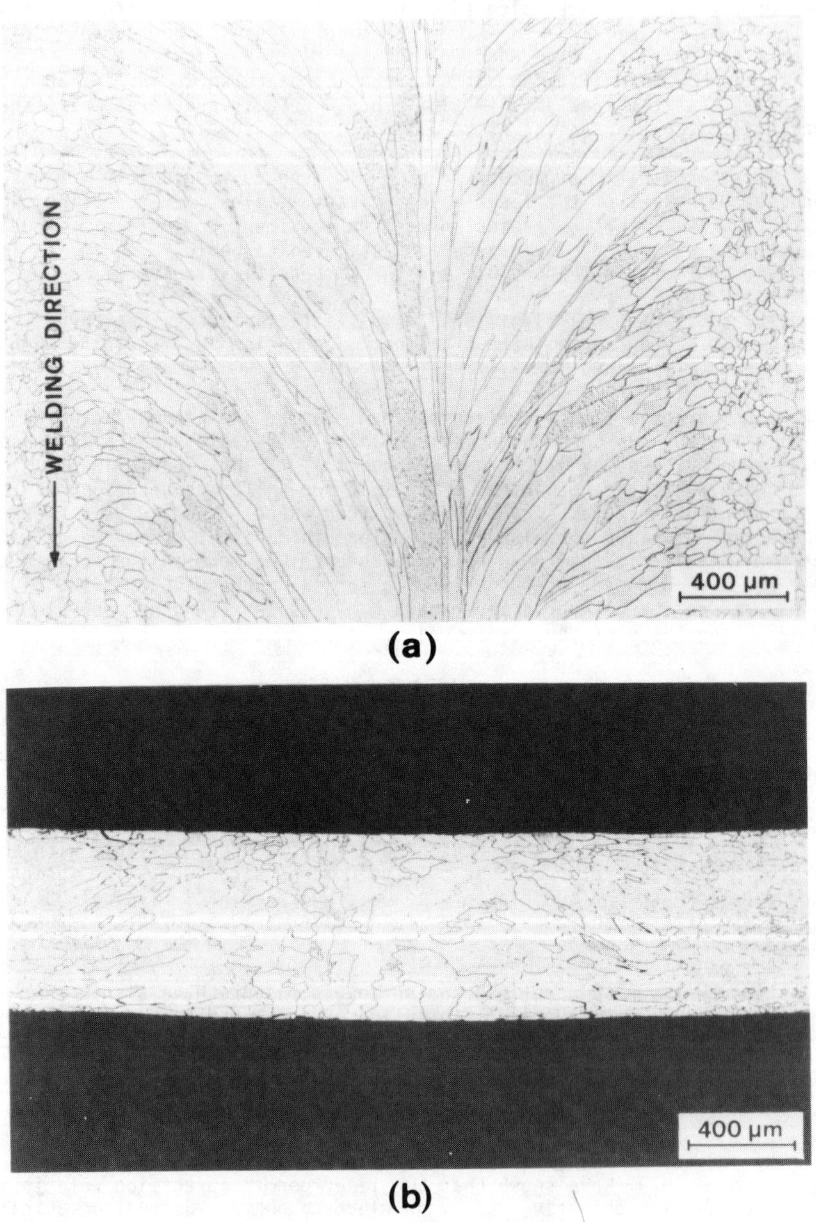

Fig. 7 Fusion zone microstructure of a laser weld.
Welding speed: 12.5 mm/s, (a) top surface;
(b) transverse section.

current welding technology and the rapid technological developments in past decades have stimulated considerable interest in laser welding. It is beginning to be accepted as an industrial production technique.

Research carried out in laser welding has already demonstrated that high quality welding can be performed for many ferrous alloys, Nickel alloys and titanium alloys. This process has also been found effective for some Iridium alloys and many dissimilar metals which are very difficult to weld by alternative processes. Some partial success is also achieved for highly reflective materials such as aluminum alloys. But more research is necessary to improve the process for aluminum alloys.

Understanding about the laser welding variables and shielding gas requirements has already been achieved to a great extent. But the plasma formation and the keyhole mechanism during laser material interaction is still a little understood process. Surface tension driven fluid flow inside the molten pool due to the characteristically high temperature gradient observed during laser welding also deserve more attention. Compared to the conventional welding processes more precise mathematical modelling is possible for laser welding due to the well defined characteristics of a laser beam. However, improvement to the predictive capabilities of mathematical models requires better understanding of plasma formation and fluid flow inside the pool.

<u>References</u>

1. Brown, C. O., "High Power CO_2-Electric Discharge Mixing," <u>Appl. Phys. Letters</u>, Vol. 17, No. 9 (1970).

2. Brown, C. O., and J. W. Davis, "Electric Discharge Convection Lasers," Paper presented at the IEEE Int. Electronic Devices Meeting, Washington, D.C. (19760).

3. Patel, C. K. N., <u>Physics Rev.</u>, Vol. 136, A1187 (1964).

4. Spalding, I. J., "Laser System Developments, "<u>Physics Bull.</u>, United Kingdom, p. 402 July (1971).

5. Arata, Y., and I. Miyamoto, <u>Technocrat</u>, p. 33, Vol 11, No. 5, May (1978).

6. Anderson, J. E., and J. E. Jackson, <u>Welding J.</u>, Vol. 44, p. 1018 (1965).

7. Duley, W. W., <u>CO_2 Lasers, Effects, and Application</u>, Academic Press, New York (1976).

8. Banas, C. M., "Laser Welding to 100 kW," United Technologies Research Center Report No. R76-912260-2 (under Navy Contract N00173-76-M-0107), Feb. (1977).

9. Meleka, A. M., ed., <u>Electron Beam Welding Principles and Practice</u>, pp. 95,96, published for the Welding Institute, McGraw-Hill, London.

10. Swifthook, D. T., and E. E. F. Gick, <u>Welding J.</u>, pp. 492S-499S (1973).

11. Klemens, P. G., <u>J. Appl. Physics</u>, Vol. 47, pp. 2165-2174 (1976).

12. Sickman, J. G., and R. Morijn, <u>Phillips Res. Rep.</u>, Vol. 23, p. 376 (1968).

13. Sickman, J. G., and R. Morijn, <u>Phillips Res. Rep.</u>, Vol. 23, p. 375 (1968).

14. Arata, Y., H. Maruo, I. Miyamoto, and Y. Inoue, "Dynamic Behavior of Laser Welding," IIW DOC, IV/222/77 (1977).

15. Locke, E. V., and R. A. Hella, <u>IEEE J. Quantum Electronics</u>, Vol. QE-10, No. 2, pp. 179-185, Feb. (1974).

16. Baardsen, E. L., D. J. Schmatz, and R. E. Bisaro, <u>Welding J.</u>, Vol. 52, pp. 227-229, April (1073).

17. Harry, J. E., <u>Industrial Application of Lasers</u>, McGraw-Hill Book Co. (U.K.), Ltd. (1974).

18. Mazumder, J., Ph.D. Thesis, London University (1978).

19. Gibson, A. F., M. H. Kimmit, and A. C. Walker, <u>Appl. Phys. Letters</u>, Vol. 17, pp. 75-77 (1970).

20. Gibson, A. F., and A. C. Walker, <u>J. Phys. C. Solid State Physics</u>, Vol. 4, pp. 2209-2219 (1971).

21. Mazumder, J., "Laser Welding," chapter in Laser Materials Procesing, M. Bass, ed., to be published by North Holland Pub. Co., The Netherlands.

22. Arata, Y., and I. Miyamoto, Laser Focus, No. 3 (1977).

23. Bramson, M. A., Infrared Radiation: A Handbook for Application, Plenum Press, New York (1968).

24. Jørgensen, M., Metal Construction, Vol. 12, No. 2, p. 88, Feb. (1980).

25. Crafer, R. C., The Weld Inst. Res. Bull. (U.K.), Vol. 17, Feb. (1976).

26. Locke, E. V., E. Hoag, and R. Hella, Weld. J., Vol. 51, pp. 245S-249S (1972).

27. Locke, E. V., E. Hoag, and R. Hella, IEEE J. of Quantum Electronics, Vol. QE8, p. 132 (1972).

28. Mazumder, J., and W. M. Steen, Metal Construction, p. 423, Vol. 12, No. 9 (1980).

29. Adams, H. J., Metal Construction and Br. Welding J., Vol. 2, No. 1, pp. 1-8 (1970).

30. Seaman, F. D., "The Role of Shielding Gas in High Power CO_2(CW) Laser Welding," SME Technical Paper No. MR77-982, Soc. of Mfg. Eng., 20501 Ford Road, Post Office Box 930, Dearborn, MI 48128, USA (1977).

31. Rein, R. M., et al., Patent 1448740, The United Kingdom Patent Office (1972).

32. Shewell, J. r., Welding Design and Fabrication, pp. 106-110, June (1977).

33. Brown, C. O., and C. M. Banas, "Deep Penetration Laser Welding," AWS Annual Meeting, San Francisco, Cal., April (1971).

34. Ready, J. F., Effect of High Power Laser Radiation, Academic Press, New York (1971).

35. Nichols, K. G., Proc., Inst. of Electronic Eng., Vol. 116, Part 12, pp. 2093-2100, Dec. (1969).

36. Schwartz, M. M., "Laser Welding and Cutting," WRC Bulletin 167, Nov. (1971).

37. Pfluger, A. R., and P. M. Mass, Welding J., Vol. 44, p. 1018 (1965).

38. Cohen, M. I., F. J. Mainwaring, and T. G. Melone, Welding J., Vol. 48, p. 191 (1969).

39. Anderson, J. E., and J. E. Jackson, Proc., Electron Laser Beam Symp., Penn-State University (ed., A. B. El-Kareh), p. 17 (1965).

40. Lebedev, V. K., and V. T. Granista, Automat. Weld., Vol. 25, p. 63 (1972).

41. Velichco, O. A., U. P. Gasshchuk, and V. E. Moravskii, <u>Auto. Weld.</u>, Vol. 25, Part 4, p. 75 (1972).

42. Miller, K. J., and J. D. Nunnikhoven, <u>Welding J.</u>, Vol. 44, p. 480 (1965).

43. Miller, K. J., <u>Weld. Eng.</u>, Vol. 51, p. 46 (1966).

44. Schmidt, A. O., I. Ham, and T. Hoski, <u>Weld. J. Suppl.</u> p. 481, Nov. (1965).

45. Crafer, R. C., Welding Institute Res. Bull. (U.K.). Vol. 17, April (1976).

46. Metzbower, E. A., and D. W. Moon, "Mechanical Properties, Fracture Toughness, and Microstructures of Laser Welds of High Strength Alloys," <u>Proc., Conf. on Applications of Lasers in Materials Processing</u>, E. A. Metzbower (ed.), published by American Soc. of Metals, Ohio 44073, pp. 83-100, Washington, D.C., 18-20 April (1979).

47. Snow, D. B., and E. M. Breinan, "Evaluation of Basic Welding Capabilities, prepared by United Technologies Res. Center, East Hartford, CT 06108, for ONR, Dept. of Navy, Report No. R78-91189-14, July (1978).

48. Breinan, E. M., C. M. Banas, and M. A. Greenfield, "Laser Welding--The Present State of the Art," 11th Annual Meeting, Tel Aviv, DOC IV-181-75, pp. 1-53, 6-12 July (1975).

49. Schwatz, M. M., <u>Metal Joining Manual</u>, McGraw-Hill Book Co., New York (1979).

50. Mazumder, J., "Laser Welding of Aluminum Alloys," unpublished research carried out at the Center for Laser Studies, University of Southern California under Contract from ALCOA Laboratories Joining Div., ALCOA Center, PA 15069.

51. Willgoss, R. A., J.H.P.C. Megaw, and J. N. Clark, <u>Welding and Metal Fabrication</u>, pp. 117-126, March (1979).

52. Crafer, R. C., "Advances in Welding Processes," <u>Proc., 4th Int. Conf. Harrogate, Yorks</u>, Paper No. 46, pp. 267-278, May 9-11 (1978).

53. Mazumder, J., and W. M. Steen, "Laser Welding of Steels used in Can Making," <u>Welding J.</u>, Vol. 60, No. 6, pp. 19-25 June (1981).

54. Seaman, F. D., and R. A. Hella, "Establishment of a Continuous Wave Laser Welding Process," IR-809-3 (1 through 10), AFML Contract F336 15-73-C5004, Oct. (1976).

55. Breinan, E. M., and C. M. Banas, "Preliminary Evaluation of Laser Welding of X-80 Arctic Pipeline Steel," WRC Bull., 201, Dec. (1971).

56. Banas, C. M., and G. T. Peters, "Study of the Feasibility of Laser Welding in Merchant Ship Construction," Contract No. 2-36214, U.S. Dept. of Commerce, Final Report to Bethlehem Steel Corp., Aug. (1974).

57. Banas, C. M., "Electron Beam, Laser Beam, and Plasma Arc Welding Studies," Contract No. NASA1-12565, NASA, March (1974).

58. Yessik, M., and D. J. Schmatz, "Laser Processing in the Automotive
    Industry," SME Paper MR74-962 (1974).

59. Banas, C. M., "Laser Welding Developments," Proc., CEGB Int. Conf. on
    Welding Res. Related to Power Plants, Southampton, England, 17-21 Sept.
    (1972).

60. Nagler, H., "Feasibility, Applicability, and Cost Effectiveness of LBW
    of Navy Ships, Structural Components, and Assemblies," Contract No.
    N00G00-76-C-1370, Vols. 1 and 2, 22 Dec. (1976).

61. Steen, W. M., and M. Eboo, "Arc Augmented Laser Welding," Metal
    Construction, Vol. 11, No. 7, pp. 332-335 (1979).

62. Blake, A., and J. Mazumder, unpublished work, University of Ill. at U-C
    (1981).

63. Mazumder, J., and W. M. Steen, " Structure and Properties of Laser
    Welded Titanium Alloy," TMS-AIME Fall Meeting, Paper No. F79-17, Sept.
    (1979).

64. Banas, C. M., "Electron Beam, Laser Beam, and Plasma Arc Welding
    Studies," NASA Contractor Report No. NASA CR-132386, March (1975).

65. Banas, C. M., United Technologies Res. Center Report No. R75-412260-1,
    July (1975).

66. Seaman, F. E., "Establishment of a CW CO_2 Laser Welding Process," USAF
    Tech. Report AFML-TR-76-158, Sept. (1978).

67. Adams, M. J., "CO_2 Laser Welding of Aero-Engine Materials," Rep.
    3335/3/73, British Welding Institute, England (1973).

68. Hochied, B., R. Klima, C. Beauvais, and C. Roux, Memoires Scientifiques
    Rev. Metallurgie LXVII, No. 9 (1970).

69. Zaidi, M. A., Master of Science Thesis, London University, Imperial
    College (1978).

70. Mazumder, J., and W. M. Steen, J. Applied Physics, p. 944 Feb. (1980).

71. David, S. A., and C. T. Liu, "High Power Laser and Arc Welding of
    Thorium Doped Tridium Alloys," Report No. ORNL/TM 7258, Oak Ridge, TN
    73830, May (1980).

SOLIDIFICATION BEHAVIOR AND MICROSTRUCTURAL ANALYSIS

OF AUSTENITIC STAINLESS STEEL LASER WELDS*

S. A. David and J. M. Vitek

Metals and Ceramics Division
Oak Ridge National Laboratory
Oak Ridge, Tennessee 37830

Solidification behavior of austenitic stainless steel laser welds has been investigated with a high-power laser system. The welds were made at speeds ranging from 13 to 60 mm/s. The welds showed a wide variety of microstructural features. The ferrite content in the 13-mm/s weld varied from less than 1% at the root of the weld to about 10% at the crown. The duplex structure at the crown of the weld was much finer than the one observed in conventional weld metal. However, the welds made at 25 and 60 mm/s contained an austenitic structure with less than 1% ferrite throughout the weld. Microstructural analysis of these welds used optical microscopy, transmission electron microscopy, and analytical electron microscopy. The austenitic stainless steel welds were free of any cracking, and the results are explained in terms of the rapid solidification conditions during laser welding.

―――――――――
*Research sponsored by the Division of Materials Sciences, U.S. Department of Energy, under contract W-7405-eng-26 with the Union Carbide Corporation.

Introduction

Austenitic stainless steels form an important class of engineering
materials in several energy systems. A significant problem in the production of fully austenitic stainless steel welds is their tendency for hot
cracking. To minimize this tendency, the compositions of welding
materials are generally modified to produce small amounts of δ-ferrite in
the as-welded microstructure. For example, in type 308 stainless steel
weld metal, ferrite contents of 5 to 10% are common. Although ferrite has
been found to effectively prevent hot cracking (1–3), it also leads to
corrosion susceptibility and embrittlement at elevated temperatures.
Hence, it would be highly desirable to produce fully austenitic stainless
steel welds without the tendency to hot crack.

Extremely high cooling rates have produced unusual microstructures in
austenitic stainless steels containing duplex structures (4). However,
neither were these microstructures characterized fully nor were their
origins understood. The purpose of our work is to characterize austenitic
stainless steel laser welds and understand the observed modifications in
microstructure.

Experimental Procedure

A multipass conventional weld overlay of type 308 stainless steel
filler metal [20.5 Cr, 10.5 Ni, 1.5 Mn, 0.44 Si, 0.065 C, 0.022 P,
0.008 S, balance Fe (wt %)] was made on a 12-mm-thick plate of type 304L
stainless steel; see Figure 1. The resultant block of type 308 stainless
steel overlay was approximately 12 mm high, 25 mm wide, and 150 mm long.
Autogenous laser welds (melt runs) were made on this overlay at welding
speeds of 13, 25, and 63 mm/s. The welds were made with an AVCO multikilowatt continuous-wave CO_2 laser system with an output of 9 kW, in the
annular beam mode. During welding, shielding was provided by helium gas
flowing through an off-axis diffuser at 5.6 m^3/h.

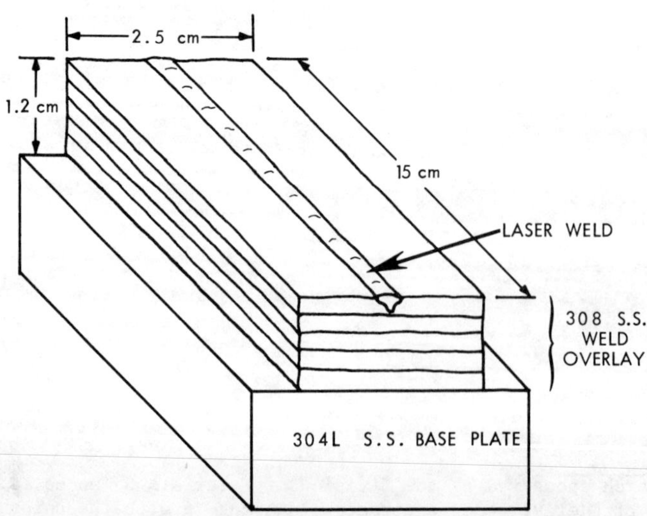

Fig. 1 – Schematic of the weld pad used for the laser weld experiment.

Microstructural analysis of these welds was performed by using optical microscopy, transmission electron microscopy, and analytical electron
microscopy. For optical microscopy the samples were etched with a solution containing five parts concentrated HCl to one part concentrated HNO_3.
Foils of weld metal for electron microscopic analysis were electro-
polished with a dual-jet polishing apparatus and a solution of sulphuric
acid in methanol (1:7). Most of the electron microscopy was performed at
120 kV.

Results and Discussion

A typical microstructure of type 308 stainless steel weld produced by
one of the conventional welding processes, namely gas tungsten arc (GTA),
is shown in Figure 2. The photomicrograph shows the vermicular and lacy

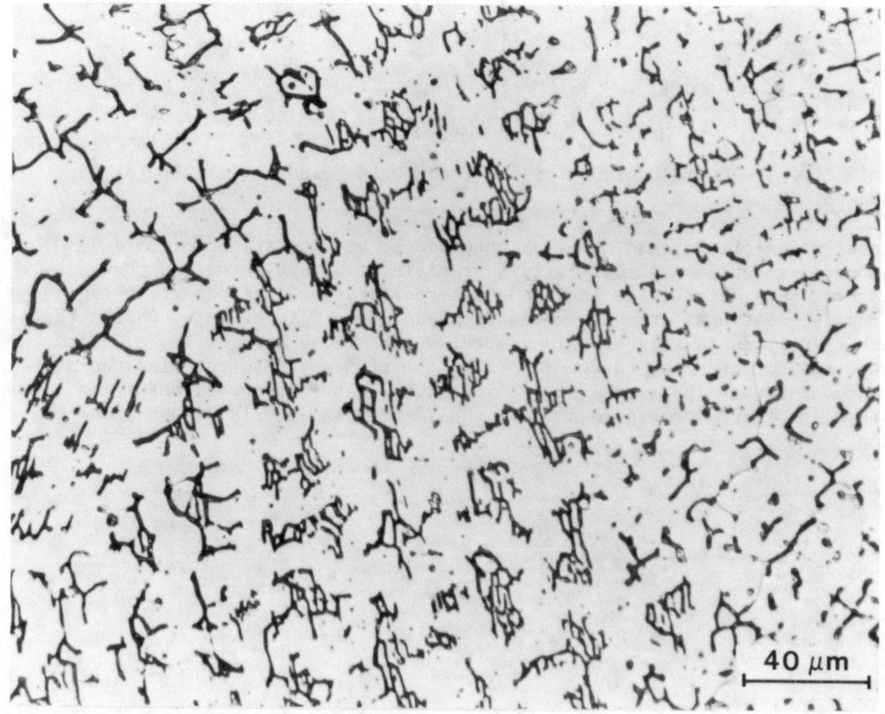

Fig. 2 - A gas tungsten arc bead-on-plate weld. Type 308 stainless
steel filler metal.

ferrite forms in austenite. An earlier study (5) revealed that the solidification sequence in type 308 stainless steel weld metal consists of the
primary crystallization of δ-ferrite with subsequent envelopment by austenite and a δ → γ transformation continuing below the solidus, leaving
behind a skeletal network of residual δ-ferrite. This particular ferrite
has been shown (6) to be located along the cores of the primary and secondary dendrite arms. Sometimes the primary δ-ferrite formed during solidification may transform to Widmanstätten austenite at lower temperatures,

leaving behind residual ferrite in acicular or lacy form. The ferrite in
this case has been shown to be located within the primary cells or
dendrites. These two modes of ferrite formation account for the origin of
δ-ferrite in austenitic stainless steel welds containing duplex
structures.

Macroscopic views of cross sections from the three laser welds are
shown in Figure 3. In contrast to the type 308 stainless steel GTA weld,

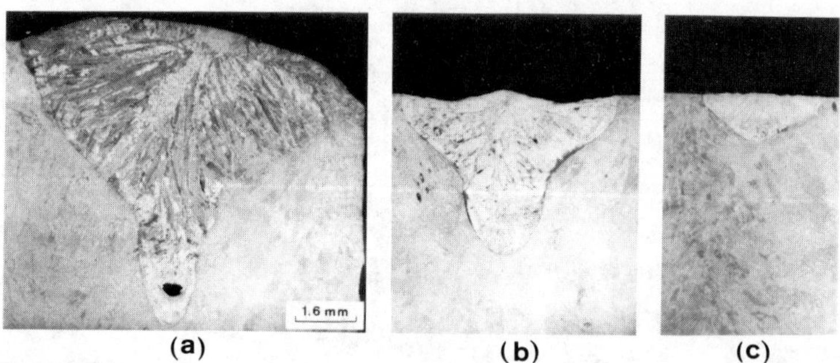

(a)   (b)   (c)

Fig. 3 - Macrostructure of the laser welds made at (a) 13, (b) 25,
and (c) 63 mm/s.

laser welds of the same material obtained at welding speeds of 25 and
63 mm/s produced an almost fully austenitic structure with less than
1% ferrite. The ferrite is in the intercellular or interdendritic regions, suggesting a primary austenitic mode of solidification. Unlike the
other fully austenitic stainless steel welds, these laser welds did not
show any hot cracking tendency. Optical microscopy also revealed the presence of a third phase uniformly distributed within the weld metal, as
shown in Figure 4. Figure 5 shows the 13-mm/s laser weld with a wide

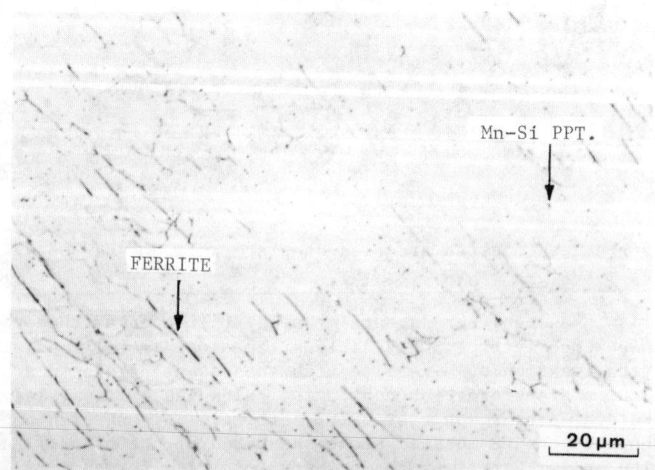

Fig. 4 - Photomicrograph showing intercellular ferrite and uniform
distribution of third-phase particle identified as MnSi. Laser weld 25 mm/s.

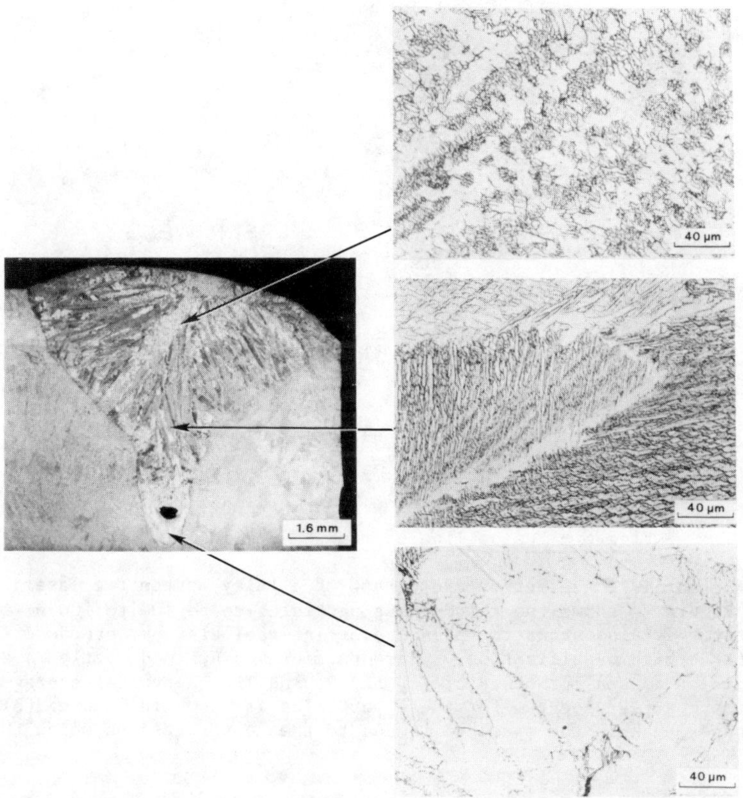

Fig. 5 - Variations in microstructure from root to the crown of the 13-mm/s laser weld.

variety of microstructures ranging from fully austenitic microstructure at the root of the weld to $\delta + \gamma$ duplex structure at the crown. The fully austenitic structure at the root of the weld may be attributed to the high cooling rates encountered in that part of the weld during welding. Midsections of the weld revealed a very fine duplex structure. The crown of the weld also contained a duplex structure of austenite and ferrite but coarser.

The origin of the fully austenitic structure in the laser welds is an interesting phenomenon that needs further investigation and understanding. Figure 6 shows the details of the microstructure along the fusion line of the laser weld made at 25 mm/s. Close examination revealed a change in the mode of solidification for type 308 stainless steel from primary ferrite, as seen commonly, to primary austenite. The presence of intercellular or interdendritic ferrite is further evidence of such a change in the mode of solidification. This may be rationalized as due to the excessive undercooling at the tip of the primary δ-ferrite cells or dendrites that could have changed the mode of solidification from primary δ-ferrite to primary austenite. A later paper will present a more extensive analysis of this result.

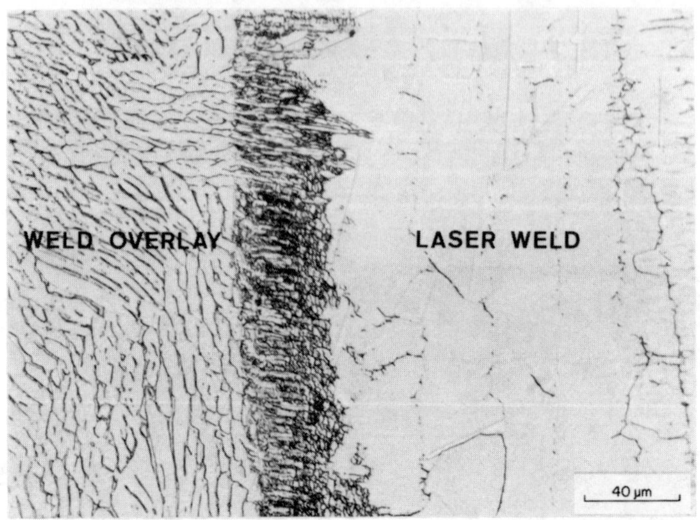

Fig. 6 - Fusion line of the 25-mm/s laser weld.

A transmission electron micrograph of a fully austenitic laser
weld, Figure 7, shows the third-phase particles to be 100 to 200 nm in
diameter. It also shows the presence of intercellular or interdendritic
ferrite at high magnification. These particles contained mostly Mn and Si
in addition to small amounts of Cr, Al, S, and Ti. A typical energy-
dispersive x-ray spectrum for these particles is shown in Figure 8(b).
The copper peak in the spectrum is due to the copper grid on which the

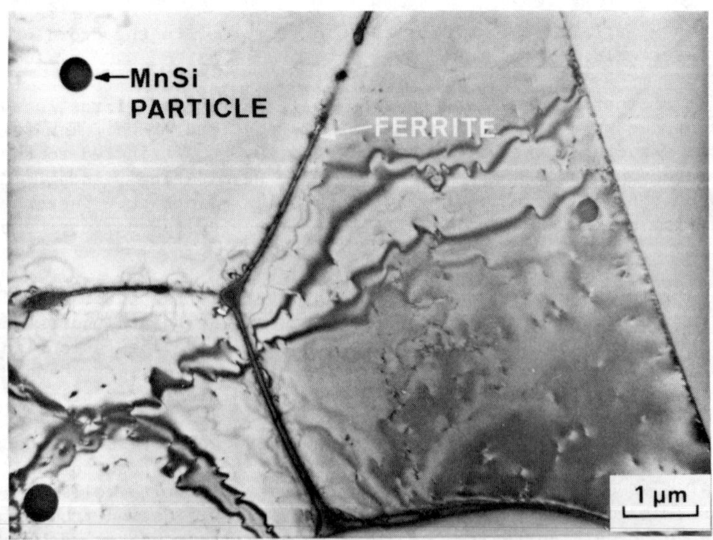

Fig. 7 - Typical transmission electron micrograph of the laser weld
showing intercellular ferrite and MnSi particles.

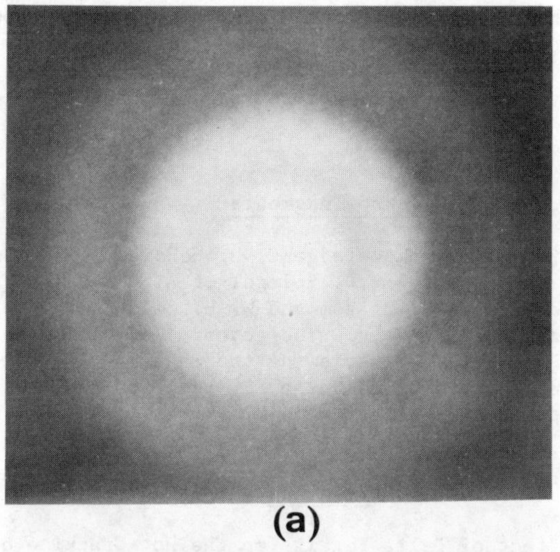

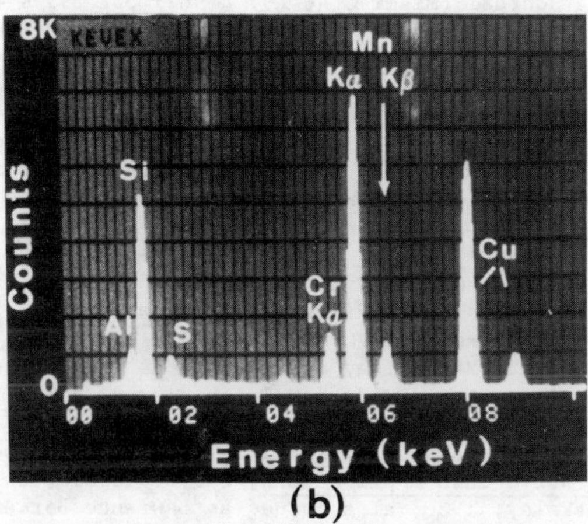

Fig. 8 - Third-phase particles observed in the laser welds made at 25 and 63 mm/s. (a) Microdiffraction pattern. (b) Typical energy spectrum.

replica was mounted for analysis. A microdiffraction technique, Figure 8(a), showed that these particles were amorphous. No evidence for these manganese- and silicon-rich particles was found in the conventional weld overlay. Spherical manganese- and silicon-rich particles have been observed elsewhere (7,8) and have been identified as oxide inclusions. In the present work, ascertaining the presence of oxygen in these particles was beyond the capability of the analytical instrument. Further work is under way to better understand the origin and nature of these particles.

Conclusions

Rapid cooling encountered during laser welding of type 308 stainless steel changed the mode of solidification from primary δ-ferrite to primary austenite. The resulting weld metal contains a very small amount of

intercellular or interdendritic ferrite and the weld metal showed no sign
of cracking. In addition to the presence of a very small amount of residual ferrite, a very fine and uniform distribution of third-phase particles
was found in the weld metal and identified as amorphous and rich in manga-
nese and silicon.

Acknowledgments

The authors gratefully acknowledge C. J. McHargue, Program Manager,
for encouragement and support, L. B. Spiegel of AVCO Metal Working Lasers
for making the welds and C. P. Haltom and W. H. Smith for metallography
and electropolishing, respectively. The authors would like to acknowledge
G. M. Goodwin and C. T. Liu for reviewing the manuscript and technical
discussion. The manuscript was edited by S. Peterson and the camera-
ready copy was prepared by A. F. Rice.

References

1. F. C. Hull, "Effect of Delta Ferrite on the Hot Cracking of Stainless
   Steel," Welding Journal (Miami), 46 (9) (1967) pp. 399-s—409-s.
2. J. C. Borland and R. N. Younger, "Some Aspects of Cracking in
   Austenitic Steels," British Welding Journal, 7 (1) (1960) pp. 22—60.
3. Y. Arata, F. Matsuda, and S. Katayama, "Solidification Crack Susceptibility in Weld Metals of Fully Austenitic Stainless Steel (Report 1) —
   Fundamental Investigation on Solidification Behavior of Fully
   Austenitic and Duplex Microstructures and Effect of Ferrite on Microsegregation," Transactions of JWRI, 5 (2) (1976) pp. 35—51.
4. W. T. DeLong, "Ferrite in Austenitic Stainless Steel Weld Metal,"
   Welding Journal (Miami), 53 (7) (1974) pp. 273-s—286-s.
5. S. A. David, G. M. Goodwin, and D. N. Braski, "Solidification Behavior
   of Austenitic Stainless Steel Filler Metals," Welding Journal (Miami),
   58 (11) (1979) pp. 330-s—336-s.
6. S. A. David, "Ferrite Morphology and Variations in Ferrite Content in
   Austenitic Stainless Steel Welds," Welding Journal (Miami), 60 (4)
   (1981) pp. 63-s—71-s.
7. K. W. Mahin, Characterization of Ferritic G.M.A. Weld Deposits in
   9% Ni Steel for Cryogenic Applications, Ph. D. Thesis, University of
   California, Berkeley, 1980; also issued as Lawrence Berkeley Laboratory Report LBL-10922.
8. S. R. Keown and R. G. Thomas, The Role of Delta Ferrite in the Thermal
   Aging of Austenitic Weld Metals, Central Electricity Generating Board
   (England) Report RD/M/R299, 1980.

LASER BEAM WELDING OF ASTM A-36 STEEL

E. A. Metzbower and D. W. Moon
Naval Research Laboratory
Washington, D. C. 20375

Laser beam weldments of ASTM A-36 steel were fabricated in thicknesses from 1/4 in (6mm) to 5/8 in (15mm). The range of welding parameters (laser power, travel speed) over which welds could be made was determined. The microstructures of the base plate, heat affected zone and the fusion zone were determined and hardness traverses across these zones at different depths were made. The mechanical properties were measured using transverse tension specimens. The ductility and the soundness of the weldments were determined by a series of bend tests. The Charpy V-notch energy was measured as a function of temperature. Additional testing was carried out in the weldments after a stress relief heat treatment. Correlations between the structure, property and process were made.

<h1 style="text-align:center">Introduction</h1>

Laser beam welding of thick plates, i.e. greater than 0.25 in (6mm) has been demonstrated in a variety of metals and alloys. At these thicknesses a carbon dioxide, continuous wave laser with power of 5 kilowatts or greater is normally used. Aluminum alloys (1), titanium alloys (1,2), steels (1,3,4,5,6), and a variety of other alloys (5) have been used to demonstrate the ability of the laser to weld. However, most of these efforts has been directed towards achieving good welds and very little effort has been made to characterize the properties of these weldments.

The most widely used weldable structural steel in this country conforms to the ASTM specification A-36. This steel is widely used in a variety of thicknesses, and has good mechanical properties. A-36 steel is welded by a variety of techniques including shielded metal arc, gas metal arc, submerged arc, and electron beam. The resulting weldments have good mechanical properties and ductility. Although not normally used in structures subjected to low temperature or impact loading, the Charpy V-notch (CVN) energy of the plate is equal to or better than that of the weldments. In large structures, the weldments are often given a stress relief heat treatment to minimize residual stresses and distortion. A program was developed to determine the parameters necessary to laser beam weld this structural steel, to measure the mechanical properties and Charpy V-notch energy of these weldments, to characterize the microstructures of these weldments, and finally, to correlate the process, properties and structures.

<h2 style="text-align:center">Experimental Procedure</h2>

Laser Beam Welding

The laser beam welding process requires that the laser beam be focused onto the joint to be welded and that the weldment be moved relative to the beam. When the laser beam interacts with the plate, the metal is vaporized and a plasma "keyhole" is made into the plate. Around the keyhole, the material is melted and a dynamic equilibrium results between the photon/plasma interaction and the forces of the liquid metal. As the keyhole is moved along the weld joint, melting occurs in the front and on the side, whereas solidification occurs in the rear. In order to produce a weld in thick plate, the plasma created by the photon/material interaction must be contained in the keyhole. If this plasma is allowed to rise above the plate surface, the laser beam will be absorbed by the plasma and the energy of the plasma will be re-radiate in all directions. This effectively decreases the energy to the keyhole and for a given laser power dictates a shallow weld. In order to control the plasma, a stream of inert gas is forced across the keyhole, moving the plasma away from the top of the keyhole. Helium, because of its high ionization energy, is usually used for plasma control, although a helium-argon mixture may be used effectively (7). In order to produce a deep weld with a high depth-to-width ratio, the focal point of the laser beam is usually (8) below the surface of the weldment by about 0.100 in (2.5mm).

For a particular thickness of plate, the relationship between the energy of the laser beam and the travel speed for welding must be determined, since there are only certain combinations of power and speed with which a keyhole can be maintained and a successful weld produced. In order to determine these parameters a series of bead on plate experiments were carried out. In these experiments for a fixed plate thickness and a fixed laser power, the welding speed was varied. At too slow of a welding speed severe undercut resulted. At too fast of a welding speed a through

thickness weld was not obtained. The results of these bead on plates welds
are shown in Fig. 1 where the travel speed is plotted as a function of
laser power for a given thickness of plate. Utilizing this data weldments
were fabricated in plate thicknesses from 1/4 in (6 mm) to 5/8 in (16 mm).

Properties

A series of laser beam weldments were fabricated according to the
parameters listed in Table 1 in order to determine the properties of the
weldments. For all of the thicknesses a series of transverse bend specimens
were tested. These side-, face- and root-bend specimens were bent about a
1.0 in (25 mm) diameter anvil. Some of the 0.5 in (12 mm) thick weldments
were machined into transverse weld tension plate specimens. When testing
was carried out on these specimens, the yield strength, ultimate strength,
and the elongation was measured and recorded. Charpy V-notch specimens
were machined from the weldments as well as the L-T and T-L (9) directions
in the base plate. These specimens were tested as a function of temperature. Additional bend, mechanical property, and Charpy V-notch testing was
done on weldments that were given a stress relief heat treatment at 1175°F
(635°C) for 1 hour.

Microstructure

Cross sections of the laser beam welds were cut, polished, and etched
in order to reveal the microstructure of the base plate (BP), heat affected
zone (HAZ), and the fusion zone (FZ). Hardness indents, using a Vickers
hardness tester, were made across the cross section of the weldments.
These traverses were made at several depths of the weldments, i.e., near the
top, mid-section and quarter sections.

Results

Structure

The microstructure of the base plate is ferrite with some pearlite.
At low magnifications the pearlite is unresolved but is easily resolvable
at higher magnification. The heat-affected zone consists of a refined
ferrite/pearlite microstructure, whereas the fusion zone consists of
bainite. These microstructures are shown in Fig. 2.

Hardness transverse across the weldment are given in Fig. 3 and
confirm the expected increased in hardness in the fusion zone, whereas the
heat-affected zone shows no appreciable change except very near the fusion
zone.

Considering the chemical composition of the base plate (0.20 w/o C,
1.00 w/o Mn), the microstructure resulting from laser beam welding is what
is expected from a rapid quench from a high temperature.

Properties

Evidence of the good ductility found in laser beam weldments was
given by the bend test. A bend test typically exposes defects in the
weldment such as undercut and interior defects (10). All of the different
thickness weldments satisfied the requirements of the bend test.

Mechanical properties were measured on the 0.5 in (12 mm) thick
plate using a transverse weld tension plate specimen. The results of

this testing is given in Table 2. All of the specimens fractured in the base plate, thus Table II is a measure of base plate properties.

Examination of both the bend and tensile specimens of the testing indicated that the fusion zone was not as ductile as the base plate. This is a result of the increased hardness of the fusion zone. Nevertheless, there is sufficient ductility in the weld to accommodate the 180° bend.

The results of testing the Charpy V-notch specimens as a function of temperature are shown in Fig. 4. The base plate specimens indicated two different levels of upper shelf toughness depending on whether the specimen orientation was L-T or T-L. The laser beam weldments had an upper shelf level of 165 ft-lb (224 joules) and a transition temperature of -4°F (-20°C). At -20°F (-30°C) the Charpy energy was 70 ft-lb (95 joules). Both base plate and laser beam weldments were given a stress relief heat treatment at 1175°F (635°C) for 1 hour. Charpy specimens were machined from these pieces. The base plate Charpy showed no change in toughness whereas the laser beam weldments had an upper shelf value of 145 ft-lb (196 joules) at -40°F (-40°C).

Analysis of Results

The microstructure of the fusion zone contained bainite which increased the hardness but still maintained good ductility and toughness. The refined structure in the heat-affected zone and the negligible increase in hardness indicate that the properties of the HAZ are comparable to the base plate.

Since all of the transverse weld tension plate specimens fractured in the base plate, very little can be said about the properties of the weldments except that they are equal to or better than that of the base plate.

The Charpy V-notch values of the base plate and laser beam welds are essentially the same. When given a post-weld stress relief heat treatment, the laser beam weld show a surprising increase in their Charpy values down to -40°F (-40°C). The Charpy requirement for an E70 weld wire is 20 ft-lb at -20°F (27 joules at -30°C). The laser beam weldments surpassed these values.

Transverse side-, face-, and root-bend specimens made from the laser beam weldments all satisfied the bead criteria for weldments of A-36 steel.

Conclusions

The objective of this program was to determine the laser beam welding parameters for different thickness of A-36 steel plate. Once these parameters were determined, weldments were fabricated and tested. Testing of the laser beam weldments showed that the mechanical properties were equal to or better than those of the base plate and that the ductility of the weldments was sufficient to satisfy the bend test criteria. Microstructural observations showed a change from the ferrite and pearlite in the base plate to a refined grain size ferrite/pearlite in the HAZ to a bainite in the fusion zone. Hardness traverses across the zone showed the correlation between hardness values and microstructures. The Charpy V-notch values of the laser beam weldments were the same as that of the base plate. When the weldments were given a stress relief heat treatment the Charpy values showed a surprising increase at lower temperatures down to -40°F (-40°C).

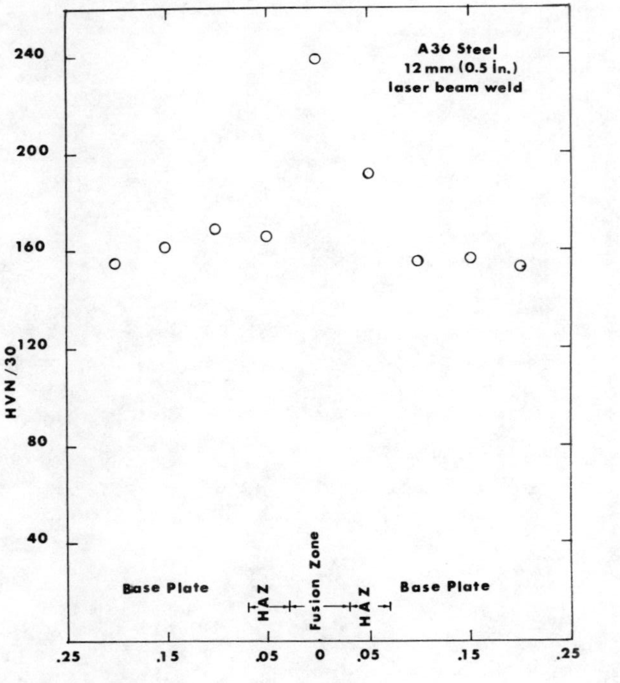

Fig. 3   Hardness traverses across the base plate, heat affected and fusion zones

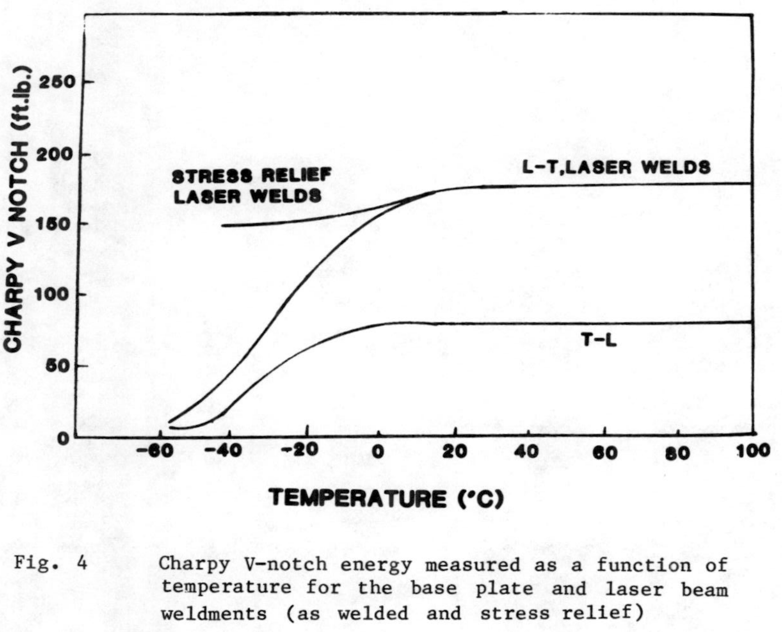

Fig. 4   Charpy V-notch energy measured as a function of temperature for the base plate and laser beam weldments (as welded and stress relief)

<u>Introduction</u>

In gas-assisted laser cutting, the gas is usually introduced coaxially with the focussed laser beam into the cutting area as shown in figure 1. The gas cools the cut area, thus lowering the Heat Affect Zone (HAZ), and also removes molten dross from the cut. In reactive fusion cutting, the cutting gas also reacts exothermically with that part of the metal which is above its kindling temperature, thereby increasing the energy input and so increasing the cutting speed.

According to Forbes (1975), nozzle design affects cutting performance in gas-assisted laser cutting. Successful coupling of the gas jet to the Kerf is an aerodynamic problem that can only be properly treated by relating the effects of variations in nozzle design and gas flow characteristics to actual performance results. By so doing, it can be shown whether, and to what extent, cutting performance can be improved by alterations to the cutting gas delivery parameters.

<u>Experimental Work</u>

The work reported here was carried out using a Control Laser 2kW CW CO_2 laser to cut various thicknesses of mild steel. The Gaussian Laser beam was focussed by a KCL lens on to the surface of the workpiece and, under normal operation, the cutting gas was introduced coaxially with the focussed beam. The gas pressure was measured as the gas line static pressure, shown in figure 1 and it was assumed that although this pressure was not equal to the total pressure behind the nozzle, yet the two pressures were related.

Nozzles of two different designs were used:-

a. those designed for uniform shock-free flow, and
b. Supersonic nozzles.

The pressure distribution at the nozzle exit was measured by means of a 1·0mm diameter pitot hole mounted in a flat plate connected to a mercury manometer. Such an arrangement would simulate the cutting situation better than the use of a pitot-static arrangement. Cutting performance was evaluated in terms of cutting speed, Kerf width, and HAZ measurements. The HAZ was measured by observation of surface oxidation marks, While the maximum cutting speed was defined as the speed V_C, (fig. 2c) at which the dross begins to reseal in the cut groove (fig. 2). "cutting speed", as used here, refers to V_C. Unless otherwise stated, the results refer to non-supersonic nozzles.

<u>Results and Discussion</u>

Figure 3 shows the relationships between cutting speed and nozzle diameter for both argon and oxygen at a constant supply pressure. When every other parameter is held constant, there is an optimum nozzle diameter to achieve a maximum cutting speed. For the nozzle design used in this work, this maximum is seen to be around 1·5mm. By comparing the maximum cutting speeds for both argon and oxygen assist using the 1·5mm nozzle, it can be estimated that about 60% to 70% of the total energy comes from the heat of reaction contribution in the case of reactive fusion cutting.

Since V_C varies with nozzle diameter, so also does cut quality (determined by Kerf width and HAZ measurements for cuts produced at the critical velocities) as shown in figure 4. This is an agreement with the general observation that cut quality increases with cutting speed (Gonsalves and Duley, 1972; Arata, et al, 1979; Kovalenko, et al, 1978).

In order to understand the effects shown in figures 3 and 4, it will be

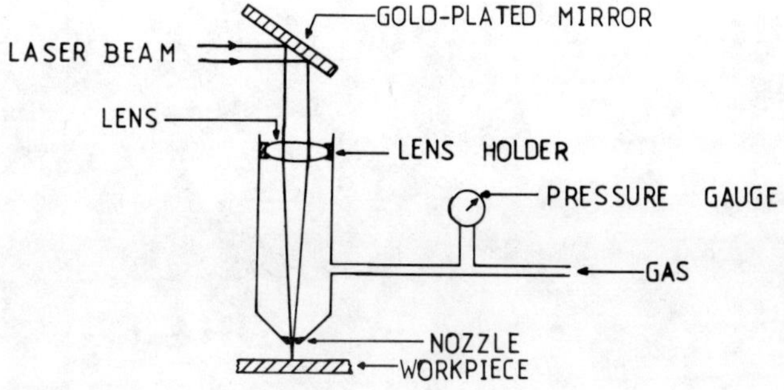

FIG. 1. - Gas-jet laser cutting

necessary to look at the nozzle exit pressure distribution as measured by
an impact hole connected to a mercury manometer (figure 5). It can be seen
from figure 5 that a 1·5mm nozzle would be more effective in dross removal
than the 3mm nozzle while the latter would give a lower HAZ because of the
wider spread of the jet. The 1·0mm nozzle suffers from two disadvantages,
namely:-

1. Its jet is confined to a narrow area of the cut, thus limiting dross
    removal at high cutting speeds.
2. It is difficult to align the 1·0mm nozzle with the focussed beam with-
    out the beam clipping the nozzle.

The variation of cutting performance with gas flow rate is shown in
figure 6. The decrease of cutting speed with gas pressure at high pressure
values is usually attributed to the cooling action of the gas jet (Adams,
1970; Babenko and Tychinskii, 1973). However, Duley and Gonsalves (1974)
have calculated that the power lost due to gas cooling at a flow rate of
2×10^3 cm^3 s^{-1} is the small quantity given by $p_c = \beta \times 12 \cdot 4W$ where, β is
the fraction of incident gas which cools the target. This calculation is
based on the assumption that the power, p_c, lost due to the gas cooling
effect is approximately:

$$p_c = \beta F^1 K (T_c - T_g)$$

where
     $F^1 = 6 \cdot 03 \times 10^{23}\ F/(60 \times 2 \cdot 24 \times 10^4)$ is the collision rate for O_2
molecules with the target, F is the gas flow rate, K is Boltzmann's constant,
T_c is the metal temperature and T_g is the gas temperature. However, the
calculation uses $(T_c - T_g)=1000^\circ K$ which is derived from the assumption that
laser cutting in the presence of oxygen proceeds at a temperature T_c which
is lower than T_n, the melting point of the metal. This assumption is not
supported by the work of Arata, et al (1979) who have determined that the
temperature of the cutting front is higher than the melting point of the
material.

The problem of the heat transfer between a plate and an impinging jet
has been studied by many workers. The heat transfer correlations of Gardon
and Cobonpue (1961) and Perry (1954) can be used to calculate the power loss
due to gas cooling.

Perry's (1954) formula gives the Nusselt number as:

Nu $= 0 \cdot 1810 (Re)^{0 \cdot 7} (Pr)^{0 \cdot 33}$, where

Nu   = Nusselt number $= \dfrac{hD}{K}$ where h = heat transfer coefficient

Re   = Reynolds number

     $= \dfrac{\rho UD}{\mu}$

Pr   = prandtl number

     $= \dfrac{c\mu}{K}$

ρ = Density of gas

u   = gas velocity

D   = Nozzle diameter

μ = viscosity of the gas

c   = thermal capacity of the gas

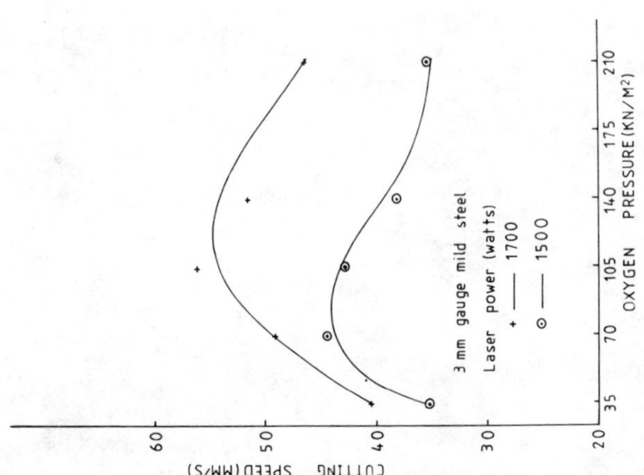

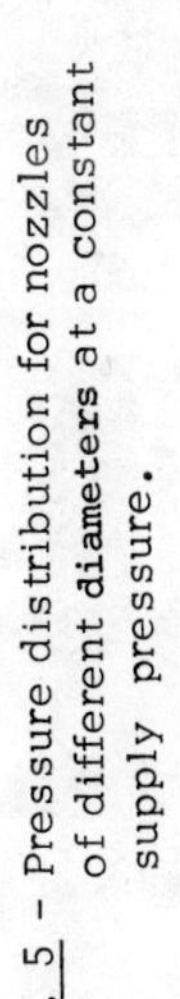

FIG. 6 - Variation of cutting speed with oxygen pressure.

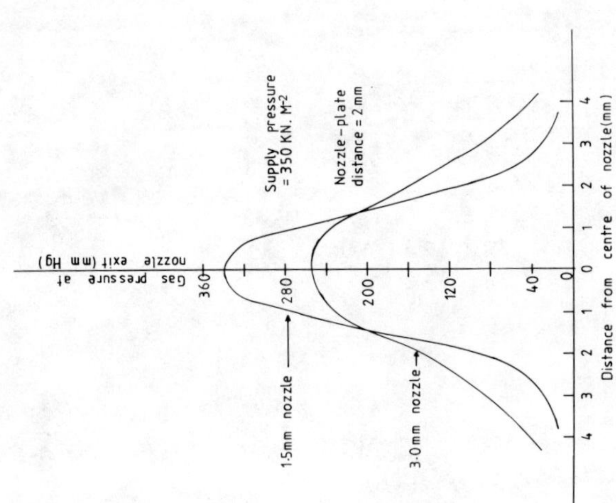

FIG. 5 - Pressure distribution for nozzles of different diameters at a constant supply pressure.

269

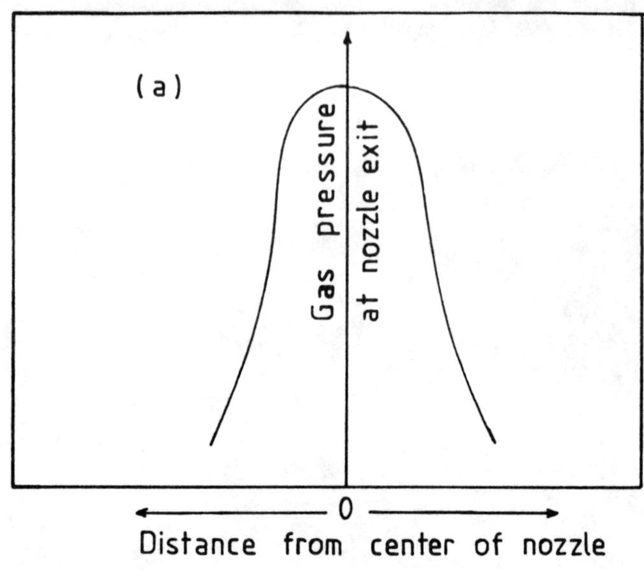

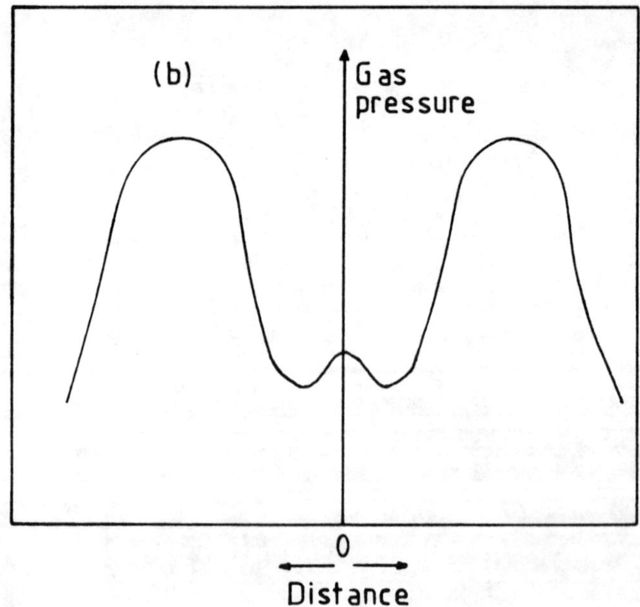

FIG. 8 Modification of pressure distribution

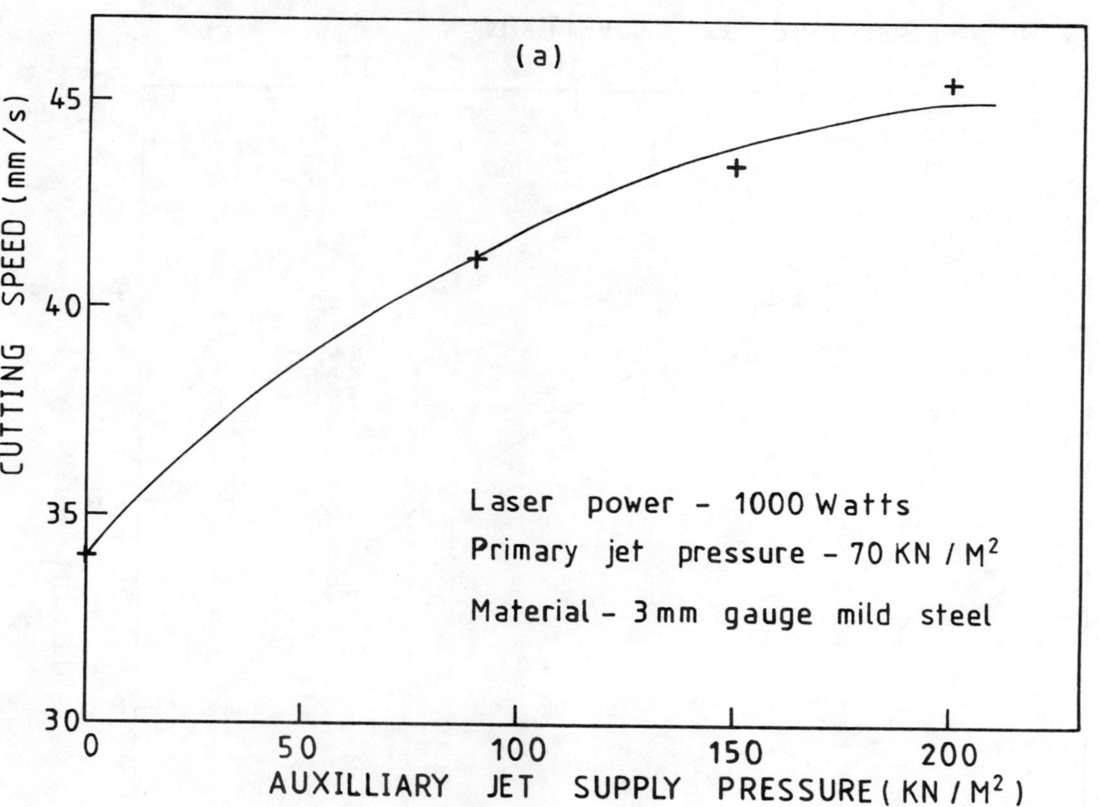

FIG. 9 Illustration of how the cutting speed increases with the auxiliary jet pressure.

273

(b)
KERF WIDTH (mm)
0.8
0.7
0.6
0.5
0
50
100
150
200
AUXILLIARY JET PRESSURE (KN / M²)

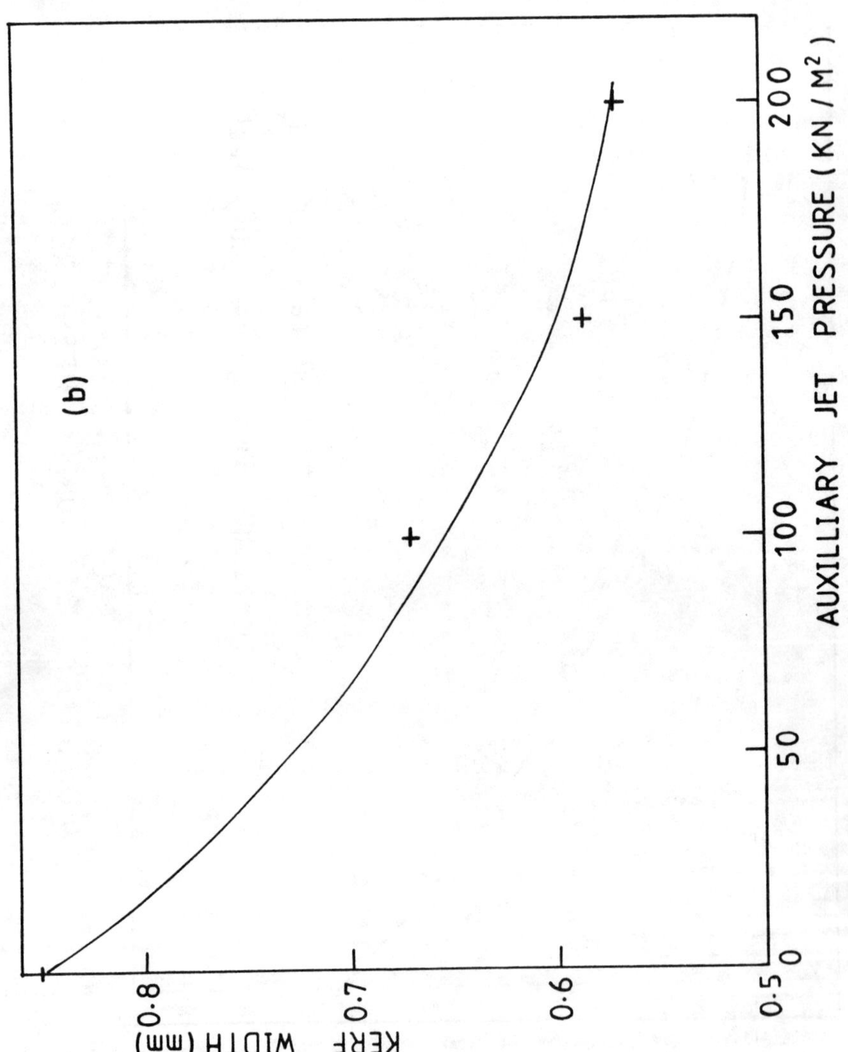

274

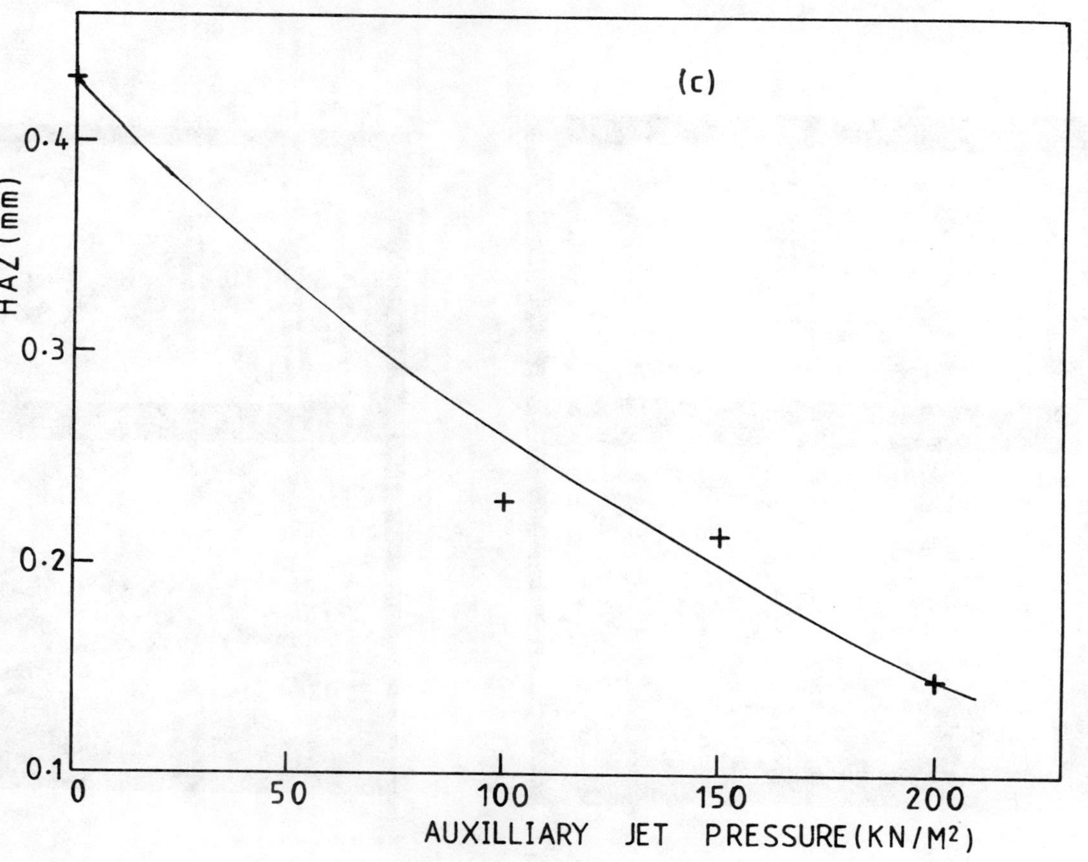

(c)
HAZ (mm)
0.4
0.3
0.2
0.1
0
50
100
150
200
AUXILLIARY JET PRESSURE (KN/M²)

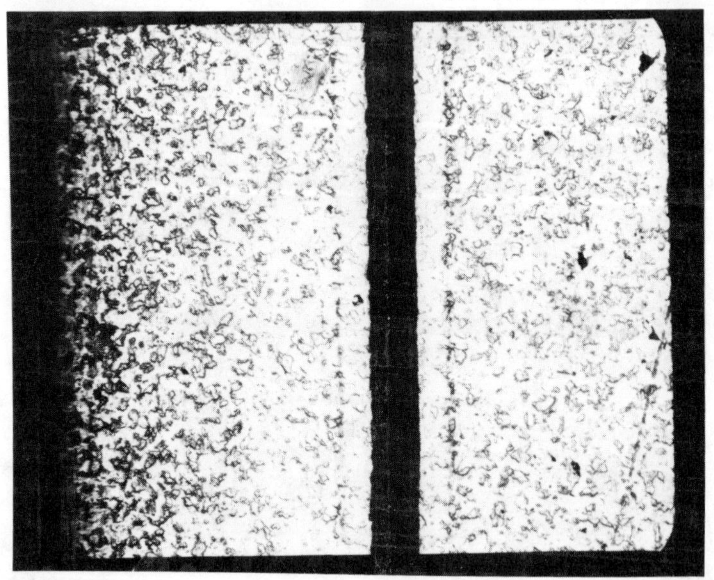

<u>FIG. 10</u> - Change of cut quality without (A) and with
(B) an auxilliary jet at the maximum
cutting speeds.

pressure region cannot interfere with the beam. This effect can be achieved
by the use of an auxiliary or side jet carefully positioned relative to the
primary jet.

Figure 9(a) illustrates how the cutting speed increases with the
auxiliary jet pressure. Because cutting speed increases with auxiliary gas
pressure, the cut quality (HAZ and Kerf width) also improves as the
auxiliary jet pressure is increased (Figures 9(b), 9(c), 10(a) and 10(b)).
A further advantage of the auxiliary jet is the fact that it can be made to
form part of the movable laser parts, thus giving no special problems in
contouring operations. The major disadvantages would be:

(a) Gas consumption may have to be increased.
(b) It may be necessary, in certain applications, to use different gases in
     both the primary and auxiliary jets.

Conclusion

It has been shown that in gas-assisted laser cutting, the nozzle
geometry affects cutting performance and that this geometry can be optimised
for a given nozzle design. At low gas pressures, cutting performance is
limited by the availability of a high enough momentum in the gas jet to overcome surface tension effects. High gas pressures result in the formation of
a density gradient field in which changes in the refractive index would
interfere with a focussed laser beam. The gas delivery system can, however,
be modified to give considerable improvements in cutting performance.

Acknowledgement

J.N. Kamalu would like to acknowledge the opportunity offered by a
Control Laser Ltd. research grant to carry out this project.

References

Adams, M.J., 1970, "Gas jet laser cutting", in Conf. on Advances in Welding
processes, Abington, Cambs.,England, April 14-16; published by the Weld.Inst.
1971, pp.140-146.

Arata,Y., Maruo,H., Miyamoto,I., and Takeuchi,S. 1979,"Dynamic behaviour in
Laser Gas Cutting of Mild Steel", Trans. JWRI, Vol.8, No. 2, pp.15-26.

Babenko, V.P., and Tychinskii, V.P., 1973, "Gas-jet Laser Cutting (Review)",
Sov. J. Quantum Electron., Vol. 2, No. 5, p.399.

Duley, W.W., and Gonsalves, J.N., 1974, "CO_2 Laser cutting of thin Metal
sheets with gas jet assist", Optics and Laser Tech., Vol. 6, (April),
pp. 78-81.

Forbes, N., 1975, "The role of the gas nozzle in metal-cutting with CO_2
Lasers", in Proceedings of Laser 75 Opto-Electronics Conf., Munich (W.Germany)
24-27 June, published by IPC Sci. & Tech. Press, 1976, pp.93-95.

Gardon, R, and Cobonpue, J. 1961, "Heat Transfer between a Flat plate and
Jets of Air impinging on it". in 1961 Int. Heat Trans. Conf., Published by
ASME, pp.454-460.

Gonsalves, J.N., and Duley, W.W. 1972, "Cutting thin metal sheets with CW
CO_2 Laser", J. Appl. Phys., Vol. 43, No. 11, pp.4684-4687.

Horing, D.F., "Shock front measurements by light reflectivity", in Physical
Measurements in gas dynamics and combustion: High speec Aerodynamics and Jet

propulsion, Vol. IX, pp.203-210.1955.

Kovalenko, V.S., et al., 1978, "Experimental study of cutting different materials with a 1·5kW CO_2 Laser", Trans. JWRI, Vol. 7, No. 2, pp.101-112.

Perry, K.P., 1954, Proc. Inst. Mech. Eng., Vol. 168, p.775.

EXPLORATORY STUDY OF LASER PROCESSING OF TITANIUM ALLOYS*

T. C. Peng, S. M. L. Sastry, and J. E. O'Neal

McDonnell Douglas Research Laboratories
St. Louis, Missouri 63166

The effectiveness of laser processing for surface alloying, rapid solidification processing, and welding of several titanium alloys was studied using a 1.5-kW CO_2 laser. The melt depth and width were correlated with laser power density and laser-solid interaction time. The dependences on process parameters of supersaturation of alloying elements, grain refinement, dispersoid formation, and weld-zone microstructures were studied, and the implications of laser-processed titanium-alloy microstructures for the development of novel Ti-alloy compositions were identified.

*This research was conducted in part under the Office of Naval Research Contract N00014-76-C-0626 and in part under the McDonnell Douglas Independent Research and Development Program.

Introduction

Laser processing is increasingly being used to increase surface hardness and improve wear, erosion, corrosion, and fracture resistance of high-strength materials (1-3). Laser processing is also an inexpensive technique for screening large numbers of candidate compositions in the development of new alloys by rapid solidification processing (RSP). To obtain rapid solidification by laser processing, the surface of a material is irradiated with a laser beam focused to a power density of 10^5 - 10^7 W/cm^2 at traverse rates of 10 - 100 cm/s. A thin, molten layer formed by laser irradiation solidifies on the alloy substrate at a rapid rate following passage of the laser beam. By varying the power density and traverse rate, melt depths of 10 - 100 μm and cooling rates of 10^3 - 10^6 °C/s can be produced. The technique of producing surface alloys by laser processing is attractive for studying the dependence of microstructures on alloying element concentration and cooling rate for a variety of alloy systems. Furthermore, laser processing offers an easily maneuverable, chemically clean, high-intensity, atmospheric welding process that can produce deep-penetration welds with narrow heat-affected zones and small distortion. In the present investigation, several titanium alloys were surface-melted, surface-alloyed, and welded using a 1.5-kW, continuous-wave, CO_2 laser with the objectives of determining the interrelation between process parameters, melt-zone geometry, and microstructures and evaluating the feasibility of producing fine, 10 - 1000 nm diameter, stable, incoherent dispersoids in rapidly solidified Ti alloys.

Experimental

The materials used for laser melting, laser-alloying, and welding are listed in Table I. The starting alloys were prepared by nonconsumable-electrode arc-melting in pure argon, and 1.6-mm and 3.0-mm thick sheets were produced by unidirectional hot rolling. The surface coatings for laser surface-alloying studies were produced by electron-beam evaporation of the coating elements onto the Ti-alloy substrates in a vacuum of 10^{-4} Pa. Laser surface-melting and surface-alloying experiments were conducted on 25 x 30 x 1.6-mm specimens with a 1.5-kW CO_2 laser using the experimental arrangement shown in Fig. 1. The specimens were surface melted by the laser radiation focused to a 500-μm beam size and traversed at 0.5 - 40.0 cm/s to obtain melt layers with cooling rates of 10^2 - 10^5 °C/s. For laser welding studies, samples of the compositions shown in Table I were used. The samples were stress-relief annealed at 720 - 790°C for 1 h and finish machined to 25 x 50 x 3 mm and 25 x 50 x 1.5 mm. Butt welds were made with the focused beam of a 1.5-kW CO_2 laser using the experimental arrangement shown in Fig. 2. The laser beam was fixed, and the specimen to be welded was traversed at controlled rates in a plane perpendicular to the laser beam. A He-gas shroud covered the specimen area where the laser beam was focused to prevent significant oxidation during welding.

The melt-zone geometry and microstructure were determined by optical and scanning electron microscopy and correlated with processing parameters.

Table I. Materials used in laser-processing experiments.

Laser-processing method	Alloy composition
Laser-melting	Ti-8Al
	Ti-8Al-0.5Y
	Ti-8Al-1.0Y
	Ti-8Al-2.0Er
Laser-alloying	Ti-RE
	(RE = Er, Y, Gd, Nd, Sc, and La)
	Ti-B
Laser-welding	Ti
	Ti-Y
	Ti-Er
	Ti-6Al-4V

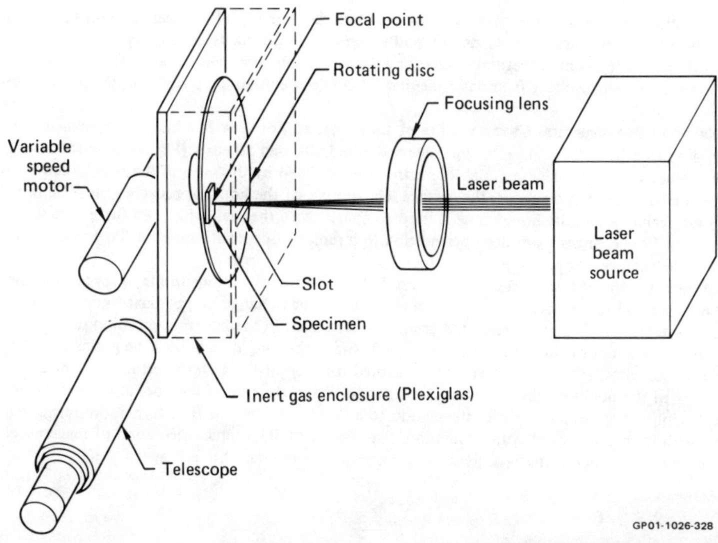

Fig. 1. — Experimental arrangement for laser-melting titanium alloys.

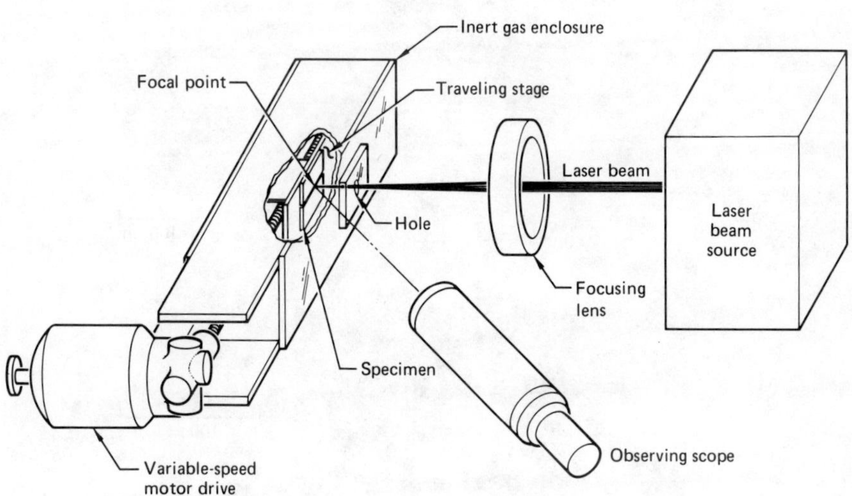

Fig. 2 — Experimental arrangement for laser-beam welding of titanium alloys.

Results and Discussion

Laser Melting

The principal objectives of the laser melting studies were to determine (1) the dependence of cooling rate and melt depth and width on the laser-beam power density and traverse rate, (2) the geometry, topology, and thermal history of laser-formed melt zones using a thermal model for a Gaussian source moving at a constant speed, and (3) the microstructural refinement resulting from laser melting of Ti alloys containing coarse constituent particles.

Topographical and Microstructural Characteristics of Laser-Melted Zones in Ti-8Al: The tranverse sections of laser-melted layers produced by a 860-W laser beam traversed at 0.5, 10, and 30 cm/s (Fig. 3) show the strong dependence of melt-zone geometry on the beam/material interaction time defined by the ratio of the beam diameter to the scanning velocity. For large interaction times, the surface temperature of the liquid approaches the boiling point, vaporization becomes appreciable, and the liquid zone is deep compared with the melt width. As the interaction time decreases, the heat flow changes from deep penetration to a radially outward flow from the beam and finally to a flow parallel to the surface.

The optical micrographs of laser-melted Ti-8Al and Ti-8Al-2.0Er alloys shown in Fig. 4 reveal that the microstructures at low beam-scanning speeds consist of columnar grains, which change into equiaxed grains with increasing scanning speeds. At the center of the melt pool, the grains are elongated in the beam propagation direction. With increasing distance from the center of the melt pool, the angle between the long direction of the grains and the beam propagation direction increases, in agreement with the predicted rotation of the growth vector from the beam-scan direction at the center of the pool to a direction nearly perpendicular to the side of the pool at the perimeter.

The microstructures also reveal that the alloys undergo a $\beta \rightarrow \alpha'$ martensitic transformation during the rapid solidification, with individual grains containing more than one crystallographic orientation of martensite. Both the martensite packet size and plate length decrease with increasing laser-beam traverse speed.

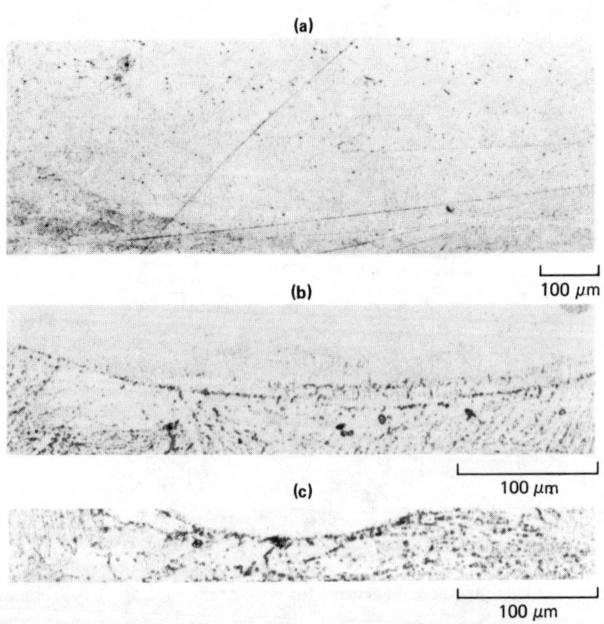

Fig. 3 — Photomicrographs of x = 0 cross sections of Ti-8Al-0.2Er laser-melted at traverse rates of (a) 0.5 cm/s, (b) 10 cm/s, and (c) 30 cm/s.

GP11-0301-4

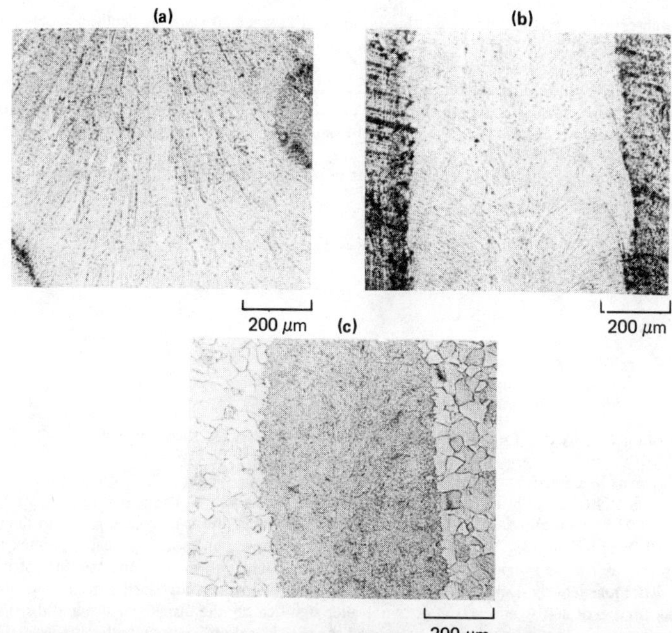

Fig. 4 — Microstructures of Ti-8Al-2.0Er laser-melted at a traverse rate of (a) 0.5 cm/s and (b) 10 cm/s, and (c) Ti-8Al at 30 cm/s with a 250-μm radius, 860-W laser beam.

GP11-0301-20

Thermal Analysis of Laser Surface Melting of Ti-8Al: When the surface of a material is scanned by a high-power laser beam, the resulting temperature distribution, depth of melting, and cooling-rate distribution are related to the laser spot size, velocity, and power level.

If the laser beam power has a Gaussian distribution over the beam area, the temperature distribution in the material being irradiated is (4):

$$T(x,y,z) = \frac{P}{C_p DR} \cdot \frac{1}{\sqrt{2\pi^3}} \int_0^\infty \frac{e^{-H(u)}}{1 + u^2} \, du, \tag{1}$$

where

$$u^2 = 2Dt'/R^2, \tag{2}$$

$$H(u) = \frac{\left[(x/R + (\rho\tau/2)u^2\right]^2}{2(1 + u^2)} + \frac{(z/R)^2}{2u^2}, \tag{3}$$

$$\rho = Rv/D, \tag{4}$$

and C_p is the specific heat per unit volume, D is the thermal diffusivity, P is the total laser power absorbed, R is the laser beam spot radius, v is the speed of traversal of the laser beam, and t' is a characteristic time. The liquidus isotherm is determined by setting Eq. (1) equal to T_m, the melting temperature of the material, and the melt depth is given by

$$z_m = z_o \ln(P/P_m), \tag{5}$$

283

where P_m is the absorbed power just before melting occurs and z_0 is a characteristic depth.

For Ti-8Al, $C_p = 2.19$ J·K^{-1}·cm^{-3} and $D = 0.048$ cm^2/s. The calculated liquidus isotherms in the y-z plane of Ti-8Al when irradiated with a laser beam having P = 860 W are shown in Fig. 5 for different laser-beam specimen-coupling efficiencies. The position of solid-liquid interfaces determined experimentally from the transverse-section photomicrographs of laser-melted Ti-8Al are shown by dotted lines in Figs. 5a and 5b. The positions of solid-liquid interfaces match the positions calculated for a specimen-beam coupling efficiency of 25%. The coupling efficiencies in other materials are typically 15-25% (5 - 7).

If the thermal gradient far from the center of the laser beam approaches that of a point source, the cooling rate can be calculated from the expression

$$\frac{\partial T}{\partial t} = -V\left[\frac{x}{r^2} + \frac{V}{2D}\left(1 + \frac{x}{r}\right)\right] T, \tag{6}$$

where

$$r^2 = x^2 + y^2 + x^2 \, .$$

Contours of constant cooling rate for different laser-beam traverse rates are shown in Fig. 6.

Dispersoid-Refinement in Laser-Melted Ti-8Al-Y and Ti-8Al-Er Alloys: Figures 7a-7f are scanning electron micrographs of conventionally cast and laser-melted Ti-8Al-Y and Ti-8Al-Er alloys. The scanning electron micrographs of conventionally cast Ti-8Al-RE alloys in Figs. 7a-7c show bimodal distributions of coarse 1 - 5 μm diameter particles, which precipitated from the liquid, and fine 10 - 20 nm diameter dispersoids, which probably formed by solid-state precipitation. In the laser-melted regions, the coarse particles are completely absent, and the microstructures contain large amounts of homogeneously dispersed, spherical, 10 - 20 nm diameter second-phase dispersoids as shown in Figs. 7d-7f. Annealing for 1 h at 800°C and 900°C does not alter significantly the dispersoid size and distribution. The dispersed-phase parameters in the laser-melted layers and the theoretical Orowan strengthening are listed in Table II. The closely spaced, fine dispersoids resulting from laser melting produce a 25-35% increase in the matrix yield stress.

The refinement of dispersoids in a laser-melted alloy results from a reduced time for coarsening during solidification and from the higher-than-equilibrium solid solubility of Er and Y in Ti-8Al with subsequent solid-state precipitation of the dispersoids. The high volume fractions of the dispersoids indicate that the dispersoids are titanium-rich Ti-Y and Ti-Er intermetallic compounds rather than pure Y or Er or oxide dispersions. These compounds are probably metastable phases because the equilibrium phase diagrams for Ti-Y and Ti-Er do not show the existence of any intermetallic compound phases (8). The precise compositions of the dispersoids is relatively unimportant in their influence on alloy mechanical behavior. The important factors are the small size of the dispersed particles and their stability at high temperatures, where they can influence and modify alloy behavior. The results of this study demonstrate the feasiblity of producing dispersion-strengthened Ti alloys by rapid solidification.

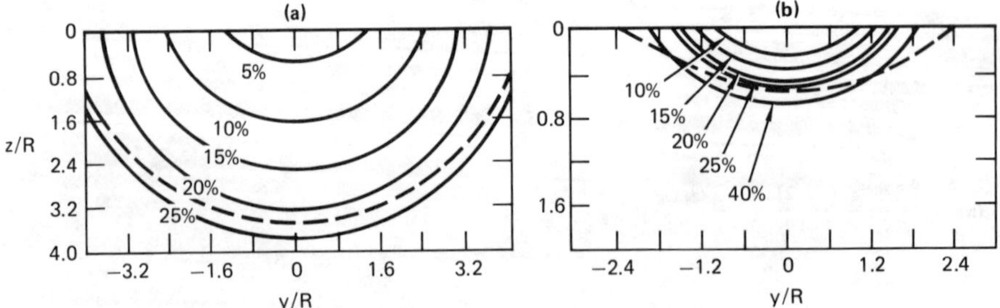

Fig. 5 — Laser-melting solid-liquid interfaces in Ti-8Al calculated for a laser-beam power of 860 W and different coupling efficiencies for laser-beam traverse rates of (a) 0.5 cm/s and (b) 5 cm/s. The experimentally determined interfaces are shown by dashed lines.

GP11-0301-19

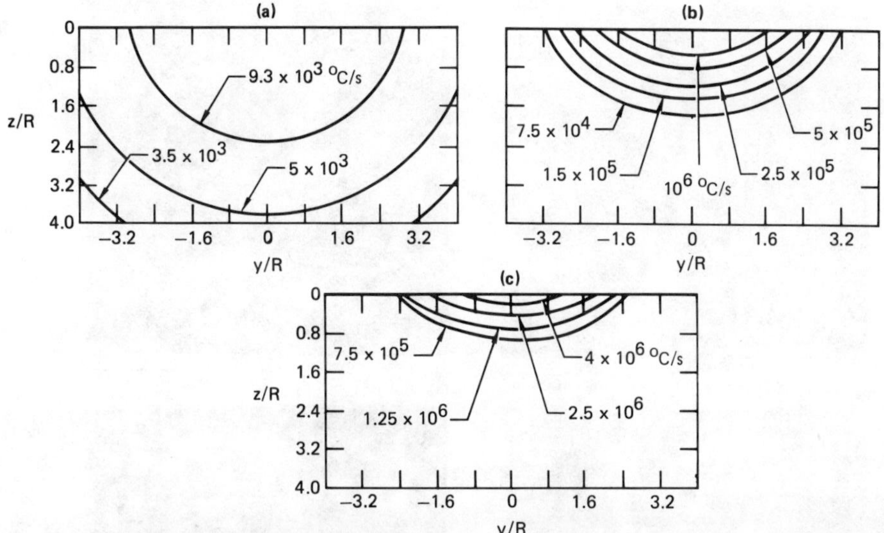

Fig. 6 — Equal cooling-rate contours in laser-melted Ti-8Al calculated for a laser-beam power of 860 W, coupling efficiency of 25%, and traverse rates of (a) 0.5 cm/s, (b) 5.0 cm/s, and (c) 10 cm/s.

GP11-0301-5

Table II. Dispersoid parameters in laser-melted Ti-8Al-0.5Y and Ti-8Al-1.0Y.

Alloy	Laser-beam traverse speed (mm/s)	Average dispersoid diameter (nm)	Average interparticle spacing (nm)	Volume fraction of dispersoids	Theoretical Orowan strengthening (MPa)
Ti-8Al-0.5Y	5	90	199	0.19	147
	400	77	180	0.20	175
Ti-8Al-1.0Y	5	94	155	0.34	192
	300	95	180	0.30	165

GP11-0301-17

285

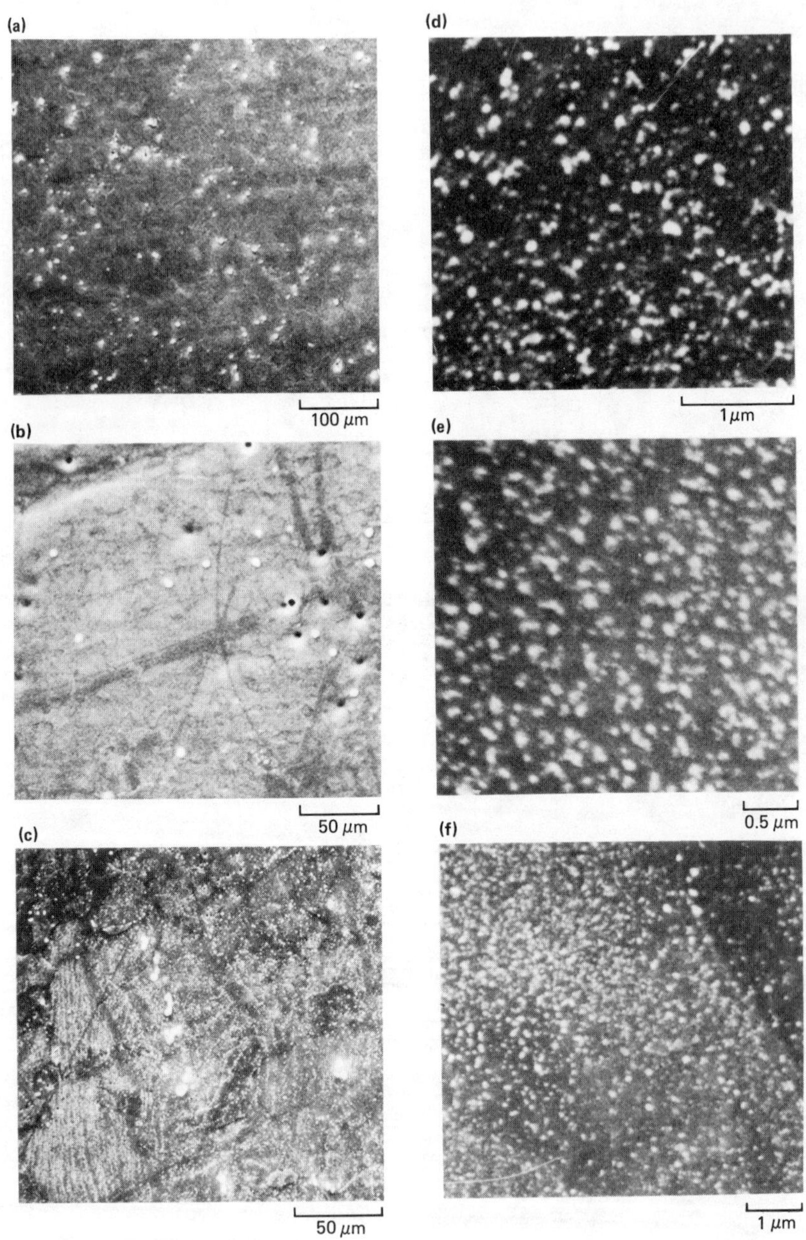

Fig. 7 — Scanning electron micrographs of conventionally melted ingots of (a) Ti-8Al-0.5Y,
(b) Ti-8Al-1.0Y, and (c) Ti-8Al-2.0Er and laser melted surfaces of (d) Ti-8Al-0.5Y, 40 cm/s,
(e) Ti-8Al-1.0Y, 30 cm/s, and (f) Ti-8Al-2.0Er, 40 cm/s with a 250-μm radius, 860-W laser
beam.

Laser Surface Alloying

Laser surface-alloying studies were performed with the objective of producing fine, stable, incoherent dispersoids in Ti-RE (RE = Er, Y, Gd, Nd, Sc, or La) and Ti-B alloys by rapid solidification. Titanium specimens, coated with 0.02 - 1.0 μm thick, electron-beam evaported, rare-earth metallic films and with 10 - 25 μm thick boron slurry, were surface melted using a 1.5-kW CO_2 laser beam at traverse rates of 0.5 to 40 cm/s.

Figures 8a, 8b, and 8c are scanning electron micrographs of transverse sections of Ti-B alloy laser melted at specimen traverse speeds of 0.5, 5, and 10 cm/s, respectively. The dependence of experimentally determined melt depths on traverse speed for pure Ti and B-coated Ti are shown in Fig. 9. The boron-slurry coating increases the laser-beam/specimen coupling efficiency and increases the penetration depth for large interaction times.

The Ti-B alloys laser-melted at low specimen traverse speeds exhibit a dendritic structure (Fig. 10a), which changes into a cellular structure at high traverse speeds (Fig. 10b-10d). The dendrite arm/cell spacing decreases from 1.82 μm at a specimen traverse rate of 0.5 cm/s (cooling rate $\approx$ 5 x 10^4 °C/s) to 0.46 μm at 40 cm/s (cooling rate $\approx$ 5 x 10^6 °C/s. The crystal structures determined by electron diffraction of the phases in laser-processed Ti-B alloy were α-Ti and TiB_2. The TiB_2 compound has a hexagonal-close-packed crystal structure with lattice spacings of a = 0.303 nm and c = 0.323 nm.

Figures 11a-11d are scanning electron micrographs of the laser-alloyed Ti-Y, Ti-Er, Ti-Nd, and Ti-Gd alloys, respectively, of Ti coated with 1-μm thick rare-earth films. In the laser-melted Ti-RE alloys with 0.2-μm and 1-μm thick rare-earth films, the microstructures consist of fine 10 - 100 nm diameter dispersoids, but the second-phase dispersoids were not observed in Ti-RE samples with 0.02 μm-thick films. The dispersoid size decreases with increasing specimen traversing speed as well as with increasing depth from the specimen surface. The volume fraction of the dispersoids increases with increasing rare-earth film thickness and melt depth. The equilibrium, coarse, constituent particles that are generally observed in conventionally cast Ti alloys containing corresponding amounts of rare-earth elements were completely absent in the laser surface-alloyed specimens. Dispersoid refinement in the laser-melted specimens results from the reduced time for coarsening during solidification as well as from the increased solid solubility of rare-earth

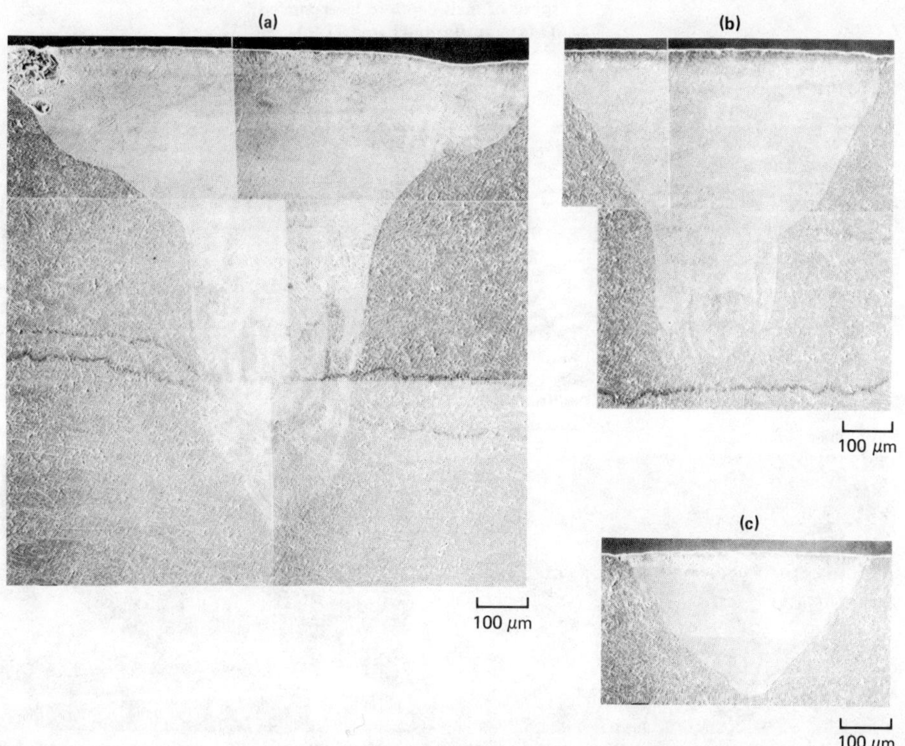

Fig. 8. — Scanning electron micrographs of traverse sections of B-coated Ti laser-surface-melted with 860-W incident power and specimen scanning speeds of (a) 0.5 cm/s, (b) 5 cm/s, and (c) 10 cm/s.

GP11-0301-10

287

elements in rapidly solidified titanium with subsequent solid-state precipitation of the dispersoids. The high volume-fraction of the dispersoids indicates that they are metastable, Ti-rich, Ti-RE intermetallic compounds rather than elemental rare-earth or oxide dispersions. Although these compounds are metastable phases, their fine size and thermal stability up to 900°C make them highly suitable for dispersion strengthening of titanium.

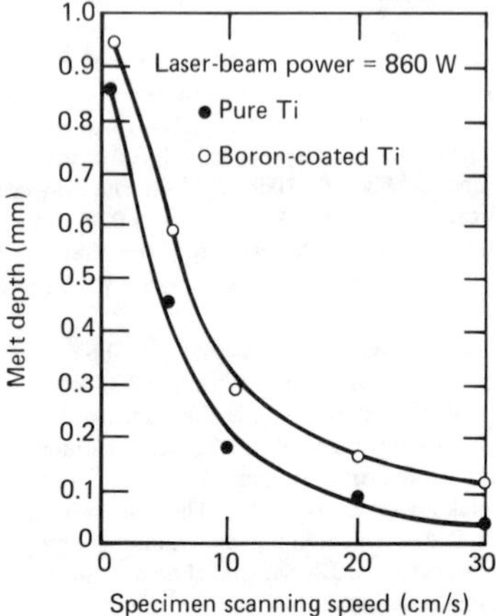

Fig. 9. — **Dependence on specimen scanning speed of melt depth in laser-melted Ti (●) and Boron-coated Ti (○).**

GP11-0301-11

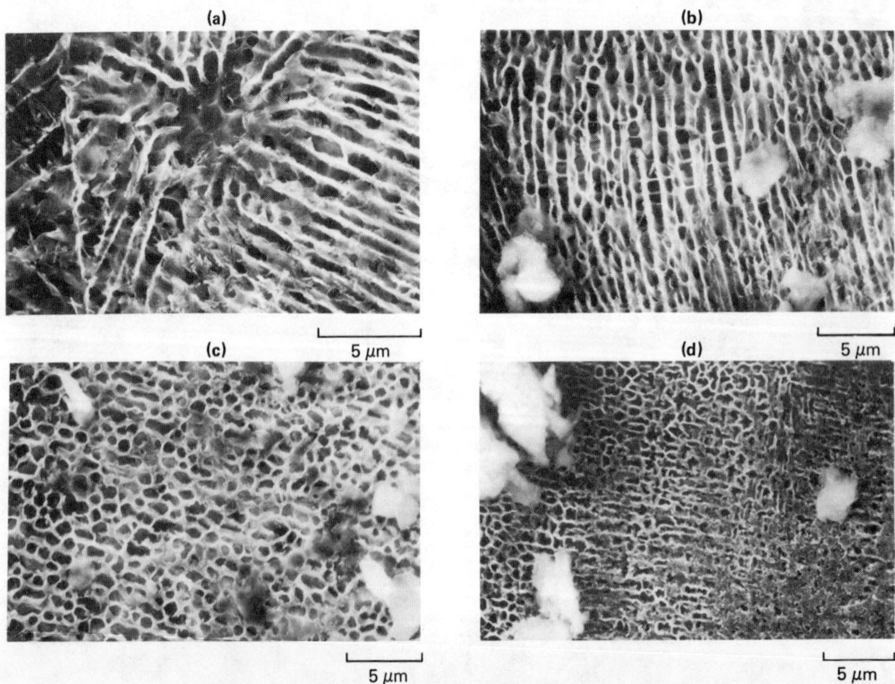

Fig. 10. — **Microstructures of B-coated Ti laser-melted at 860-W beam power and specimen traverse rates of (a) 0.5 cm/s, (b) 5 cm/s, (c) 10 cm/s and (d) 40 cm/s.**

GP11-0301-12

288

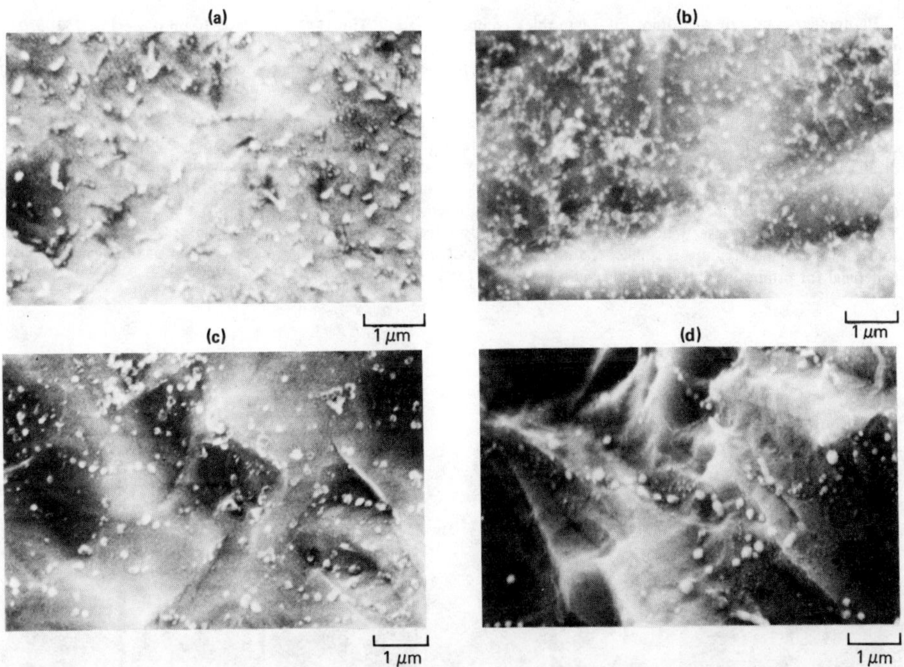

Fig. 11 — Scanning electron micrographs of dispersoids in laser-surface-alloyed (a) Ti-Y, (b) Ti-Er, (c) Ti-Nd, and (d) Ti-Gd alloys with a 250-μm 860-W laser beam at a traverse rate of 5 cm/s.

GP11-0301-1

Laser Welding

Titanium alloys containing 50 - 500 nm diameter Er and Y dispersoids and Ti-6Al-4V were chosen for the laser welding studies with the objectives of (1) determining the interrelation between process parameters and weld-zone geometry, (2) comparing the microstructures and tensile properties of laser-welded specimens with those of electron-beam-welded specimens, and (3) determining the modifications in the size and distribution of dispersoids as a result of laser welding. The laser-beam-welding parameters are listed in Table III.

Table III. Laser-Beam welding parameters.

Alloy	Plate thickness (mm)	Laser-beam traverse rate (mm/s)	Laser-beam power (kW)	Laser-beam penetration
Ti-0.27Er	1.5	0.254	0.580	Full
Ti-0.05Er	1.5	0.254	0.580	Full
Pure Ti	1.5	0.254	0.435	90%
Ti-0.02Y	1.5	0.254	0.290	Partial
Ti-0.014Y	1.5	0.254	0.377	Partial
Ti-0.09Y	1.5	0.254	0.508	Partial
Ti-6Al-4V	3.0	0.254	0.870	Full
Ti-6Al-4V	3.0	0.254	0.986	Full
Ti-6Al-4V	3.0	0.333	1.102	Full
Ti-6Al-4V	3.0	0.254	1.101	Full
Ti-6Al-4V	3.0	0.254	0.754	Partial

GP11-0301-14

Photomicrographs of weld joints of 3-mm thick commerical-grade Ti-6Al-4V with laser beam powers of 750, 870, 986, and 1102 W are shown in Figs. 12a-12g, which show the increases in weld width and depth with increasing laser beam power. The microstructure at the center of a weld zone consists of columnar grains, and away from the center, the grains are equiaxed.

The fusion zone in laser-welded Ti-6Al-4V and Ti-Er alloys consists mainly of acicular martensitic α (α'), whereas the heat-affected zone consists of a mixture of martensitic alpha and primary alpha. The martensitic structure in Ti indicates that the cooling rates were 10^2 - 10^4 ° C/s. The weld width and microstructures are similar to those for electron-beam-welded specimens: the weld-depth dependence on laser power and feed speed for commercial-grade Ti-6Al-4V is summarized in Table IV. The room-temperature tensile properties of laser-welded alloys, summarized in Table V, indicate that laser welding does not significantly affect the tensile properties. The average size of Er dispersoids in laser-welded regions is smaller than that in the base metal (Fig. 13), and this refinement of the dispersoids is beneficial for promoting ductile fracture and increasing the fracture toughness of the welds.

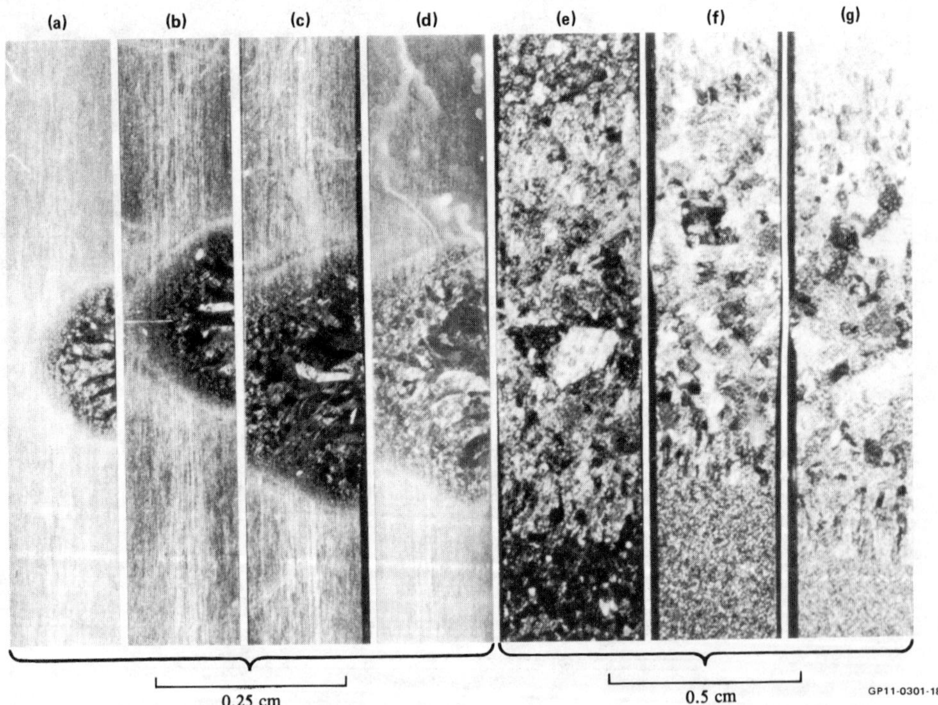

Fig. 12 — Cross sections of laser-welded butt joints of 3-mm thick commercial-grade Ti-6Al-4V for laser-beam powers of (a) 750 W, (b) 870 W, (c) 990 W, and (d) 1100 W, and 1.5-mm thick pure Ti with (e) no additive, 440 W, (f) 0.05 wt% Er additive, 580 W, and (g) 0.27 wt% Er, 580 W.

**Table IV. Dependence of weld depth
on laser-beam power and traverse rate
for 3.0-mm thick Ti-6Al-4V plates.**

Laser-beam power (kW)	Laser-beam traverse rate (mm/s)	Weld depth (mm)
0.75	0.25	1.00
0.87	0.25	1.12
1.10	0.25	1.70
1.10	0.33	1.37

GP11-0301-16

**Table V. Room-temperature tensile properties of laser-welded
titanium alloys.**

Alloy	Yield stress at 0.2% offset (MPa [ksi])	Ultimate tensile stress (MPa [ksi])	Total elongation (%)
Ti-0.27Er	247 [35.8]	366 [53.1]	27
Ti-0.05Er	279 [40.4]	375 [54.3]	28
Commercial-grade Ti-6Al-4V	901 [131]	979 [142]	12
Commercial-grade Ti-6Al-4V	909 [132]	971 [141]	12
Electron-beam-welded Ti-6Al-4V	965 [140]	1000 [145]	11

GP11-0301-15

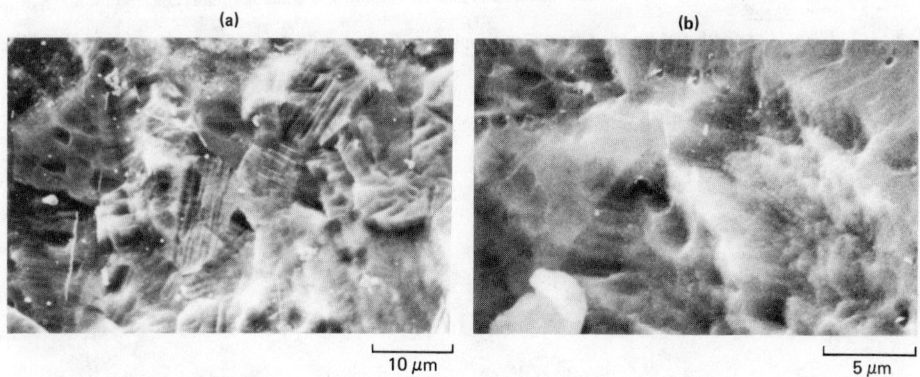

Fig. 13 — Scanning electron micrographs of polished and etched surface of Ti-0.27Er (a) before and (b) after laser welding.

GP11-0301-13

291

Conclusions

1. Cooling rates of 10^4 - 10^7 °C/s in titanium alloys can be achieved by laser surface melting with a 1.5-kW CO_2 laser beam at specimen scanning speeds of 5 - 40 cm/s.
2. The geometry, topology, and thermal history of melt zones formed by laser surface melting can be adequately described by a heat transfer model for a Gaussian source moving at a constant velocity.
3. The laser-beam coupling efficiency for titanium alloys is $\approx$ 25%.
4. Laser melting results in a hundredfold refinement of insoluble second-phase particles in Ti-8Al-Y and Ti-8Al-Er alloys.
5. The laser-surface-melted microstructures in Ti-8Al-Y and Ti-8Al- Er alloys indicate the feasibility of increasing by 25 - 35% the yield stress of Ti alloys by the incorporation of second-phase dispersoids by rapid solidification processing.
6. Rapid solidification processing of Ti-RE (RE = Er, Y, Gd, Nd, or Sc) produces fine, stable, incoherent dispersoids in the Ti matrix.
7. The weld microstructures and mechanical properties of laser-welded titanium alloys are comparable to those of electron-beam-welded alloys.
8. Laser welding of Ti alloys containing Er and Y dispersoids results in dispersoid refinement in the weld zone.

Acknowledgement

The assistance of G. M Niemeyer and D. L. McGinnis of the McDonnell Aircraft Co. in the laser processing experiments is gratefully acknowledged.

References

1. E. M. Breinan, B. H. Kear and C. M. Banas, "Processing Materials with Lasers," Physics Today **29**, 44 (1976).
2. E. M. Breinan, B. H. Kear, C. M. Banas, and L. G. Greenwald, "Surface Treatment of Superalloys by Laser Skin Melting," In **Proc. of the Third Int. Sym. on Superalloys: Metallurgy and Manufacture** (Claitor's Publ. Div., Baton Rouge, LA, 1976), p. 435.
3. L. S. Weinman and C. Kim, "Rapid Solidification of Aluminum and Titanium Alloy Surfaces," In **Proc. of the Int. Conf. on Rapid Solidification Processing** (Claitor's Publ. Div., Baton Rouge, LA, 1977), p. 117.
4. R. Mehrabian, "Relationship of Heat Flow to Structure in Rapid Solidification Processing," In **Proc. of the Int. Conf. on Rapid Solidification Processing** (Claitor's Publ. Div., Baton Rouge, LA, 1977), p. 9.
5. H. E. Cline and T. R. Anthony, "Heat Treating and Melting Material with a Scanning Laser or Electron Beam," J. Appl. Phys. **48**, 895 (1977).
6. O. Esquivel, J. Mazumder, M. Bass, and S. Copley, "Shape and Surface Relief of Continuous Laser-Melted Trails in UDIMET 700," In **Rapid Solidification Processing: Principles and Technologies, II** (Claitor's Publ. Div., Baton Rouge, LA, 1980), p. 180.
7. P. R. Strutt, M. Kurup, and D. A. Gilbert, "Comparative Study of Electron Beam and Laser Melting of M2 Tool Sheet," In **Rapid Solidification Processing: Principles and Technologies, V** (Claitor's Publ. Div., Baton Rouge, LA, 1980), p. 225.
8. R. P. Elliott, **Constitution of Binary Alloys, First Supplement** (McGraw Hill Publishing Company, New York, NY, 1965) p. 407, 854.

Subject Index

Author Index